Organic
Hobby Farming

A Practical Guide
to Earth-Friendly Farming in Any Space

Andy Tomolonis

I·5
·PRESS·

Organic Hobby Farming

Project Team
Editor: Dolores York
Photo Editor: Elizabeth Tunnicliffe
Design: Mary Ann Kahn
Indexer: Elizabeth Walker

I-5 PUBLISHING, LLC™
Chief Executive Officer: Mark Harris
Chief Financial Officer: Nicole Fabian
Vice President, Chief Content Officer: June Kikuchi
General Manager, I-5 Press: Christopher Reggio
Editorial Director, I-5 Press: Andrew DePrisco
Art Director, I-5 Press: Mary Ann Kahn
Digital General Manager: Melissa Kauffman
Production Director: Laurie Panaggio
Production Manager: Jessica Jaensch
Marketing Director: Lisa MacDonald

Library of Congress Cataloging-in-Publication Data
Tomolonis, Andy, 1955-
  Organic hobby farming : a practical guide to earth-friendly farming in any space / by Andy Tomolonis.
      p. cm.
  Includes bibliographical references and index.
  ISBN 978-1-933958-58-3 (alk. paper)
1. Organic farming. 2. Farms, Small. I. Title.
  S605.5.T66 2014
  631.5'84--dc23
                      2013044506

This book has been published with the intent to provide accurate and authoritative information in regard to the subject matter
within. While every precaution has been taken in the preparation of this book, the author and publisher expressly disclaim any
responsibility for any errors, omissions, or adverse effects arising from the use or application of the information contained herein.

I-5 Publishing, LLC™
3 Burroughs, Irvine, CA 92618
www.facebook.com/i5press
www.i5publishing.com

Printed and bound in China
14 15 16 17   1 3 5 7 9 8 6 4 2

To Valerie, my wife and farming partner,
whose patience, understanding, and indefatigable optimism
helped make our farming dream—and this book—possible.

Contents

Sometimes the actions of a child can inspire you to move mountains—or at least to till a patch of soil.

I was picking sugar snap peas one Saturday morning in June when my wife, Valerie, returned from errands with our two young daughters in tow. After unharnessing the girls from their car seats, she pointed down to the garden.

"Look," she announced, "Daddy's picking peas."

In an instant, Alyssa—still at the tender age of 4—tossed her box of cookies to the ground like last year's toy on Christmas morning.

"Peas!" she squealed, as she ran full tilt to the garden.

Tiny fingers wrapped around a pod at the bottom of the trellis, and with a quick tug it was hers. In seconds, she was stuffing fresh green vegetables into her mouth and reaching for more.

I grabbed a pea pod for my own taste buds. It was crisp and sweet, with a delicate earthy finish—too perfect to resist another. So Alyssa and I both snacked as we plucked the remaining peas— one for the bowl, one for the mouth...two for the bowl, another for the mouth.

Fat bumblebees bury their faces in squash blossoms.

The choice my daughter made that morning—to toss aside her box of packaged, store-bought cookies for a fistful of organically grown snap peas—stirred my heart. Months earlier she and I had planted the seeds—presoaked overnight and dusted with natural soil organisms. The seeds sprouted like little green corkscrews, then developed leaves and clung with threadlike tendrils to the trellis I had built. We watched as the vines climbed higher and blossomed, finally bearing crisp, green pods. The moment came full circle, as we picked and ate our perfect harvest, fresh from the vine.

There was no price tag. No bar code. No plastic bag or carbon footprint. No fossil-fuel–consuming drive to the grocery store. And most important, there were no worries about environmental damage or toxic chemicals. The peas were flawless and 100 percent guilt-free.

As our daughters grew older, Valerie and I strengthened our commitment to organic growing, knowing we were producing healthy food for our children and teaching them life-long lessons. We witnessed paper wasps descend and fly away with garden pests. We watched ladybugs arrive en masse to devour aphids on a Stanley prune plum tree. We handled snakes and toads and worms and giant praying mantises. The backyard became not just a place to grow vegetables but also a thriving ecosystem buzzing with life. Carolina wrens scolded noisily from an overgrown apple tree. Fat bumblebees buried their faces in yellow squash blossoms. And emerald dragonflies buzzed along the pepper plants like tiny Black Hawk helicopters.

Every plant and animal—from microscopic soil bacteria to bug-eating birds—played a role not only in creating guilt-free peas but also broccoli, pole beans, tomatoes, Swiss chard, jalapeños, Asian greens, fingerling potatoes, and dozens of other vegetables.

When the harvest came, our bounty was often greater than the refrigerator could hold. So we shared our crops with neighbors, coworkers, family, and friends—all of whom were eager to receive fresh organic tomatoes, cucumbers, zucchini, and greens.

But that wasn't enough. I knew I wanted to be a farmer.

My earliest childhood memories were of visiting my Uncle Joe's dairy farm in southern New Hampshire, watching the cows cross the street on a warm summer evening. As a teenager staying on my brother-in-law's family farm in Le Sueur, Minnesota, I marveled at the rows of corn that seemed to reach the horizon in every direction. Later, while working as a produce manager unpacking crates of lettuce, I wondered about the farmers who grew such round and perfect heads.

So when our daughters left home for college, Valerie and I turned our food-growing hobby into a small-scale agricultural business, never leaving our suburban backyard in southern New England. I shaped gardens into efficient raised beds, equipped with hoops for floating row covers that stretched the season and guarded crops from insects. My wife added beehives to the operation. We built a chicken coop and ordered layer hens. I converted lawn to make room for more raised beds, then cleared space to plant berry bushes and fruit trees.

Today, we operate a small-scale farming business on our 1-acre (0.4 ha) suburban plot, selling vegetables to neighbors, coworkers, and friends. We use intensive gardening techniques borrowed from such vegetable-growing gurus as John Jeavons, Eliot Coleman,

Why Go Organic?

The best reason for growing organic is that it's simply healthier for everyone involved—you, your family, farm workers, and customers—thanks to a lack of chemicals and an abundance of nutrients.

It's also economical. While organic production takes more time, effort, and understanding at the onset, sustainable practices can save money in the long run by recycling farm waste and avoiding expensive inputs. And if you're a USDA certified organic farmer, your crops will command a higher price.

The Organic Center, a Washington–based research and education group, lists twelve reasons to go organic. Among them are the following:

- Reduces your risk to harmful synthetic pesticides.

- Avoids unknown genetically engineered food risks.

- Decreases your intake of unnecessary hormones and antibiotics.

- Preserves local crop varieties for future generations.

- Improves water quality and the safety of drinking water. (Chemical pesticides and fertilizers can flow with rainwater into aquifers and rivers.)

- Maintains healthy soil. (Organic practices restore nitrogen and nutrients and help sequester carbon to help fight global warming.)

Organic Farming: Its Origins and Evolution

Before the advent of chemical pesticides and synthetic fertilizers, farming was essentially an "organic" operation. Growers kept livestock and recycled their animal wastes to fertilize fields. They conserved their soil, rotated crops, and practiced long-term, sustainable agriculture. However, that all changed dramatically after World War II. Farmers moved away from raising mixed crops and livestock. Thanks to rising market prices, larger machinery, and high-yield varieties, they turned toward large-scale production of grain. Without animal manures, farmers opted for chemical fertilizers. The North American factories once used to produce wartime explosives began churning out synthetic high-nitrogen fertilizers, and the nerve agents developed and stockpiled for warfare led to the production of organophosphate pesticides.

The organic movement didn't spring up to confront the use of chemicals overnight. It grew from studies and observations around the world by prominent people on both sides of the organic debate who laid the groundwork.

Father of Synthetic Fertilizers

The shift away from natural growing took root in the mid-1800s, when German scientist Justus von Liebig and agronomist Carl Sprengel independently preached the "Law of the Minimum": that plant growth was limited by whatever single nutrient was lacking in the soil. The science is still used today and is a key part of soil testing and plant health. But the reductionist thinking led to an over-reliance on synthetic nitrogen, phosphorus, and potassium (N-P-K) to feed plants and boost production. For that reason, Liebig is sometimes referred to as the father of the fertilizer industry.

Advocate for Spiritual Growing

As chemical fertilizers gained popularity in Europe, scientist, clairvoyant, and philosopher Rudolf Steiner lectured on the importance of balancing animals, plants, and soil in a more holistic approach to farming. His talks in 1924 led to the publication of Spiritual Foundations for the Renewal of Agriculture. Steiner, who advocated an understanding of the spiritual connection to farming and nature, pioneered the biodynamic farming movement.

Observer from India

British botanist Sir Albert Howard, an adviser and farming researcher in India from the early 1900s to 1931, observed the healthy relationships among India's farmers, their animals, and the crops they grew. His books, *An Agricultural Testament* and *The Soil and Health,* noted the importance of the farm as an ecosystem and the recycling of compost and animal waste to

The shift away from natural growing took root in the mid-1800s.

J. I. Rodale is the first in the U.S. to describe natural, sustainable agriculture as "organic" in 1940.

Rudolf Steiner pioneered the biodynamic farming movement in 1924.

British botanist Sir Albert Howard's *An Agricultural Testament* is published in 1940.

feed soil organisms and build humus. Howard is often described as the founder of the organic movement, although it wasn't called organic at the time.

Spreading Howard's Word

Sir Howard inspired the work of British scientist Lady Eve Balfour, who wrote *The Living Soil* in 1943. Howard's teachings also helped spark the research of American J. I. Rodale, who was the first in the United States to use the word *organic* to describe sustainable and natural agriculture in an article in *Fact Digest* in 1940. Rodale later launched *Organic Gardening* magazine and established the Rodale Institute in Pennsylvania, which became a prime source of information for the organic farming movement.

Pesticide Whistle-Blower

Scientist and author Rachel Carson sounded the alarm on chemical pesticides in 1962 with her groundbreaking book *Silent Spring,* which illuminated the dangers of DDT and acted as a call to action against the widespread use of chemical pesticides. Environmental concerns about DDT helped turn consumers to the fledgling organic farming movement, whose followers grew steadily in the 1960s and 1970s, increasing the demand for naturally produced pesticide-free food.

Apples and Alar

In 1989, a *60 Minutes* report on research linking the apple growth-regulator Alar to increased cancer risk sparked a backlash against chemicals in food. Actress Meryl Streep became a famous face for the Alar alert, and growers looked for ways to salvage the reputation of apples, some turning to the label "organic." At the time, there were only regional organic certification groups with varying standards for defining the term organic. The industry needed national standards.

Organic Foods Production Act

In 1990 Congress approved the Organic Foods Production Act "to establish national standards governing the marketing of certain agricultural products as organically produced products; to assure consumers that organically produced products meet a consistent standard; and to facilitate interstate commerce in fresh and processed food that is organically produced." The program, implemented in 2002, uses independent certifiers to ensure that farmers who call their products organic adhere to sustainable agricultural practices and avoid using chemical pesticides and fertilizers derived from synthetic means or human sewage. Certified organic farmers are also prohibited from using genetically engineered organisms.

Organic farming followers grow steadily in the 1960s and 1970s.

Congress approves the Organic Foods Production Act in 1990; law takes effect in 2002.

Scientist Rachel Carson raises awareness of DDT with publication of *Silent Spring* in 1962.

Organic industry grows rapidly in new millennium with global sales surpassing $63 billion in 2013.

Dick Raymond, Ed Smith, and the late Bob Thomson. We became members of the Northeast Organic Farming Association, and we attend workshops to hone our growing skills. To supplement our income, we keep a small group of subscribers supplied with fresh groceries all season long, using a Community-Supported Agriculture (CSA) business model.

It's not always easy, but it's enjoyable, profitable, and 100 percent possible. You can do it, too. This book will get you started!

- **Chapter 1** will help you assess the land you live on to determine whether it meets your plans for organic growing. If you need to lease more land or buy an organic farm, you'll learn what to look for—factoring in the climate, soil, water, and location.

- **Chapter 2** can assist you in selecting the right implements for digging, weeding, seeding, and raking—starting slowly with quality hand tools and determining whether you need to invest in power equipment.

- **Chapter 3** will help you understand the importance of organic soil. Cultivating the relationship between plants and underground microbes is the key to successful organic agriculture. You'll also find information on nutrients, testing your soil, making compost, and planting cover crops.

- **Chapter 4** shows you how to develop a farm plan with an organized schedule for growing successive crops all season long. Whether you plan to farm for a growing family or start a small-scale agribusiness, you'll learn to start your own seeds, schedule crops for production, maximize growing space, and prevent incursion from weeds and pests.

- **Chapter 5** covers the best vegetable and herb crops for organic production, with advice on varieties, heirlooms, seed starting, soil requirements, and harvesting.

- **Chapter 6** suggests ways to branch out with fruit trees and berries, with advice on choosing varieties, preparing the soil, and dealing with pests and diseases naturally.

- **Chapter 7** describes the basics of adding chickens to the farm— for wholesome organic eggs or pastured meat. There are suggestions on the right breeds, tips for raising a flock from day-old chicks, and protecting the birds from predators.

- **Chapter 8** will help you add to your organic system with bees or small livestock. Learn the basics involved with keeping honeybees, rabbits, and dairy goats.

- **Chapter 9** includes details on marketing your farm-fresh products. Learn what you need to know to sell your goods at restaurants, farmers' markets, or through a CSA program.

Grow healthy, delicious scallions and other fresh vegetables without the use of harmful chemicals; learn how in chapter 5.

- **Chapter 10** covers the things to think about if you want to make a lasting business out of your hobby. Here's where you'll learn the basics of crafting a mission statement, setting goals, and creating a budget. You'll also find information to help you weigh the pros and cons of becoming USDA certified organic.

- Finally, in the **"Resources"** section at the back of the book, you'll find information on soil testing, organic certification, local organic farming organizations, publications, and more.

Whether you're a large-scale gardener or a fledgling organic farmer looking to share the earth's bounty and earn a second income in the process, it's time to get started. You'll eat safer, more nutritious food; learn about the bond between soil organisms and your plants; discover new ways to join the local food movement; and help save the planet—1 acre at a time.

Dig in.

The Good Earth

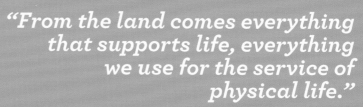

1

"From the land comes everything that supports life, everything we use for the service of physical life."

—Henry Ford

Whatever your dream farm encompasses—peach grove, poultry farm, or mixed-vegetable market garden—nothing is more important than its location. The right property forms the foundation for everything else you do on a farm. You can tear down small buildings; you can clear and plow fields; you can grade roads and string fences. But features such as clean well water, shelter from prevailing winds, soil that drains well, a southern exposure, and proximity to a population center with potential customers are all attributes that can't be altered once you settle in and begin farming. So before you even pick up a shovel or place your first seed order, you need to do some research.

Assessing Your Own Property

Dreaming of an organic farm on land you already own? You may be able to make it happen. Just consider some basic information about the property. You might find that the land is perfectly suited to an organic farm, you may need to make a few adjustments to make it work, or you may decide that you're better off purchasing more suitable land. Start your assessment by asking the same questions you might ponder when scouting a new property:

Issues beyond the Land

It's true that land issues are the primary concerns when looking for an organic farm property. But there are other things to think about, too—the farmhouse, the community, the neighborhood, and your personal preferences. Some things to consider:

- Is the farmhouse up to your standards? Many farm homes are older and may need upgrades to heating, electricity, plumbing, and structure. The cost of fixes in the first few years could add significantly to your overall price.

- Is the house's foundation strong? Does the basement leak? Is the roof in good shape? Hire a good home inspector.

- Are the outbuildings in need of repair? A barn, shed, greenhouse, and any other structures that need work will add to your financial burden.

- Is the property fenced and is the fencing adequate? If you're keeping livestock on a large piece of land, fencing can become a big expense.

- Does the climate suit you? If it's snowy and you're located at the end of a private road, your most valuable piece of equipment may be a four-wheel-drive truck with a plow.

- Beyond irrigation needs, how is the water? Is it potable or are there too many minerals? Have the wells been tainted by fertilizer or chemical runoff? Check municipal water records.

- Is the community to your liking? Consider your proximity to neighbors, shopping, good schools, hospital, and your place of worship.

- Is there enough level land with a sunny exposure to provide growing space for all of your vegetables?

- Is the earth beneath your lawn, backyard, or field deep and loamy? Or is it lean, thin, compacted, and riddled with rocks?

- What is the climate? Is the season long enough to grow the crops you want to produce? Is the seasonal rainfall adequate?

- If you need to irrigate, will there be ample water, either from a public supply or a pond and private wells?

- If you want to farm commercially, are there enough customers living within your geographical area? When you answer this question, consider whether you will farm for yourself, sell to restaurants and farmers' markets, run a farm stand, or sell CSA (Community-Supported Agriculture) shares to customers who buy a season's worth of vegetables and to pick up their goods each week.

- Is the property zoned for agriculture? Can you keep chickens, goats, cows, or other animals? If so, how many can you have?

- If you want to keep animals, is there enough land to grow hay? Or are there nearby growers who can supply it?

Your land may have everything you need for an organic hobby farm. Check zoning regulations to find out if you can build a chicken coop and raise layer hens.

- What kinds of shelters are allowed on your land? Can you have a barn, stable, chicken coop, greenhouse, and storage shed for tools and equipment?

- Is your land safe for growing fruits and vegetables? Has it been contaminated with lead, heavy metals, or other pollutants that will make it difficult to grow and market food?

- Does it border a farm that uses pesticides? If you want to be certified organic, you will need to create buffer areas that stop pesticide drift.

Chances are you already know the answer to some of these questions, but it pays to look at your own property with the same critical eye that you would use on other land. The conditions could be enough to warrant a move.

Assessing New Property

One sure way to get information on a piece of farm property is to interview previous owners, who should be able to tell you everything from the approximate date of the first frost to the spot-on location of each submerged boulder in the fields.

Organic farmer John Mitchell, owner of Heirloom Harvest CSA in Westborough, Massachusetts, says there is no substitute for the information you can glean from a farmer who has worked the land before you. Mitchell leases his

Is the Acreage Right?

This is more a question about you and your ambitions than the actual size of the property. Many new farmers find that too much land turns into too much work. Maybe you should look for 5 acres (2 ha) instead of 20 (8 ha).

farmland from a church parish and says he is fortunate that the previous lessee was also a certified organic farmer. To maintain their certification, organic farmers must keep records of crop rotations, soil tests, and types of fertilizers used, along with information on weeds, pests, and diseases and how each problem encountered during the growing season was solved organically. Such meticulous records are like having a two-way crystal ball that looks into a property's past and can help you divine its future.

Even without records, the previous owner should have answers to the following critical questions.

How Was the Land Farmed?

If the land wasn't farmed organically and you want to market your produce as organic, you'll need a three-year transitional period before you can legally advertise your goods as USDA

When looking at a new piece of land, find out how it was farmed. Repeated use of machinery could mean compacted soils. If chemicals were used to control insects, weeds, or diseases, you will have to farm organically for three years before you can legally call your goods organic.

Creative Ways to Find Farmland

Want to farm but can't afford to buy land? One solution may be to hook up with a farmer who wants to sell or lease his land with a creative financing option. Some retiring farmers will offer flexible terms and financing options just to ensure that their properties are passed on to others who want to continue the farm.

Land Link programs in many states look to pair young and old farmers for just such arrangements. In many instances, a retiring farmer can work with a would-be successor and learn about the operation as the transfer of ownership progresses. (More information is available at the International Farm Transition Network website: www.farmtransition.org)

Another option is buying land under an agricultural easement. An agricultural easement is a legal restriction voluntarily placed on a piece of farmland or ranch land ensuring that it will never be used for anything but farming. No strip malls, no housing developments.

Sometimes a farmer will have donated or sold those rights to a conservation land trust. The benefits for farmers are the immediate tax advantages and the long-term comfort of knowing the property will never become a shopping center. The benefit for the community is the preservation of open space.

New farmers can benefit because once an agricultural easement is in place, the property can only be resold at its agricultural value. That means new farmers can occasionally pick up an old farm or ranch in a pricey location for well below the fair market value. The drawback is there are restrictions on the types of buildings allowed on the property—possibly including the house you want to live in. And, if you decide to resell the property, you'll be restricted to the agricultural value, too. Buyers generally have to bid on a property and provide a farm plan that demonstrates continued productive use of the land in agriculture.

Canadian growers who need a place to farm can meet up with landowners to accommodate them via Landshare Canada, www.landsharecanada.com, an online agri-networking website that "brings together people who have a passion for home-grown food." Users start by logging in and creating a profile. Then they post listings as either growers in a specific area looking for land to farm or as landowners in a specific area who will allow farming on their properties in exchange for maybe a cut of the harvest. The website also has a tutorial, sample legal agreements, blog postings, and maps showing where landowners and farmers want to get growing.

To find out about opportunities in your area, contact farm advocates and land conservation groups in local communities. (For additional information, see "Resources" on page 356.)

Certified Organic. In addition, land that has been farmed with conventional practices may need cover cropping and repeated applications of manure or compost to eliminate weeds and regenerate the soil microbes that play such a crucial role in sustainable agriculture.

Has Anyone Plowed the Land Before?

A field that has been used for pasture may not have been plowed or renovated. If so, boulders or even ledge may be lurking just beneath the surface. One indication that rocks have already been cleared is the presence of stone walls. The picturesque walls around New England that date back to colonial times were built more out of necessity than for aesthetics. Rocks in the field were lugged to the perimeter and stacked in rows to create those now-scenic walls.

How Frequently Was the Land Tilled?

There are good and bad answers to this question. Land that has been cleared, plowed, and tilled should be relatively free of rocks. But a long history of being worked by farm machinery could mean compacted soil, erosion, and the loss of nutrients and humus. The ideal field is one that has been used in a regular crop rotation, with minimal tilling performed to incorporate cover crops and organic matter into the topsoil.

Are Any Diseases Lingering in the Soil?

Knowing about past diseases will help you prevent them in the future—or could give you pause about whether the land is right for your purpose. For example, one of the worst diseases for a vegetable farmer is clubroot, which affects cabbage, broccoli, cauliflower, mustards, and other Brassicas, severely stunting their growth and even wiping out the crop. Once clubroot is in the soil, its spores can live for a decade. And if it's in one field, a farmer must clean and disinfect equipment before working another field to prevent the disease from spreading. The bottom line, says Mitchell, "You don't want soil that has been contaminated with clubroot."

What Kinds of Weeds Are Present?

Perennial weeds like bindweed, quackgrass, and yellow nutsedge are difficult to eradicate because broken bits of roots and plant material can sprout new weeds. One of the most pernicious

Weeds that go to flower and seed near your vegetable fields can create a nuisance for years to come. But a well-planned meadow garden somewhere on the property will increase diversity and encourage beneficial insects.

of annual weeds, Galinsoga, can spread thousands of seeds that germinate, grow to maturity, and release more seeds in a matter of weeks. If the previous farmer has controlled the weed-seed banks (eliminated weeds before they went to seed), crops will be easier to grow. And if weed-killing cover crops were planted and the fields were mowed around the perimeter to prevent incursions into growing areas, fewer unwanted seeds will be waiting to be raked to the surface where they can germinate.

Did the Previous Owner Sell the Topsoil?

I once looked at farm property that had been stripped of its topsoil. The farmer or previous owner had sold the loam from his unused fields to a developer, leaving mostly sandy subsoil with little organic matter. A farm that is missing much of its topsoil would create a serious setback for future farmers, potentially requiring seasons of cover cropping and tons of manure, compost, or other organic matter. Finding out general information on a property is helpful, but a site inspection should reveal more details. (For more information, see page 19.)

After you've received all the information you can get from the previous landowner, public records, and local historians, you can look for more answers online.

Gathering Additional Information

The previous sections illustrate just a few of the questions that need answers before you invest in an agricultural business. Some of the information, as discussed, is easy to glean from knowledge of your own property or the previous owner's knowledge of his or her property. But thankfully, when you can't get the answers you need, you'll find plenty of reliable sources for information. If the former property owner is no longer available, you may be able to get information from a real-estate agent, local historian, community's assessor office, or neighboring property owners. If the property was farmed before, state or county agricultural officials should be able to help you learn about its past.

Going Online

What do you do if you can't find the information you need from farmers, neighbors, and others? You can search online.

For detailed technical reports on a property's geology, land slope, drainage, average temperature, and rainfall, along with such valuable information as its suitability for crops, consult the national database that has been more than a century in the making. The Natural Resources Conservation Service (NRCS) keeps detailed records on the physical characteristics of land across the United States, updating the information as new reports and surveys are completed. (See "Soil Surveys" on page 24 for information on how to access the database.)

According to David Hvizdak, now retired soil scientist for the NRCS, soil surveys began in the late 19th century. Until recently, information was collected and recorded in hardbound reports that were used by farmers when buying or leasing land. Today, the information is posted online, where it can

Soil scientists take samples and make observances, then update the information online after others at the NRCS have approved it.

Soil Surveys

The NRCS (Natural Resources Conservation Service) is actively surveying land—visiting sites to view the terrain, observing and documenting soil properties and morphology, and then posting the data online periodically upon undergoing a quality-assurance process by the agency. That makes their online database the most complete and up-to-date source of land information available.

To get started, visit the NRCS Web Soil Survey home page at websoilsurvey.nrcs.usda.gov, and follow instructions on the page. If you're more adept at maneuvering a tractor than a computer mouse, the home page contains a help link and instructions to walk you through the basic steps.

With a little practice you can learn to home in on a particular area that is displayed in a satellite image map on the screen. With tools at the top of the map, you can draw a rectangle or polygon around a block of land and designate that specific property as an "Area of Interest." Once the land is highlighted on the map, click on tabs to access reports about the property. After completing the exercise and collecting pertinent information, you can save it for future use.

be updated by scientists and used by anyone with a computer and access to the Internet. The depth and accuracy of information has improved with recent onsite land surveys.

For farmers, the information will help determine things such as mean temperature, rainfall, length of the growing season, the soil's capacity to drain or retain water, the depth of the topsoil and subsoil, and whether the land is stony, hilly, or prone to flooding. The survey can help steer you away from land that would be better left unfarmed.

Canada's National Soil Database

Canadian growers have an online resource for soil information recorded by the Canadian Soil Information Service, CanSIS for short, which operates under the umbrella of Agriculture and Agri-Food Canada. To find soil data, go to sis.agr.gc.ca/cansis/ and follow the quick link to Soil Maps, then Soils of Canada link. You can view a map, click on the information in a text box in the left margin, and find such attributes as the land's drainage, surface material (down to 1 yd. [1 m]), and the amount of organic and mineral content.

Other links on the page will take you to print maps that have been scanned and uploaded to the website. These maps have areas that are color coded and delineated with numbers, with corresponding information contained in documents that you can download.

Testing the Soil

Ask for the farmer's record of soil tests to determine whether amendments are needed to increase soil nutrients, improve tilth and structure, or alter the pH (acidity/alkalinity). Also take your own soil samples around the property and send them to a reliable soil-testing laboratory for a thorough analysis. For your produce to be legally certified as organic, the land it came from must have been free from prohibited chemicals and non-organic produce for three years.

Keeping this in mind, it's worth checking soil to make sure the land has not been damaged by recent years of chemical farming. You might find residual pesticides, poor quantity of organic material in the soil, erosion, or compacted soil. If the previous farmer used an integrated pest management (IPM) program on the property, the impact should be less severe. In an IPM program, farmers monitor their crops and use physical traps to determine an optimum time for action. Physical barriers, natural pests, and biological controls are used before spraying pesticides. It's a practice that is more environmentally conscious than repeated chemical sprayings throughout the year.

These considerations are also important for farmers who are leasing land. Using organic amendments to increase soil fertility can be an expensive investment, often required over a period of years. You should only make such a costly soil investment if you can work out an acceptable long-term lease for the property. Otherwise, all of your hard work might just benefit the next person who rents the property.

Looking for Water

One of the most important considerations in locating viable farm property is making sure there's enough water to keep the crops irrigated and farm animals happy and healthy. It's a bonus if the property has a pond or river within its boundaries. If it doesn't have any obvious

A soil scientist conducts a soil test on a farmer's field in Virginia.

water sources, find out if there is a well, and ask whether the water supply is reliable year round. Talk to someone in the community's conservation district or water department about the long-term health of the aquifer. If the area is drought-prone, there may be restrictions on use. Make sure there's plenty of water available during the summer, when you'll need it most.

It's a good idea to have a second option for water, in case one source fails. If your farm is supplied with public water, you may want to install a private well for irrigating crops. Many communities enforce public watering bans during the summer months, and other communities charge high fees for water use. Check with the cooperative extension office or with other farmers in the area for a reputable well installer who can give you more information and cost estimates.

In addition to looking at groundwater sources, check the rainfall totals for your region. (For more information on climate conditions, see opposite.)

Just as inadequate water can be problematic, too much water can harm the farm—especially when you want to get started with the season in the spring. A low-lying field that's covered with puddles until June won't be ready to till, rake out, and prepare for planting until

A landowner and conservationist with the NRCS uses a soil probe to take a soil sample on a farm near Iowa City, Iowa. Information from around the country is available at the agency's website; search for Web Soil Survey.

late in the growing season, wasting valuable time. So keep in mind that while land near a rising river may be fertile and flat and provide a steady source of water, a rainy spring and early summer could shorten the growing season.

Finding the Right Climate

Climate counts, too. Locating your property's position on a USDA or Agriculture Canada Plant Hardiness Zone Map (see page 384 and inside back cover) will tell you what the regional climate will allow you to grow (without the aid of season-extending greenhouses, high tunnels, or other weather protection).

Most seed and plant companies make recommendations based on the latest spring frost dates and average minimum winter temperatures listed on government climate maps. Before trying to grow okra, sweet potatoes, or melons—all of which require hot summers and a long growing season—consult the number of frost-free days to determine whether your season is long enough for vegetables to reach maturity. The average minimum winter temperature for your region will tell you whether cold-sensitive plants such as artichokes, rosemary, or lemongrass can survive outdoors in the winter.

For those who are buying new property, the maps can help you choose an area suitable for the farm you envision. For instance, if you grew up on a farm in northwestern Pennsylvania, which is located in Zone 6, the map will show you that the Oklahoma Panhandle has close to the same growing season, even though it's located much farther south. Many factors determine climate, which is why it's hard to predict without the maps. For example, land along the Pacific Coast in British Columbia can have the same hardiness zone as some areas of Florida. That's because land near the

Dealing with Lead Contamination

Organic farmer John Mitchell, owner of Heirloom Harvest CSA in Westborough, Massachusetts, also recommends testing the soil for the possibility of lead or other potentially harmful materials. Lead and other soil contaminants aren't confined to urban areas. Land that was used for agriculture before the EPA began banning the most dangerous agricultural chemicals could still be contaminated with toxic materials. Some areas that were used for apple orchards in the early to mid-1900s, for example, could have been repeatedly exposed to lead arsenate, which was widely used as a fruit-tree pesticide.

Lead remains in the soil for decades and can be harmful in several ways. It's most likely to contaminate the hands of anyone working in the soil, but it can also be inhaled in dirt and dust that's kicked up by tools or machinery. Although leafy vegetables can accumulate some amounts of lead, fruiting crops cannot. The most likely way to actually ingest lead is by eating root crops to which small amounts of contaminated soil have clung.

Most university labs will check for lead content as part of a routine soil test. A natural level of lead in soil is roughly 7 to 20 parts per million (ppm) (7 to 20 mg/kg), but some urban areas with lead contamination may contain more than 1,000 ppm (1,000 mg/kg), most of it in the top few inches. Because lead isn't easily transmitted from roots to leaves and fruit, you can still grow garden crops in soil with a concentration of less than 300 ppm (300 mg/kg).

You can make soil safer for growing by adding phosphorous-rich compost or other organic material that will keep plants from taking in lead. Ensuring that soil pH is kept to at least 6.5 or higher will help keep vegetable crops safe.

Before investing in a property for vegetable farming, it would certainly be worth the time to get a lead test, especially in an urban area or on a site where an old apple orchard once stood.

ocean or another large body of water can stay warmer into the winter. It can also take longer to warm up in the spring, since onshore winds affect temperatures. Elevation also plays a role in temperature, although it's not always what you think it will be. Read on to find out how the rise and fall of your land may create microclimates.

Understanding Microclimates

Not only will you have to think about the general climate of the area in which you'll farm, you'll also need to understand and account for the effects of microclimates (climates that differ from those of the surrounding areas). Features of the land are responsible for numerous microclimate changes. For example, sometimes lowlands can collect and hold cold air, creating early and late frosts. This is known as lowland chill.

Have you ever stood at the top of a mountain and looked down at fog blanketing the valley below? It's an illustration of simple physics: Hot air rises and cold air sinks. And fog in the valley forms because water vapor condenses in the cold air that settles against the ground. You can sometimes see these effects in early fall, when driving through areas with hills and valleys: Trees in the lowest-lying areas show the first foliage colors. That's because cool air sinks and flows like a slow-moving river downhill, settling into valleys—especially on calm,

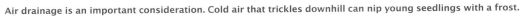

Air drainage is an important consideration. Cold air that trickles downhill can nip young seedlings with a frost.

The Effects of Slope

The slope of a property can also enhance the sun's power. Land that is slightly graded toward the south and southwest catches more direct rays from the sunlight, which warms the earth faster, allowing an earlier planting time. Too much slope, however, will create problems. At a grade of 3 percent or more, some drip irrigation systems don't perform at their optimum efficiency. And at grades of 5 percent or more, water runoff can be fast enough to cause soil erosion.

Take note of the land's slope when planning your beds. Sunlight collects and warms land that has a grade facing south. But erosion may be a problem if the grade is too steep. Slow erosion by planting in rows that run perpendicular to the slope.

clear nights. Understanding this downward flow of cool air, called air drainage, will help you take advantage of microclimates caused by physical attributes of the land.

Farmers who are tending fields at the bottom of a valley might find their plants nipped by an unexpected frost, while plants growing at a higher elevation remain unharmed. Even if you're not at the bottom of a hill, buildings, hedges, a row of trees, or other physical obstructions can block the streams of cold air. This can be a problem for vegetable gardeners who are stretching the season in spring or fall. Fruit growers, who tend orchards on sloping land, can suffer crop losses when trees on south-facing slopes bloom early after a streak of sunny days, then get hit by a late spring frost. The cold can damage blossoms and reduce fruit set.

Prime Farmland

The NRCS (Natural Resources Conservation Service) classifies prime farmland as "land that has the best combination of physical and chemical characteristics for producing food, feed, forage, fiber, and oilseed crops. It has the right soil properties, growing season, and moisture supply needed to produce sustained high yields of crops in an economic manner if it is treated and managed according to acceptable farming methods."

Land that has been classified as prime farmland must meet the following conditions:

- It has an adequate and dependable water supply from precipitation or irrigation.
- It has a favorable temperature and growing season.
- It has an acceptable soil pH: (acidity or alkalinity), acceptable content of salt or sodium, and few or no rocks.
- It's permeable to water and air.
- The land is not excessively eroded or saturated with water for long periods of time, and it doesn't flood during the growing season.
- The land can't be built up in an urban area.

Air drainage isn't just a desirable thing in fall and spring. During the summer, a good flow of air between rows of tomatoes will help deter fungal diseases, which multiply under warm, moist conditions and stagnant air.

If you're buying a piece of property in a northern climate, it pays to make sure the topography provides ample air drainage. If you already own the land and want to take advantage of air drainage, look around your property for obstacles that prohibit airflow. The best time to detect such an impediment is at dawn on a morning when ground fog is apparent. Go to the highest point on your property and take note of spots where the fog collects. Then walk through the fog banks looking for barriers. If there are no obstacles and the fog is moving, make a mental note of its path so you can avoid creating future blockades.

Air drainage is important in the summertime as well as during cold weather. Tomato vines can succumb to midsummer fungal diseases if they aren't given enough space between rows. Remember, those tiny transplants will eventually grow into big, bushy plants.

Planting Trees for Energy Savings

Windbreaks or shelterbelts that separate fields can help control erosion by interrupting and reducing the wind speed. In small growing areas, trees should be kept far enough from the growing areas to prevent unwanted shade. Carefully planted windbreaks around your home and buildings can help cut energy costs and create a more comfortable living space, too.

One energy-saving landscape design is to plant deciduous trees on the south side of your home and evergreens on the north. In summer, when the deciduous trees are covered with leaves, they will shade your home from the hot sun. And during the winter months, when you want the sun to shine in and warm your home, the leafless trees will let the sun through. Meanwhile, the thick evergreens on the north side of your home will keep their foliage all winter long, blocking the icy gales.

When you find an obstacle to air drainage, remove it, create a larger area for the air to collect, or create a different route for the cold air to flow downhill.

Taming the Wind

Airflow is good, but too much can be a problem. A large area of flat land without windbreaks can lead to erosion, which is one reason why so much valuable farmland has disappeared. Soil that's left unplanted or that has been recently plowed or tilled can become airborne under dry, windy conditions. And most of the material that gets picked up and carried away is organic matter or fine silt, both of which help conserve moisture and retain nutrients in the soil.

In most of the country, the prevailing wind blows west to east, but storms and other weather patterns can temporarily alter wind direction. Winter winds tend to blow from the northwest, and summer winds from the southwest. Consider the prevailing directions when scouting out a property. To avoid wind erosion, choose a farm site that receives some shelter from the prevailing winds. A ridge, hill, or line of trees can slow down the steady wind. If you've already chosen or purchased your site, plant trees and shrubs yourself to disrupt the continuous flow of wind. Just beware of wind funnels, which can occur when prevailing winds are channeled between hills, ridges, buildings, or wooded areas.

Windbreaks can also come in handy if properties adjacent to yours are or will be used for non-organic farming. To be certified as organic with non-organic farming next door, you'll need to create buffers between your farming operation and any property that is sprayed with chemicals. You could, for example, plant a windbreak or create a roadway around the perimeter. Keeping a row of trees, some tall grasses, or a meadow between you and your neighbor are a few other ways of creating buffers that will minimize chemical drift. Talk with an organic certifier to generate more ideas.

2

"One $20 digging bar will save a dozen $30 shovels."

—Installer at Mr. Fence,
Bellingham, Massachusetts

Since the very first farmers planted grains in the Fertile Crescent 10,000 to 12,000 years ago, mankind has witnessed an evolution in agricultural hardware. Stones were lashed to sticks to make primitive digging tools. Metal was forged into blades for plowing and slicing. Larger equipment harnessed the brawn of horses and oxen. Eventually, the internal combustion engine brought the power of fossil fuels into the farming equation. Today, you'll find far more specialized tools in the world than there are farmers to use them.

Choosing and Using the Right Tool

From clearing land and starting seeds to harvesting, cleaning, and packing produce, the tool and farm-equipment trades have responded to every imaginable task with an array of hand tools, better hand tools, machines that make the hand tools seem archaic, and newer machines that render the old machines obsolete.

Beyond the tools are structures—greenhouses, cold frames, hoop houses, root cellars, packing sheds, raised beds, trellises, henhouses, barns, pole barns, stables, and storage sheds (to name more than a few). There are also myriad mechanic's tools for repairing farm machinery as well as carpentry tools for building, fixing, and modifying those many structures just mentioned.

There's a balancing act to follow when buying tools for a small-scale organic farm. It's not complicated, but it meanders delicately between two very sensible rules:

1. Always use the right tool for the job.

2. Never buy a tool that you don't need.

Small hand weeders and forks are helpful tools for small-scale growers, complementing the array of shovels, rakes, and full-sized hoes used for larger jobs.

Organics and Hand Tools: Peas in a Pod

A synergy forms between manual tools and organic farmers. Hand rakes and shovels don't consume fossil fuels or pollute the air, and they don't leak gasoline, diesel fuel, oil, or hydraulic fluid onto the soil. Once you cradle a well-made digging fork or grub hoe in a calloused palm, it becomes an extension of your own arms, hands, body, and mind. I feel a kinship with my hand tools every time they help me strip away sod, carry compost from bins to beds, and smooth out soil to prepare for seeding. I'd be lost without them. Try using hand tools before bolting for the heavy-machinery. You won't regret it.

So how do you figure out which tools you really need and which ones you can covet now and buy later?

Start with rule number one: Always use the right tool for the job. Not following this rule will cost you more in the long run. For example, if you use a shovel to pry 90-lb. (40 kg) boulders from a future corn field, the handle is likely to snap off at the fulcrum point, dropping the rock back into its hole with the shovel blade stuck underneath. (Here's where the fence installer would say, "I told you so.") Likewise, you wouldn't use a shovel to unearth potatoes; chances are you'll slice through some tubers and damage the crop. A digging fork is a much better option. The list goes on.

To avoid misusing tools—and possibly breaking them, ruining your crops, or injuring yourself—you will need a well-selected array of high-purpose implements that perform all the basic farming functions. Once you have these starter tools, consider each new piece of specialized equipment on its individual merits, carefully weighing the cost against the amount of time and money it will save.

For small-scale vegetable growers, the basic tasks include digging, moving, and shaping earth; planting; watering; and weeding—with a strong emphasis on digging and weeding. Fortunately, you can perform most of these chores with a reasonable assortment of manual tools. Some are available at hardware stores and mass-market garden centers; others you'll have to order from specialty catalogs. You might be able to craft or copy a few with your own hands.

I consider most of the following hand tools as must-have items in my shed. Together, they do 90 percent of my work. While hand tools are not as fast as machinery, the effort I put into using them gives me my daily exercise. (After all, what's the point of eating healthily if you don't follow it up with a little exercise?)

The Diggers

Shovels, spades, forks, and mattocks do the digging, scooping, and earth-moving chores, shaping the earth for growing and keeping beds in shape for years to come. They are the first tools to reach for when looking to expand your planting area.

Long-Handled Round-Point Shovel

Long-handled round-point shovels are specifically designed for digging and swinging a load of soil from one side of your body to the other without putting too much strain on your back. The spoon-shaped blade holds a heavy mound of soil, and the handle—usually at least 4 ft. (1.2 m) long—allows leverage for flipping the dirt a short distance. The longer handle also helps when digging a deep hole, providing easy reach with minimum stooping. This tool is well suited for planting fruit trees and blueberry bushes, making raised beds, cutting compacted soil, transferring finished compost into the sifter, and filling those holes after transplanting. It's one of the most useful small-farm tools you'll buy.

It's always good to have at least one high-quality shovel on hand. Look for one with a 14-gauge forged-steel blade and a sturdy-but-lightweight handle made of ash or fiberglass. This is the tool for heavy jobs, such as digging through new soil that sometimes involves cutting through compacted earth, rocks, and tree roots. Cheaper shovels with thinner pressed-steel blades can loosen on the handle after continued use in hard or rocky soil. I have one high-quality shovel that is sturdy, dependable, and built to last.

A good shovel, fork, pick mattock, rake, and hoe will cover most of the basic chores in your small farm or garden. Power equipment will be a time–saver for larger properties.

Long-handled shovels are useful for myriad chores—including digging large holes for planting your squash or pumpkin seedlings.

But the good shovel isn't enough. Whenever it's time for digging loose soil or piling compost into a wheelbarrow, I often reach for the cheaper shovel with the thinner blade and lightweight fiberglass handle. The price for these lightweight shovels is usually around $10 to $15, and the handles are strong enough for most tasks.

Wood or fiberglass? Just because a lot of inexpensive shovels have fiberglass handles doesn't mean that fiberglass is lower in quality. Many well-made shovels have handles made of fiberglass, which is actually stronger than wood.

Short-Handled Round-Point Shovel

Short-handled garden shovels have blades that are similar in design but have a smaller shaft (under 2 ft. [60 cm]) with a closed handle in a D or Y (wishbone) shape for easy grasping. As with long-handled shovels, look for heavier-gauge steel, noted by number (14-gauge steel is thicker than 16-gauge). While you might want two or three long-handled shovels around for various jobs, you can almost certainly get by with just one short-handled model. The compact size makes it useful for stepping down and digging straight in front of you with a good grip and firm control. And the short handle makes it handy for loading or unloading soil or compost with a wheelbarrow. While short-handled shovels are handy for working in tighter spaces and small enough to load into a wheelbarrow or garden cart, they require bending or stooping and can be tiring to work with all day.

Easy Does It!

Don't throw away a broken shovel. I have a blade that I saved from a shovel that broke many years ago (before I bought a digging bar). I find it useful for scooping compost out of a wheelbarrow and top-dressing crops. It's also good for stirring compost, perlite, and soil amendments to make potting mixes and for any chore that needs an oversized scoop. A hoe or weeder with a broken handle can be fitted with another wooden handle, or simply cut shorter and turned into a large-bladed hand tool. I have one that I use for hilling potatoes. As I pull up the stalks with one hand, I use the mini-hoe to scoop soil with the other.

Both long- and short-handled shovels can sport specialty blades with serrated edges for cutting through heavy soil and small roots and/or larger steps on the blade's shoulders that add strength and stability when stepping onto the blade.

Garden Spade

Spades differ from shovels in that the handles are almost always shorter (about 3 ft. [1 m] at maximum), and the blades are thinner and flatter. Most spades have straight-edged blades for cutting into soil and digging neat well-shaped holes. They won't carry much dirt on those flat blades, so they're not as versatile as round-point shovels. But spades are indispensable when digging in heavy soil. Strong, flat blades with sharp tips cut

If your budget is tight, you might be able to find some old tools at a garage sale. Stainless steel blades, as seen on this spade, have the benefit of resisting rust.

through sod, penetrate heavy clay, and make quick work of smaller tree roots; shovels are prone to jerking sideways out of your grip in these conditions. (I'm sure you'll come to know the feeling of stomping down on a shovel's treads and having the blade twist under your feet as the handle wrenches out of your grip—after striking a stone or root.)

Spades also do a better job of digging straight lines—for example, when you're cutting a line along the edge of a bed. And if you're using plastic mulch to heat up the soil and block out weeds, the spade's flat blade is perfect for slicing into the soil and tucking in the plastic at the ends and sides.

For long-lasting usefulness, choose a garden spade with a forged-steel head and sturdy steps. A strong handle and comfortable grip are also important. Less-expensive spades with pressed-steel blades don't hold a sharp edge for long, and they can flex and bend in heavy soil. Strong steps allow you to insert the blade straight into the soil and step down, letting the weight of your body do most of the work.

A digging fork with sturdy tines will break up heavy soil. It's also good for digging potatoes—without damaging many tubers.

You'll find numerous variations on the size and shape of a spade's blade, some of them extremely narrow. The trenching spade and similarly shaped rabbiting spade, which was originally designed for digging rabbits from their burrows, has a long, narrow blade with a rounded point. I find it useful in transplanting—digging small holes and holding plants with their root balls intact during a short walk to a new growing spot. It's also good for planting through black plastic mulch without ripping wide holes in the material. And when you need to pull dandelions or other deep-rooted weeds, one well-placed plunge of the blade cuts the taproot (the main root from which smaller roots grow) so deeply that it rarely grows back.

Digging Fork

Shovels and spades will handle most digging chores, but a digging (or garden) fork will help with more specialized tasks, such as lifting and loosening soil, unearthing root crops

Easy Does It!

without damaging the harvest, busting up soil clumps, and mixing compost, manure, or another organic material into a garden bed. Digging forks usually have four sturdy tines, which are easy to push into compacted ground. The spaces between the tines make this tool ideal for slipping past small stones without twisting or turning away.

A digging fork has a shorter handle, and its tines are usually shorter, flatter, thicker, and more closely spaced than those of a pitchfork. Pitchforks, hay forks, and manure forks have thin, gently curved tines designed for scooping and tossing a load of lighter material. (Don't make the mistake of trying to dig into heavy soil with a fork that's made for tossing hay. You'll quickly bend the tines.)

Broadfork

When you want to loosen soil without the use of noisy power equipment, reach for the broadfork. It's an oversized fork with two handles—one on each side—and a wide row of long, sturdy tines in the middle. You push the fork into the ground, step onto the top (between the two handles), and rock the tines deep into the earth (12 in. [30 cm] or more). Then you pull back on the handles an inch or two until "the earth moves." Rock the fork forward a little, pull it out, and step back 6 in. (15 cm), taking care not to tread on the newly loosened soil. Repeat the process and continue until the bed is finished. Tilling with a broadfork is a low-impact way to let air, water, and organic matter penetrate the soil. Unlike a rotary tiller, which can damage soil structure, kill earthworms, and break up fungal hyphae, a broadfork does the job without damaging the soil ecosystem.

Pick Mattock

There are very few digging chores that can't be handled by a determined farmer with a long-handled shovel and a mattock at his hands. Mattocks, which combine two tools in one—most commonly a pick with a blade—have sturdy wooden or reinforced fiberglass handles (3 to 4 ft. [1 to 1.2 m] long) and heavy solid-steel heads (2½ to 6 lb. [1 to 3 kg]). They're terrific for digging new beds, peeling back sod, and taming compacted or rocky soil.

I like the versatility of a pick mattock, which will penetrate hard-packed soil and loosen it with ease. The pick can work around rocks and undermine the roots of a bush or small tree. The blade side of the tool cuts into the earth and pulls it toward the user, much like a deep-digging heavy-duty hoe—perfect for cutting long trenches to lay irrigation or drainage pipes.

The heavy head of a mattock can quickly get your heart pounding if you put too much effort into each swing. But if you concentrate on using your muscles to lift the tool, then let the weight of the mattock do the work on the way down—much like swinging a sledgehammer—it's less exhausting.

Grubbing Hoe

Halfway between the mattock and an ordinary garden hoe is the grubbing (or grub) hoe. It's heavier and thicker than the hoe you might use for scraping weeds and digging furrows (it looks like a tool from medieval times). The thickness of the blade

The trusty pick mattock has a sturdy blade that can cut into heavy soils, dig trenches, break through roots, and loosen compacted, rocky soils.

allows it to hold a strong, sharp edge that digs and moves earth quickly and efficiently—whether cutting through sod, carving trenches, shaping mounds of soil into raised beds, or busting up hard-packed dirt.

I find that I actually dig faster and with less effort when using a grubbing hoe instead of a shovel, although the hoe won't dig as deep a hole. (It also won't load soil into a wheelbarrow or toss it aside.) This tool also levels the soil like almost no other. With shallow chops, followed by a long dragging motion, the heavy blade shaves off bumps and fills holes, quickly leveling the hard-packed space between raised beds.

Grub-hoe heads attach to their heavy-duty 5-ft. (1.5-m) -long tapered handles by eyehole and screw and are therefore interchangeable. You'll find several sizes, starting with 4-in. (10-cm) -wide heads that weigh a little more than 2 lb. (1 kg) and are suited to lighter garden work or narrow trenches. I like a head that's about 6 in. (15 cm) wide and weighs about 3 lb. (1.5 kg), because it cuts a wider path and moves more soil. (I leave the narrow trenching to my mattock blade.)

You'll also find named variations. The Italian grape hoe has an even heavier blade that's as big as 8 in. (20.5 cm) wide. A fork hoe, also called a Canterbury hoe, has three prongs instead of a blade and is well suited to busting up sod or pulling out the heavy roots of perennial weeds. It can also rake potatoes out of a hill without too much damage to the tubers.

The blade of a grubbing hoe is thicker and heavier than the business end of a weeding hoe. It chops into soil and is ideal for creating raised beds.

Digging Bar

The last of my must-have manual digging tools is the often-overlooked and incredibly versatile digging bar. Once you use one for the first time, you'll wonder how you got by without it for so long. At about $25, a digging bar will quickly pay for itself in shovel handles you don't have to replace.

There are several variations, but my favorite is a 6-ft. (1.8-m) -long solid-steel bar, about 1½ in. (4 cm) in diameter, with a pointed tip on one end and a 2-in. (5-cm) -wide chiseling edge on the other. When I run into large rocks with my mattock, grub hoe, or shovel—and there are plenty of big rocks in southeastern New England—I plunge the pointed end of my digging bar around the perimeter of the rock to judge its size. If it's movable, I thrust the chisel end of the bar underneath until it catches a ridge. I then push a piece of wood or another rock close to the boulder to use as a fulcrum. With a quick pry (or maybe a few), most 50- to 100-lb. (22.5- to 45-kg) rocks can be shifted and lifted.

A digging bar is also indispensable for creating pilot holes to pound in stakes, pipes, or posts for trellis netting. In addition, you can wedge the bar beneath the main roots of a small tree or large shrub and wiggle it back and forth to loosen the soil before prying out the stump.

Hand tools

Sometimes you need to get close to your work, and that calls for using hand tools. There are trowels for transplanting, hand weeders for working between rows, and hand rakes for light work.

Trowels

Keep several of these handy for transplanting, seeding, scooping fertilizers, and occasionally for squashing a nasty bug. Two big considerations are the size of the blade and ergonomics. Wide blades scoop more soil and make the work go faster, but narrow blades are easier for digging small transplanting holes. A stainless steel blade resists rust, and a comfortable handle is essential when planting on hundred tomato, pepper, and eggplant seedlings.

Hand Weeders

Hand weeders come in as many varieties as upright weeders. Tools with small blades are great for tight places. One good choice is the Cape Cod weeder, which has a small teardrop-shaped blade for precision.

Hand Rakes

A small hand rake is sometimes useful for smoothing out areas of soil before direct-seeding, especially on raised beds.

Other Choices

I like using a mason's pointing trowel, which works two jobs: first as a dibble for transplanting small lettuce seedlings and secondly as a weeder, deftly scraping away tiny weeds just below the soil surface.

The Weeders

The majority of organic growers will tell you their most tiresome task is weeding. Tools aside, the best method for fighting weeds is a comprehensive strategy that controls the weed seed bank, thereby keeping weed seeds from getting started at all. Mowing the area around beds keeps weed seeds from dispersing onto fields, mulches help smother weeds before they get a foothold, and keeping soil loose and friable makes it easier to pull weeds once they germinate. However, no matter how much you prepare, some dormant seeds will always make their way to the soil's surface. Once they germinate, you'll need to deal with them before they develop deep roots and tough stems. An arsenal of weed-eating tools is available to you for fighting this battle—just remember to do your weeding in the early morning so that the weeds shrivel and disappear in the midday sun.

Basic Hoe

The ordinary garden hoe is a multi-duty tool that eliminates weeds but also digs, moves earth, shapes beds, and cuts planting furrows in loose soil. Even when confined to the single task of weeding, it has multiple purposes. With light pressure, it scrapes tiny weed-lings from the surface of the soil—without unearthing new seeds to germinate. Add a little muscle, and it cuts deeper to uproot larger weeds. Finally, it can chop the ground and roots around stubborn weeds, then pull them out with a firm tug.

It's also useful for shaping raised beds. Standing on one side of a raised bed, use the hoe to reach across and pull up soil from the other side. You can use a steel rake to smooth out the bed later. When you top-dress an unplanted bed with compost, dump a wheelbarrow load in one area and spread it around with a hoe first, then smooth it with a steel rake. Hoes will also help you pile soil around potato plants and cut clean, straight rows through finished beds, allowing you to quickly drop in seeds or transplants.

Look for a comfortable, well-made hoe, since it's a tool that you'll turn to for many tasks. One of the things I require is a handle that's long enough to accommodate my large frame. (I'm over 6 ft. [1.8 m] tall, and I don't like tools with short handles that make me bend at the waist.) Before buying a hoe, stand in the store and test it, going through the motions. Make sure the handle is long enough for you to do the job without tiring.

Although the basic hoe is versatile on its own, there are a number of specialty blades with different features: The triangular hoe, also called a Warren hoe, has a pointed blade that's ideal for cutting neat furrows. It also picks weeds out of tight spots. Some other specialty hoes are designed to make the chore of weeding more efficient.

The versatile garden hoe will move the earth—or if you prefer, it can shape beds and eradicate weeds.

Collinear Hoe

This better-built hoe is small, lightweight, and crafted to prevent fatigue, with the blade designed at a sharp angle to the handle so users can stand completely upright, gripping the handle with thumbs up and cutting weeds with a gentle sweeping motion. The collinear hoe is the design of organic farming guru Eliot Coleman, who is also a tool consultant at Johnny's Selected Seeds.

The hoe has a thin, narrow 4-in. (10-cm) blade that is easy to control and suited for working through established beds, scratching in and around heads of lettuce, salad greens, garlic, and other delicate vegetables. A larger 6-in. (15-cm) blade is available for wider areas, but it can still be maneuvered into tight places. Another version has a removable blade for easy sharpening. There's also a hand-sized collinear hoe for close-up work.

Tool Care

Tools left in the sun, wind, and rain won't last nearly as long as tools that have been well maintained. Here are some tips to make your manual tools perform for years and years:

- **Put them away.** That may sound simple, but it's easy to leave them lying around. (I've done it myself.) Locating a small shed near the garden or fields is perfect. If not, carry the tools back to a garage, barn, or shed, and keep them out of the elements.

- **Clean off dirt.** Mud and dirt that is caked onto the blades of tools can cause rust. Brush it off before storing tools.

- **Clean and oil tools before winter.** Brush soil off the blades and file the ends to a working edge (not sharp enough to cut you, but sharp enough to feel an edge). Finally, spray or rub the steel with cooking oil, and hang up the tools—don't lean them blade–down.

- **Wooden handles should be prepped for winter, too.** Sand any rough spots and rub with linseed oil.

Hoes with smaller blades, such as the collinear hoe, the diamond hoe, and this small pull hoe, have smaller blades that can be used for weeding close to leaf lettuce and other garden crops.

Oscillating Hoe

Call it a hula hoe, a scuffle hoe, an oscillating hoe, a two-way hoe, or a stirrup hoe. Any way you spell it, it's a great time-saver when weeding. I bought my first oscillating hoe more than twenty years ago, after the late Bob Thomson, former PBS host and author of *The New Victory Garden*, recommended it. I still have that hoe, and it still works miracles. It has a blade made from a loop of steel that is bent into a D shape. The flat side of the D faces down and is sharpened at both its front and back; the entire loop wobbles back and forth about ¼ in. (6 mm) on its handle, allowing it to remain in contact with the soil whether you push or pull the handle. The result is two-way hoeing that clears open areas quickly and efficiently. Oscillating hoes work best on small weeds, smartly cutting them just below the soil surface.

The blade of an oscillating hoe wobbles slightly, enabling it to hold the proper angle to the ground whether you push or pull it across the soil.

Diamond Hoe

Yet another improvement on the old-fashioned garden hoe is the diamond hoe, which has a diamond-shaped blade that is sharpened on all four edges. It also has a 6-ft. (1.8-m) -long handle that allows users to stand upright, reducing fatigue.

The diamond hoe's four-way cutting edge makes it easy to maneuver, allowing the user to pull, push, and slide sideways—cutting weeds just beneath the surface without moving a lot of soil. Because you face forward when working the hoe, it causes less neck strain. And, of course, it's devastating on young weeds.

Wire Weeder

This precision weeding tool looks so simple you might try to make one yourself. Or you could buy one created by inventor Eliot Coleman, whose design is sold through Johnny's Selected Seeds (see "Resources" on page 356 for more information). The tool is essentially a bent piece of wire that's flattened at the business end, creating a thin blade that slices along the soil around onion plants, carrots, and other closely planted crops. It's not for hacking through tall thick-growing weeds, but it's made to do the job in close quarters. There's also a short-handled version for close-up, down-on-your-knees weeding.

Wheel Hoe

This is one of those purchases that you might think about for a while—wheel hoes can cost several hundred dollars, but they make quick work of bigger weeding jobs between row crops. A wheel hoe looks a little like half a bicycle, with handlebars leading to a shaft that has a wheel at the bottom. Behind the wheel is a wide weeding blade, designed like a stirrup hoe. You walk behind the wheel hoe, pushing it along at an angle that cuts weeds just below the soil surface.

Models vary among manufacturers, with blade sizes ranging from 5 in. (12.5 cm) wide all the way to 18 in. (45 cm) wide. Some models have pneumatic tires while others roll across the soil on a wide metal wheel. There are also attachments for cultivating, digging furrows, and shaping beds.

Wheel hoes are a good choice for larger gardens and small farms. Attachments will cut weeds, cultivate the soil, or dispense seeds.

Seeding Machines

If your small farm or garden is too big for hand weeding, it might be too big for stooping over to seed by hand as well. Consider a garden seeder, which looks a bit like a wheel hoe. You push it along in front of you, but instead of scraping up weeds, it cuts a furrow and drops seeds in at preset intervals. Models vary, but there's usually a hopper with a disc that has holes or slots to dispense seeds one by one. Also look for a handy row marker, which extends from the unit to scratch a groove in the soil at a preset distance. You follow the groove as you put down your next row.

Seeders are fairly inexpensive, starting at about $100, and can save considerable time—especially for farmers with long, straight rows. At least one model is sold as an attachment to a wheel hoe.

Another type of seeding machine is a stab seeder, which looks like a 3-ft. (1-m) section of pipe that's pointed on one end with a trigger attachment on the side. You plunge the pointed end into the soil, pull the trigger to drop a seed, and move on. Stab seeders are useful for sowing larger seeds like corn, squash, and beans. They can also be used for transplanting seedlings, especially through black plastic mulch.

A lawn spreader can be used to spread minerals, limestone, and granular fertilizers or to sow cover crops.

Other Essentials

Rakes, wheelbarrows, and other cleanup tools have myriad uses beyond season-ending cleanup tasks. Rakes spread compost or mulch, scratch rocks out of delicate planting areas, and remove weeds and sod from newly turned soil. Wheelbarrows and garden carts tote debris to the compost bin and carry compost back to the planting areas. And small hand tools are a necessity for working on raised beds.

Metal Rake

The metal rake is one of the organic grower's key tools—simple in design but useful in multiple ways. When used with the tines down, it pulls rocks, sticks, leaves, and clumps of sod out of the soil. It also drags wide patches of loose soil into mounds for raised beds. And it can flatten the tops of mounded earth to create level planting areas where seeds receive an even distribution of water and sunlight. When flipped over (tines up), a metal rake will smooth a seedbed to finish it for black plastic mulch or the delicate job of sowing carrots, greens, or mesclun seeds.

Road/stone metal rakes, which connect to the handle at the center of the rake's business end, are best for smoothing soil but aren't as strong as bow rakes, which connect via a steel bow that ties each end of the rake to its handle. The bow is a sturdier connection, but it can get in the way of smoothing out a seedbed when you're using the rake in its upside-down position (tines up).

With either version, I prefer lightweight fiberglass handles for my metal rakes—the longer the better. I can't think of a time when I've ever stressed a rake handle to the breaking point, and the lighter weight of a fiberglass handle makes using the tool easier. The extra length lets me stand upright and work the rake at any angle almost parallel with or nearly perpendicular to the soil, teasing out small rocks before sowing carrots or salad greens. A long handle also lets me stoop low and reach farther when smoothing out a wide planting bed.

Another handy thing about metal rakes is that you can improvise your own attachments to make straight, even rows for sowing beans, beets, and other direct-seed vegetables. Here's how: Buy a section of stiff PVC tubing that's just wide enough to fit tightly over the tines of your rake. Cut sections of the tubing into lengths that are an inch or two longer than the tines. Push the tubing onto the tines at 4-, 6-, or 8-in. (10-, 15-, or 20-cm) intervals, depending on the distance you want between your crops. When finished, the tubing should stick out an inch or two beyond the other tines on the rake. Drag the rake down the center of your prepped planting area, and sow seeds in the grooves left behind. You may need to run a

Don't leave your rakes lying on the ground where they can cause an accident.

triangular hoe through the grooves to plant larger seeds like beans, but the marked rows will be evenly spaced.

Wheelbarrow

I get a lot of use from my wheelbarrow. No matter what task I'm tackling, I invariably load up the wheelbarrow with all the tools, supplies, fertilizers, compost, manure, buckets, and even seedlings that I'll need.

I've learned that not all wheelbarrows are created equal. In my opinion, the best choice is a 6-cu.-ft. wheelbarrow with sturdy handles, a pneumatic tire, and a deep bucket made of heavy-duty steel. Buckets that hold only 4 cu. ft. are just too small for most organic farm chores, and the price difference for a larger wheelbarrow is only $10 to $15—it's well worth the extra money.

Although the newer lightweight plastic buckets are enticing, I prefer the metal ones. My wheelbarrows carry boulders and firewood just as frequently as they tote bales of straw. And no matter what the advertising says about the "tough poly tub," I don't trust it to hold up to a dropped boulder, tossed pieces of oak firewood, or other heavy debris—especially in below-

Wheelbarrows are handy for carrying compost, mixing a potting medium, or just carrying an assortment of tools.

freezing weather. (I'm speaking from experience: I have a large two-wheel model with a plastic tub. The extra wheel is great for added stability, but the tub now has some strips of duct tape where the plastic has cracked after just two seasons.)

If you're living in a warm climate and you want a large-capacity wheelbarrow for carrying wood chips, brush, yard debris, and other lightweight materials, a poly tub will do the trick. But make sure you have a metal one handy for heavier jobs—especially during winter weather.

Garden Cart

Large-capacity garden carts are also great for moving lightweight supplies and planting material to and from the fields. The old-style garden cart with plywood sides, bicycle tires, and a handle made from tubular steel is perfectly functional, but it has some flashier competition these days. New four-wheeled carts built like oversized kids' wagons have fat pneumatic tires, a sturdy ergonomic handle, and steel-mesh sides that fold down and lock into position for carrying bulky loads. One big advantage that garden carts have over wheelbarrows is that the load stays fairly level when you're carrying it. (When you lift a wheelbarrow, you tilt the payload forward, and contents can bounce around or topple.)

Bigger Tools

Most small-scale organic farmers face a bigger battle than rain, wind, insects, weeds, and fungal diseases: Their foremost enemy is time. And to save time, it's often necessary to spend serious money for motorized tools that help them finish work faster and more easily.

All these purchases should be weighed carefully, since they can be big investments. Consult with other farmers and growers at farmers' markets for advice, and keep in

Garden carts are handy for moving lightweight materials or for carrying seedlings to the planting area and crops back to the cleaning station.

mind that you can rent some tools to try them out before you buy them. Buying well-maintained secondhand machinery is another option that will keep prices down. (But, as with all things, beware of purchasing someone else's problems.) Also consider that machinery comes with costs beyond the original price tag—maintenance, gasoline, repairs, and so on. And you'll need a shed or barn to safely store your motorized tools.

As with manual tools for the small organic farm, motorized tools should cover the basics first. So before you rush out to buy a tractor with attachments for every farm task you can think of, look to buy some entry-level tools—a heavy-duty rotary tiller, a mower, or a gas-powered weed-whacker.

Rotary Tiller

One of the first motorized tools that many small-scale organic farmers and market gardeners choose is a rotary tiller. It cuts new beds, churns through hard-packed soil, and turns lawn and meadows into vegetable-ready growing areas. Tillers fall into three basic categories: small garden tillers, medium front-tine tillers, and rear-tine tillers.

Small Garden Tillers

With two-cycle engines, these tillers buzz like mosquitoes, and they shred through the top 2 to 3 in. (5 to 8 cm) of soil. I own one of these, and I use it to mix compost and integrate organic matter into topsoil. But I try to run it slowly, and I'm careful not to use it too frequently. At full speed, the blades spin quickly and can pulverize soil and leave it looking like finely milled powder—damaging the structure created by soil organisms.

Medium Front-Tine Tillers

With handles at the rear and circular blades in the front, these tillers usually have about a 3- to 5-horsepower engine. They cut deep into the soil, but they can wrench your shoulders when they hit roots, rocks, and other underground debris. If you're tilling any sizable plot of land, you will quickly tire of your front-tine tiller.

Rear-Tine Tillers

Although these tillers can be small and are made for home use, they are generally larger and more productive than front-tine models. If your farm has any size to it, or if you're considering expanding operations in the near future, you're better off spending money on a rear-tine tiller that is built to last. Below are some features to consider.

Rear–tine rotary tillers clear the land for a new planting area or mix cover crops into the soil.

Powerful Engine. Rear-tine tillers typically have at least a 5-horsepower engine. For heavy farm use, an 8-horsepower engine might be a better option. If you're considering a smaller tiller, an industrial/commercial engine is more durable and less likely to break down under heavy use.

Tine Rotation. Rear-tine tillers are built with tines that move either in forward direction (Standard Rotating Tines), reverse direction (Counter-Rotating Tines), or have settings that allow you to operate the machine with tines that move either backward or forward (Dual Rotating Tines). The forward rotation keeps tines moving in the same direction as the wheels, kicking out rocks and smoothing the bed. Tines moving counter-rotationally (in reverse as the tiller moves forward) cut into sod and dig deep into the soil to mix in cover crops or organic matter. But counter-rotating tines can present a serious safety hazard in walk-behind tillers.

If the tines get snagged on an underground object when they are rotating in reverse, the tiller can lurch back toward you, possibly catching your feet or legs. If you're working along a fence or building, the tiller could pin you against the structure. Forward-rotating tines may not dig in as aggressively, but they are much safer because the tiller will pull away from you if the blades become stuck on an underground object.

Sprayers

Organic growers don't rely on pesticides to solve all their garden problems, but sprayers are still useful for foliar feeds, compost tea, and well-timed and sometimes necessary organic pest controls.

Pressure Sprayers

When spraying organic controls, I like to use the smallest container possible because I don't want to mix up any more than I need. That means my 1-gal. sprayer gets the most action (sometimes only half-filled).

A sprayer should have a strap to help you carry it, a pump to provide internal pressure, a relief valve, and a wand with a trigger and adjustable nozzle (from stream to fine mist). Brass wands and nozzles are more durable than plastic nozzles.

For larger areas, consider a backpack sprayer, which will keep both hands free. Most hold 3 to 4 gal. Many models have a hand lever on the side to pump the sprayer and build pressure without removing the backpack. Try a few on and choose a size that's best suited for your farm or garden, keeping in mind that the larger-capacity sprayers will be heavier when full.

Some liquids—like fish-emulsion fertilizers and compost tea—can contain particles that will clog the nozzle of a pressure sprayer. They may be better applied with a hose sprayer or watering can.

Hose Sprayers

These sprayers typically have a half-liter poly container, which screws onto a head with a nozzle at one end and a hose connector at the other. The better models use a feeder straw that draws an even supply of fertilizer into the stream of water exiting the hose. They are not as precise as a pressure sprayer with a brass wand, but they work well for applying liquid fertilizers.

Adjustable Tilling Depth. A bar in the back of the tiller will set the tilling depth. Make sure your tiller adjusts for deep or shallow settings. Shallow tilling is desirable when stripping sod or turning organic matter into the top few inches of soil. A deeper setting is handy for creating a new bed (after stripping the sod). It will also pull up small rocks and incorporate organic matter deeper into the soil.

Adjustable Handle. A handle that adjusts up or down to the user's height will make the tiller operation safer and more comfortable, and one that swings left or right will allow you to work the tiller from its side and prepare beds without having to walk on them. This isn't just a cosmetic consideration—stepping on freshly tilled, fluffed-up soil can compress the soil and destroy its structure.

More Land = Bigger Tools

Growers who are looking to tame more than an acre or two of farmland may set their eyes on bigger and more expensive machinery. But they'll have decisions to ponder before making such a large investment as a compact or full-size farm tractor. Heavy-duty motorized equipment can create sizable debt—sometimes too big for a fledgling small-scale farmer to handle. Consider the following before purchasing a tractor or other major farm tool.

Return on Investment. Think about how long it will take to pay back the expense. Will the savings in time and effort be worth the purchase? (Don't forget to calculate the cost of repairs, maintenance, and fuel.)

Risk of Injury. Buying and using heavy motorized equipment carries with it the increased risk of injury. Make sure you understand how to use the equipment and that you read and follow all instructions and safety precautions.

Used Equipment. Purchasing used equipment is a cost-saving option, but caution is important here, too. In the case of a tractor, a used piece of equipment won't do you any good if it's broken when you need it most. There's nothing more frustrating than beginning one job with a short window—such as prepping a field when the soil is dry and planting time is at hand— only to find that the equipment is broken and the repairman busy for the next two weeks.

However, if you're mechanically inclined and you have the workshop space, you may want to pick up a secondhand tractor. Make sure it's a make and model that will still have parts available if you need to perform repairs. The good news is tractors aren't used every day, and

You can save money by purchasing a second–hand rotary tiller or tractor. Just make sure the equipment is in good condition and won't break down when you need it most.

well-maintained machines will last for decades, so a twenty-year-old tractor may have plenty of quality life ahead of it.

There are garden tractors, subcompact tractors, compact tractors, and full-sized tractors, many of them with a wide range of attachments for tasks as varied as baling hay, cultivating row crops, chipping brush, and digging fence postholes.

It's a good idea to make a list of all the work on the farm that you expect a tractor will help you accomplish and talk over the options with a reputable tractor dealer. Also ask a dealer about used tractors; some may have reliable used equipment to sell.

If you do choose a small tractor and are planning to use it for multiple tasks, look for a three-point hitch and PTO (power takeoff)—features that allow versatility in attachments. The three-point hitch has three connecting points from tractor to attachment, with one link above and two below in a triangle. The three-point hitch is the standard in farm tractors, allowing hydraulics on the lower attachments to lift the equipment up and down. There are four categories/sizes of three-point hitches for different sized tractors: 0, I, II, and III. Make sure your attachments fit your tractor's hitch.

Walk–behind tractors may be the solution for small–scale growers who need more power and versatility than a rotary tiller. The machines can handle multiple attachments.

The PTO is a shaft that couples with the attachment to provide power from the tractor. It's what makes a posthole auger turn, a wood chipper chip, and a tiller's blades rotate. It's also one of the most frequent causes of serious farm injuries, usually from items of clothing that become entangled in an exposed shaft.

Two-Wheeled Tractor

A less-expensive but still versatile option for small-scale farmers is a walk-behind tractor, also called a two-wheeled tractor. Walk-behind tractors look like the handlebars and engine section of a large rear-tine rotary tiller, but they hook up to multiple attachments, from snowblowers and plows to seeders, tillers, mowers, and even wood chippers. Not only are the engine sections smaller and less expensive than a ride-on tractor, you can also add attachments as you expand your operation.

Cleaning Station

Large-scale farming operations have packing sheds where harvested vegetables are washed, trimmed, and packaged for market. You might set up an outdoor station of your own for small-scale production. We took a double-well sink from one of our CSA (Community-Supported Agriculture) customers who was remodeling his kitchen and mounted it onto a sturdy frame with an old kitchen countertop. Then we attached a hose to the sink's rinse-sprayer.

It has ample counter space for cleaning and trimming lettuce or cooking greens and two wells for rinsing and stacking the veggies. We keep a small compost bin nearby for any trimmings that don't go to the chickens. I've seen other small farms with similar arrangements for cleaning their crops.

Other Large Expenses

Tractors aren't the only major farm expenses. Additions such as a greenhouse should be given the same consideration and weighed against other major purchases. New greenhouses—constructed of tubular steel with heavy-duty stretched plastic sheeting for a roof—have been reduced in cost to a few thousand dollars, depending on size and such features as lights, fans, ventilation, and heating. They can stretch the growing seasons, allowing farmers to produce salad greens into winter and start tomato plants in early spring. A quick look at your customers' needs and the potential for market expansion could lead you to spend money on a greenhouse before you invest in a tractor.

Somewhere between owning a greenhouse and not owning a greenhouse there's an option called the high tunnel. A high tunnel is a low-cost movable greenhouse that can be placed over a planting bed to extend the season for the equivalent of one full USDA hardiness zone. Extra protection can be gained by placing floating row covers over crops on cold nights.

High tunnels—often called hoop houses—can be made from inexpensive materials available at building supply stores or purchased as kits from specialty suppliers. Some include sideboards for a more permanent structure. (See the "Resources" section on page 356 for more information.)

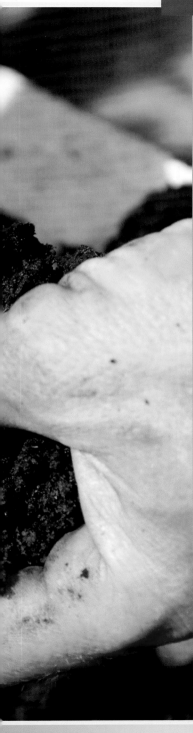

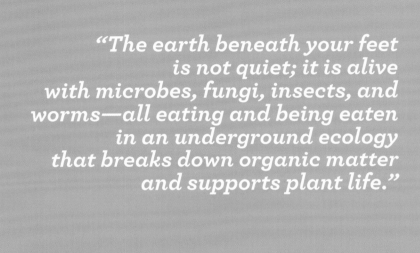

"The earth beneath your feet is not quiet; it is alive with microbes, fungi, insects, and worms—all eating and being eaten in an underground ecology that breaks down organic matter and supports plant life."

There's something about the smell of earth on a late April morning that stirs my urge to plant. It's not just a whiff of sand or silt or clay. And it's more than the smell of new grass and old leaves; it's the sweet aroma of living, breathing soil— earth that is rich in organic matter and teeming with activity.

The Soil Food Web

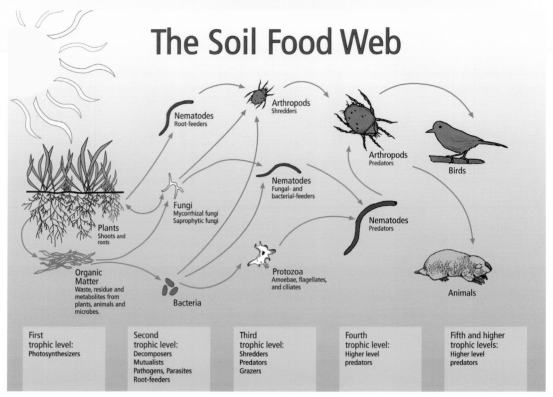

First trophic level:	Second trophic level:	Third trophic level:	Fourth trophic level:	Fifth and higher trophic levels:
Photosynthesizers	Decomposers Mutualists Pathogens, Parasites Root-feeders	Shredders Predators Grazers	Higher level predators	Higher level predators

Relationships between soil food web, plants, organic matter, and birds and mammals. Image courtesy of USDA Natural Resources Conservation Service http://soils.usda.gov/sqi/soil_quality/soil_biology/soil_food_web.html

Healthy Soil—The Keystone to Success

Just 1 teaspoonful (5 ml) of well-maintained organic topsoil holds up to a billion bacteria along with miles of hyphae (probing filaments of underground fungi). A shovelful of that same organic topsoil contains more species of living creatures than are known to exist aboveground in the Amazon rain forest, from bacteria and single-celled protozoa to nematodes, earthworms, and burrowing insects. And they all get busy in springtime.

Sunshine warms the landscape and dries the earth, allowing air to penetrate the pores and pockets between soil particles. Oxygen in the air stirs microbes to action. Soon, these legions of tiny organisms begin feeding on old roots, plant litter, and other organic material softened by the months of ice and snow. And as they feed, they multiply. Like crazy. In the process, microbes called *actinomycetes* release geosmin, a sweet-smelling compound that farmers, gardeners, and anyone in tune with the changing seasons recognize as the pleasant aroma of fertile soil.

So what does all this have to do with organic farming? Everything. Healthy soil is the backbone of any successful organic farming operation. And to be a good grower—whether you're farming 50 acres (20 ha) in row crops or using intensive methods on raised beds in a suburban backyard plot—you need to understand that soil is more than just a layer of crumbled rock. The earth beneath your feet is a living ecosystem, and each organism has a role to play in the cycle of growth and decay.

The Fruits of Their Labor

Birds, small mammals, and insects shred fallen leaves that break the material into tiny pieces. Burrowing earthworms then ingest the litter and pull it deep into the ground with other organic matter that has seeped into cracks in the soil with rain and melting snow. Once underground, bacteria and fungi consume the litter and convert it into nutrients that can be absorbed by plants. The microbes also bond to plant roots to fight diseases and help plants reach farther for water. Some bacteria even team up with legumes to pull nitrogen (the key to all things green) from the atmosphere and store it in the ground. As the microbes die off or are eaten by other soil creatures, the protein in their cells adds nitrogen to the soil.

All this bustling biological activity is called the *soil food web,* and it does more than break down the residue of last year's field crops. Soil structure improves as carbon turns to humus and the underground armies secrete gluey substances that bind particles of sand, silt, and clay into water-absorbing aggregates. The end result is rich organic topsoil that drains well, holds moisture during dry spells, and stores essential nutrients where they can be absorbed by plant roots.

The Underground Army

Populations in the soil are subject to a natural ebb and flow, with underground residents rising in population and then subsiding as other creatures grow. After cover crops are plowed into the soil, bacteria populations grow quickly, feeding on the soft green material. As the bacterial

Horizons—The Layers in Your Soil

Soil is formed over time and develops sections, like layers in a cake. Topsoil, subsoil, and substratum are the major layers found in most soil profiles.

Topsoil

The topsoil is a mineral layer that is typically dark brown to black due to an accumulation of decayed organic matter called humus. The topsoil may be covered with a relatively thin layer of organic material from leaves, grasses, bark, and twigs that have undergone some degree of decomposition.

Subsoil

Typically, the dark topsoil will quickly fade with depth to a lighter layer. This is the subsoil, which is made up of mostly mineral material, with the exception of deep roots. It is typically more yellow or red in hue than the underlying material due to soil weathering processes.

Weathered Parent Material

Next is the parent material—the rock that was broken down to make the topsoil. Weathered means it's partly deteriorated and on its way to becoming soil.

Parent Material

Next is the substratum, the relatively unweathered geologic material from which the overlying subsoil developed. Weathered means it's partly deteriorated and on its way to becoming soil.

Bacteria dot the surface of these fungal hyphae. When microscopic protozoa feed on bacteria, they release ammonium, which is a plant-usable form of nitrogen.

population grows, soil acidity rises, and then the population starts to fade. Next come the fungi—which thrive in the more acidic soil—with hyphae that release acids and then grow into the woody material that breaks down.

Adding synthetic fertilizers or using insecticides, herbicides, or fungicides can kill off soil organisms, upsetting the natural balance and ultimately harming both the soil and the crops that you plant. What follows is a quick guide to the microbes you'll be sharing your soil with.

Bacteria

Basics. Bacteria are the smallest and most plentiful of soil organisms; a single teaspoon (5 ml) of garden topsoil can contain 100 million to 1 billion of them. Their population rises and falls with soil temperature, acidity, and available food. Bacteria live in the top layers of soil, where they consume and recycle organic material, as well as around the root zone, where they feed on simple sugars given off by plant roots. Some bacteria, called actinomycetes, tolerate more acidic conditions and help break down tough, high-cellulose material.

Benefits. Most bacteria are decomposers that play a role in nutrient cycling, converting organic matter into usable plant nutrients. Many feed on simple sugars given off by plants, blocking pathogenic (disease-causing) bacteria from contacting plants. Some help legumes fix nitrogen from the air and store it in the soil.

Potential Problems. Some bacteria cause plant and animal diseases, but in a healthy soil with a diverse microbe population, beneficial bacteria and other soil organisms will crowd out harmful bacteria.

Fungi

Basics. Fungi are plants that can't make their own food, so they live off the organic material around them. Many soil fungi form tiny strands of cells called *hyphae,* which snake between soil particles and give off acids that break down woody materials. Networks of intertwined hyphae can form a vast underground mat, called a *mycelium,* which can span across acres. Mushrooms, which sprout from the mycelium, are the reproductive parts of some fungi.

Benefits. Saprophytic fungi (those that feed off decaying organic matter) attack tough woody material in soil, converting lignin, cellulose, and other complex carbon molecules into nutrients that plant roots can absorb. They also create humus, long lasting organic matter that holds dissolved nutrients until they can be absorbed by plant roots.

While bacteria work rapidly through freshly turned soil and compost piles, fungi do their work at a slow and steady pace, spreading out and decomposing wood chips, rye-grass stems, corn stubble, and straw. (You can sometimes see mycelium as a white weblike network in a pile of matted straw or wood chips.)

Mycorrhizal fungi (root-zone fungi) colonize the roots of nearly every crop except spinach, beets, and cabbage-family plants. The fungi develop a partnership with their hosts, feeding on simple sugars given off by the roots and using their far-reaching hyphae to draw in phosphorus, nitrogen, zinc, and other nutrients. Mycorrhizal fungi perform another helpful function in farmland, coating plant roots and protecting them from some parasitic nematodes.

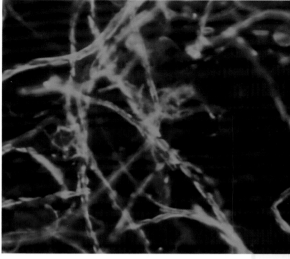

Thin strands of mycorrhizal fungi develop a relationship with plants, feeding on simple sugars exuded by the roots and helping to draw water and minerals into the root zone

Potential problems. Not all fungi are good for your plants. Pathogenic fungi might be termed the evil cousins of the fungal world, feeding on living plants and destroying valuable farm crops. They cause such diseases as Verticillium wilt, damping off, and root rot. But diverse and active colonies of beneficial fungi, along with a disciplined regimen of crop hygiene and rotation, will help prevent most fungal diseases.

Protozoa

Basics. You may remember looking at amoebas and paramecium under a microscope in biology class. They're among the many protozoa that swim through water among soil particles, feeding on bacteria, fungi, and other single-celled organisms.

Benefits. While they're happily munching on bacteria, protozoa release ammonium, a cation, which is taken up by bacteria or absorbed by plants. Ammonium and nitrate are the two common forms of plant-usable nitrogen. Feeding protozoa also keep the bacteria herds young and productive, resulting in more bacteria and more organic matter being converted into usable plant nutrients. Some parasitic protozoa also feed on pathogenic fungi, making them valuable in controlling plant diseases.

Potential Problems. Protozoa do not present a threat to vegetable crops in the soil but can cause human diseases. Researchers have found that when protozoa consume some pathogenic bacteria, they can help the bacteria survive longer on leafy vegetables. (Always wash your greens.)

Nematodes

Basics. These are microscopic roundworms that wriggle through the soil and feed on bacteria, fungi, protozoa, smaller nematodes, and other soil life. Scientists estimate there may be up to half-a-million different species of nematodes.

Benefits. As they consume bacteria, nematodes give off ammonium, which is used by bacteria and taken up by plants as nitrogen. And as with protozoa, when nematodes feed on bacteria, the bacteria reproduce faster and take up more of the waste. A few predatory nematodes are extremely effective as organic pest controls.

Potential Problems. Some types of nematodes are known for the damage they cause. Root-knot nematodes, for example, attack the roots of many vegetable crops, causing galls that stunt growth.

Earthworms

Basics. Most people think of worms as squiggly fish bait or as the perennial losers of so many early morning battles with hungry robins. But for organic growers, worms are prized as the oxen of the underground world. They till the soil, they lug organic material and minerals from layer to layer, and they clear underground roadways by burrowing through the

Squiggling earthworms perform multiple chores in your field or garden. They ingest material and expel microbe-rich castings. And their tunnels help to aerate the soil.

soil. There are at least 7,000 different species of earthworms, including some that can reach 6½ feet (2 m) long.

Benefits. The tunnels that earthworms make in their travels allow water and air into the soil, making it more porous. Worms also take crop residue and organic material from the surface layers and pull it into the subsoil, providing nutrients that encourage deeper plant roots. And when worms take in material, their guts become microbe incubators, and they release microbe-rich castings. In many ways, they work like natural tillers and aerators, but without breaking aggregates and without compacting the soil. Worms even improve soil structure by secreting a slimy coating that binds soil particles together.

Pay Dirt

The long-term effect of soil microbe activity is the accumulation of humus—stable, high-carbon material that holds nutrients, builds soil structure, and helps the earth retain water. Most humus is created by soil fungus breaking down tough, high-cellulose organic material.

Humus remains in the soil for hundreds of years without breaking down, effectively keeping carbon sequestered in the soil and lowering greenhouse gases.

Potential Problems. Earthworms in the garden are signs of good soil health. But some species imported for indoor composting have become invasive in sensitive forest ecologies after being released outdoors.

Arthropods

Basics. Insects, spiders, millipedes, and mites all hang out in the soil for at least part of their lives—digging, eating, and being eaten, all the while adding nutrients to the soil food web. You might think of all these wriggling creatures as insects, but many have more than six legs.

Benefits. Most are shredders, feeding on dead plant parts, increasing the organic material's surface area and providing more feeding places for fungi and bacteria. Earwigs, millipedes, sow bugs, and pill bugs (those little gray prehistoric-looking critters that roll up into balls when you touch them) all feed on dying plant parts. Other tiny arthropods—such as springtails and a few mites—feed on fungi and soil microbes.

Some are predators of the bug world—spiders, centipedes, ants, and beetles. All together, the arthropods break down large pieces of organic matter into smaller pieces that can become food for bacteria, fungi, and other microorganisms. The larger arthropods move material and microbes around in the soil, and the predators keep the grazing population in check.

Potential Problems. Some arthropods such as wire worms (click-beetle larvae), mole crickets, grubs, and earwigs can become pests when they eat living plants.

Pay Heed to pH

Whether you're an organic farmer in California, Canada, or the Carolinas, you need to know the pH of your soil. The pH is a reading of a soil's acidity or alkalinity, measured on a scale that ranges from 0 to 14. A reading of 7.0 is neutral, with higher numbers indicating alkalinity and lower numbers showing acidity.

All plants have their own preferred soil pH; blueberries thrive in acidic soil, while asparagus, onions, and okra like neutral to slightly alkaline conditions. Most veggies perform best in soils that are slightly acidic, ranging from 6.0 to 7.0.

When soil is too acidic (lower than 6.0), plants have difficulty taking up the major nutrients—nitrogen, phosphorus, magnesium, and calcium. When soil is too alkaline, plants can suffer from iron, magnesium, phosphorous, and zinc deficiencies.

To raise the pH or make it less acidic (sweeter), add ground or pelletized limestone or a small amount of wood ashes. To lower the pH or make it less alkaline, add elemental sulfur. This is an expensive amendment, and it may not be practical for a large farm with highly alkaline soil (above 8.0). Organic materials such as wood chips and peat moss can also lower pH.

Naturally acidic soil occurs in the northeastern United States and Canada, partly due to years of acid rain. Acid soils also occur in forests, the Southeast, and in areas where heavy rains can wash silt and calcium out of soils. Soils in the arid western states have higher alkalinity because calcium has accumulated due to a lack of rain. In western Canada, most soils are alkaline or neutral, but large areas of soil with a pH of 6.0 or less are found in Alberta, northeast British Columbia, Ontario, and Saskatchewan.

In case you were wondering: The pH scale is logarithmic, which means a reading that is 1 point lower (5.5 instead of 6.5) is 10 times more acidic. Also, the "p" and the "H" in pH refer to potential hydrogen in water, which is the number of positively charged hydrogen ions.

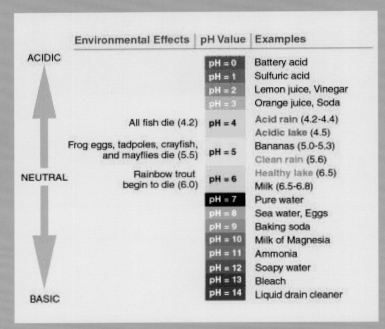

	Environmental Effects	pH Value	Examples
ACIDIC		pH = 0	Battery acid
		pH = 1	Sulfuric acid
		pH = 2	Lemon juice, Vinegar
		pH = 3	Orange juice, Soda
	All fish die (4.2)	pH = 4	Acid rain (4.2-4.4) Acidic lake (4.5)
	Frog eggs, tadpoles, crayfish, and mayflies die (5.5)	pH = 5	Bananas (5.0-5.3) Clean rain (5.6)
NEUTRAL	Rainbow trout begin to die (6.0)	pH = 6	Healthy lake (6.5) Milk (6.5-6.8)
		pH = 7	Pure water
		pH = 8	Sea water, Eggs
		pH = 9	Baking soda
		pH = 10	Milk of Magnesia
		pH = 11	Ammonia
		pH = 12	Soapy water
		pH = 13	Bleach
BASIC		pH = 14	Liquid drain cleaner

Building Your Own Soil

Once a grower understands how each living creature—no matter how small—fits into the larger ecology, his or her goal becomes not just planting and harvesting, but also finding a way to nurture relationships between beneficial soil organisms and food crops to make the harvest more successful. That partnership is at the heart of organic agriculture and is the reason why good growers treat their soil as a living habitat—not some sterile medium that anchors roots and holds a monthly dose of chemical fertilizer. Making the underground world more hospitable is one part of what growers mean when they say they are "building the soil."

Not all vegetable growers are blessed with ideal soil, which varies slightly from crop to crop but falls into the general range of sandy loam (see "Soil Texture" on page 69) with high levels of humus and a slightly acidic reading on the pH scale of about 6.7. So how do you get there? It's a slow process—more a journey than a destination, considering you only finish when you've retired from farming.

Even the best organic farmers work diligently to keep their soil fertile, friable, and biologically active: They plant cover crops, add organic matter, and monitor progress with soil tests. When the test results come back, they add the proper nutrients to replace what plants have used. Farmers analyze the results of these tests over years to determine whether their efforts are producing the desired effect.

The best organic growers actually keep problems in check by managing the underground ecology. When farmers are successful in cultivating their soil, the microbes break down nutrients and plants grow stronger cell walls that resist diseases and insects. Wherever you are in your own soil-building journey, consider these five steps to help Mother Nature create the perfect underground environment for growing food crops.

1. Do no harm.

2. Understand your soil.

3. Feed your soil.

4. Mark your progress and adjust accordingly.

5. Protect your investment.

Step 1: Do No Harm

As discussed earlier, your soil is its own delicate ecosystem, and it's easier than you think to cause it harm. Chemicals are an obvious red flag in organic farming, but you'll also see repercussions from too much or too little physical intervention.

The effects of all this soil abuse are especially damaging over an extended period. Every time you apply an artificial fertilizer, chemical pesticide, herbicide, or fungicide, you can inflict some degree of harm to soil life. And after years of heavy plowing, chemical applications, and neglect, soil can become lifeless, compacted, low in organic material, and inhospitable to anything but the spindliest of weeds. Repairing that damage and building the soil back into a living ecosystem can be a difficult and expensive process; the more soil that needs to be restored, the greater the cost and effort. So the best course of action is to avoid creating that damage in the first place, which you can only do if you're aware of what causes it.

Chemicals
Before you read another word in this chapter, take a pledge to stop using harmful chemicals.

Simply put, synthetic fertilizers and chemicals that are designed to kill weeds, fungal diseases, and insects can all cause harm to the delicate soil organisms that perform so many vital functions.

Fungicides. Even though fungicides may help you battle blight on your tomatoes, using them is shortsighted. Agents that kill fungal diseases—even a few fungicides approved for organic use—also carry the potential to destroy beneficial soil fungi.

Herbicides. Herbicides can make quick work of weeds, which is why they're so popular in conventional agriculture. But in addition to killing aboveground growth, they can harm beneficial microorganisms beneath the soil.

Insecticides. An obvious danger to the bugs they target, insecticides can sometimes hurt the people who apply them. They're also harmful to beneficial organisms above and below your soil. Aboveground, broad-spectrum insecticides can kill bees and other insects that pollinate your crops. And the pesticides that seep into your soil can kill beneficial arthropods, earthworms, and microorganisms living inside the root zone.

Physical Abuse

Occasional shallow-depth tilling benefits the soil by mixing compost, cover crops, and organic amendments into the top layers of the soil, where they can be broken down by hungry microbes. It also shreds any remaining crop residue and disturbs hibernating pests in early spring. But the more you till, the more you risk crushing earthworm tunnels and churning through the delicate network of fungal hyphae. As a tiller's blades cut through the soil, they also pulverize the soil aggregates created by underground organisms. The tiller fluffs up the soil, but the effect is only temporary if the tunnels and structures created by underground organisms have been destroyed. Heavy rains combined with the weight of machinery or even foot traffic will pack the soil back down again. Once the soil packs back down, it becomes more condensed than it was before tilling. The damage is exacerbated when you till wet soil.

Repeated tilling also harms the subsoil. The downward force of a tiller's blades on the soil beneath them creates a densely packed sublayer called *tiller pan*. This compressed layer can slow root growth and impede drainage. Some tiller damage is mitigated in an organic farming system; organic growers plant cover crops and mix in organic matter, encouraging soil life and building back structure.

Lack of Crop Rotation

Planting the same crops in the same place, year after year, can also harm your soil. The practice, called *monoculture cropping,* allows insect pests to overwinter in the soil and crop residue and then return the following year. Beetles that feed only on potatoes, for instance, find it easy to repopulate if their favorite meal is within crawling distance. They emerge to mate in the spring and head straight for the new-grown potato leaves.

In addition, crop-specific diseases that remain dormant during winter can come back when new crops are planted the following spring. Monoculture also effectively mines important minerals from the soil; plants draw the nutrients up through the roots, and the results are harvested. But by alternating crops that draw different minerals from the soil, you can help conserve nutrients.

Neglect

The final soil killer is neglect. In this instance, neglect doesn't mean complete inattention; it means neglecting to do what is necessary to prevent harm. For example, failing to plant cover crops or spread mulch over fallow beds can result in the loss of topsoil to erosion. Sunshine bleaches and dries out the organic matter, allowing rain and wind to whisk it away.

Neglect can also include failing to follow the basic rules of soil care that I've already listed: understand your soil; feed it; mark your progress; and protect your investment.

Crop rotation and companion planting will help fight insect pests and diseases, while helping to conserve soil nutrients.

Step 2: Understand Your Soil

Before you can start improving your soil, it would be helpful to understand the basic geology of dirt. While the ground beneath your feet may appear firm, only a portion is solid material. Gaps and pores between the individual particles make up the rest of the volume, with water and air consuming that pore space. That makes your soil more like a giant sponge, with varying amounts of solid, liquid, and gas.

Soil Texture

The solid part of a typical soil is made up of mostly mineral material—pieces of rock fragments that have been broken down into tiny particles by time and nature. In places, that material is similar to the bedrock beneath the soil, but some of it may have been transported thousands of miles by wind and water and deposited over the centuries. The size of these individual particles determines the texture of your soil, which affects its ability to retain water in dry times or drain water away when the landscape is flooded. The texture class of the soil is determined by its proportions of sand, silt, and clay.

The finest particles are clay, which are so small that individual grains can be seen only with an electron microscope. The slightly larger particles make up silt. Finally, the largest granules are sand and are big enough to see with the naked eye. The ideal soils for growing

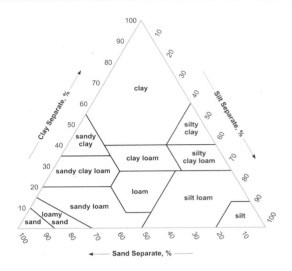

crops are loam or silty loam, which have a fairly equal balance of sand, silt, and clay; the worst soil is a mix that is sharply skewed toward clay or sand, each of which creates its own problems.

Sand. Sandy soil is easy to dig, allows roots to spread freely, and typically drains well, unless the water table is high. But the coarseness of sand particles allows water to seep through quickly, leaching nutrients with it. Sandy soils do not have the capacity to hold water and make it available to plants. In extremely sandy soils, plants don't have much time to absorb water and nutrients before the material disappears from the root zone.

Silt. Silty soil is usually generally good for farming because it holds water and nutrients, but the smaller particles have a tendency to nestle together. That means silty soil is easily compacted. Water can then run off the surface without sinking in. In addition, roots may have difficulty spreading out, and any foot traffic will add to the compaction. Soils high in silt are also those that are most readily eroded by wind and water. Silt particles are smaller than sand, so they are more easily transported. Clay particles adhere to each other, so they are more difficult to detach.

Clay. Clay soil often holds nutrients, but the fine particles leave almost no space for water to escape. So once clay soil is wet, it remains soggy and unworkable for some time. When it finally dries out, the surface can harden. Plowing or tilling clay soils often creates clods and clumps that make it difficult to plant crops.

Putting a Name to Soil Types

Soil types around the world are classified by twelve orders, with such names as Mollisol, Alfisol, Ultisol, and Oxisol—each determined by its overriding characteristics. For instance, Mollisols (*mollis* is Latin for soft) are soft soils with high amounts of organic matter and are found in the fertile grasslands of central North America. Oxisols refer to the weathered (oxidized) soils of the tropics. Beyond that, soils are classed by great groups, groups, orders, families, associations, and series. The most local categories are series with labels such as Montauk Series or Sudbury Series, named for the geographic area where they were first logged. Under that name, you'll find information on that particular soil's texture and characteristics. The soil survey also includes data on drainage, organic content, depth to impervious material, and an indication of how deep the roots stretch. (For more information on the soil survey, see page 24.)

How to Determine Your Soil's Texture

Luckily, you don't need complicated equipment to figure out your soil's texture. First, pick up a fistful of moist soil and squeeze it into a ball. Then open your fist and let the sticky ball stand on your open palm. If it breaks apart easily, you have sandy soil. If it sticks together but breaks apart when you tap it firmly with your finger, you have a mix of sand, silt, and clay. And if it stays in a ball after repeated thumps with your finger, there's more silt and clay than sand.

Another way to tell if your soil has high clay or silt content is to make a muddy paste from the soil and rub it between your fingers. If it feels sticky smooth, it has higher clay content; if it's slightly gritty, you probably have fine sand and silt; if smooth it has more silt.

You can also get an idea of your soil's texture—its ratio of sand to silt to clay—by performing the following simple test with a canning jar, hardware cloth, and a few household ingredients:

Determining the texture of your soil can help you understand its limitations. While it's not practical to improve soil texture by adding sand, silt, or clay, organic matter can help compensate for problems.

1. Scoop up about 2 cups (475 ml) of topsoil, taking samples from several spots in your field or garden: Scratch away the loose mix of mulch and soil at the immediate surface—about the first inch. Then dig a hole that's 6 to 8 inches (15 to 20 cm) deep. Use a hand trowel to slice a layer of soil from the side of the hole so you get a sampling of several different depths within the root zone.

2. Let the soil dry in a planting tray lined with newspaper and then sift it to remove rocks, pebbles, wood, and larger bits of organic matter.

3. Place 1 cup (240 ml) of soil into the jar and fill jar with warm water; shake it for a few minutes to ensure the soil is completely mixed. Look at the jar from the bottom to check for soil clumps and shake again if needed.

4. Place the jar on a table and wait for a day or two. Once all the particles have settled, you should see three layers of soil: from the bottom up—sand, silt, and clay—topped by water.

5. If the three layers are even, your soil is loamy. If the bottom layer is thicker than the others, you have sandy soil. If the middle layer is thickest, you have higher silt content. (Note: Adding a tablespoon of nonsudsing dishwasher detergent will help the soil particles separate and settle.)

This test is not perfect, but you can use it in combination with information from the Natural Resources Conservation Services' soil survey to better understand your soil's texture. Visit the Natural Resources Conservation Service website, websoilsurvey.nrcs.usda.gov, and consult the survey as described in chapter 1. In addition to details about topography, climate, and rainfall, the survey lists the dominant and secondary soil characteristics for specific areas.

Knowing what type of soil you have will help you decide which crops are best suited for the land. This information will also help you better understand how to improve your growing conditions.

If you're using the soil survey as a guide to buy farmland, make a physical check, too. Visit the property and take core samples or just dig a hole a few feet deep to see whether the layers or "horizon" match the description. There should be a visible section of darker topsoil (roughly a foot deep), usually with a lighter-colored subsoil that should stretch downward. The land should also drain well.

Step 3: Feed Your Soil (Read: Composting!)

How do the best organic farmers maintain soil? They keep the underground creatures happy and healthy. New York organic farmer Steve Gilman, who authored the booklet *Organic Soil Fertility and Weed Management,* writes: "The organic grower's primary task is to provide high-quality room and board for the legions of microbes and soil animals who do the actual work of growing the crops." Being a good microbial host involves feeding soil organisms by planting cover crops (see page 96), adding manure and natural soil amendments, and regularly top-dressing (adding a thin layer on the surface) with compost.

The best thing about compost is that it serves two purposes. First, it replenishes the soil by adding some nutrients, as well as microbes and organic material to feed the organisms. Second, it reduces the amount of solid waste generated on a small farm.

In a home composting operation, you might add kitchen scraps, coffee grounds, lawn clippings, and fallen leaves to a pile or bin in the backyard to make small amounts of compost. On an organic farm, the volume of available waste for composting is far more abundant:

The Importance of Organic Matter

Healthy soil contains organic matter—anything that is still alive or was once living. This includes plant roots, compost, mulch, leaves, fungi, bacteria, worms, insects, and any other organisms that live and eventually die in the soil.

Organic matter, while making up only 5 to 10 percent or less of the topsoil in most gardens, is important because living organisms convert material into plant-ready nutrients. As mentioned earlier, stabilized (decomposed) organic matter is called humus, and it helps bind particles into crumbs called aggregates that improve soil structure. It also creates places where ions can collect, improving the soil's ability to hold nutrients and transfer them to plants. Loamy soil with lots of organic matter holds moisture, stores nutrients, and feeds creatures in the soil food web.

Organic material also helps overcome the limitations of soil texture that is not inherently suited for farming. Once you've determined that your soil is either too sandy or too high in clay content, the best way to amend it is by adding organic matter.

What kind of organic matter? Compost (a deliberately decayed mixture of organic items) is the universal cure-all. It loosens clay soils and helps sandy soil hold water. It brings more living material into the soil to help build structure, and it feeds the organisms that are already present.

What Doesn't Belong in Your Compost Pile

Avoid adding the following items to your pile:

- Cat, dog, and pig manure often carry parasites and diseases that may be harmful to humans. They should never be added to compost that is going to be used on food crops.
- Meat, bones, fish waste, dairy products, and eggs can all create foul odors and attract flies and larger animals. Eggshells, however, are usable.
- Leaves or grass clippings that were removed from a non-organic lawn might contain residual herbicides, insecticides, and chemical fertilizers as well as dog manure.
- Weed seeds and diseased plants should be avoided. It's a tricky call. Hot aerobic composting kills many of the weed seeds and plant pathogens, but if you're not very experienced in composting, parts of your pile may not heat fully, and some weed seeds and diseases may survive. If your plants are obviously diseased, burn them or throw them away.

livestock manure (including poultry droppings); straw bedding; crop residue and greenhouse waste; spoiled hay; fallen apples, peaches, or pears; and overripe vegetables—oh, and whatever leaves, lawn clippings, and household waste the daily routine generates. Depending on how many animals you keep, the farm by-products might generate all the compost you will need.

Some Like It Cold

Producing compost can be as technical or as low-key as you make it. *Cold composting* is the easiest way to recycle farm and yard waste, mimicking Mother Nature's own method of breaking down material. Simply gather leaves, hay, manure, and other material, place it into a pile, and wait. In time, the material will break down into fine, crumbly black compost that you can add to soil—but it may take a year or more to get there.

Time isn't the only drawback to cold, or *passive,* composting. Parts of the pile may not break down completely. And because cold composting doesn't generate much heat, the finished product can contain weed seeds or harmful plant diseases. But you can avoid some of these problems if you watch what you throw into the compost pile. Never add diseased plants to the pile, and add weeds only if they are immature—before the plants have begun to produce seeds.

Some Like It Hot

To produce quick compost that is free of weed seeds and diseases, you need to enhance Nature's process by creating the ideal conditions for an explosion of aerobic (oxygen-

Small compost containers located near the house are ideal for collecting kitchen scraps. Add brown leaves to balance the nitrogen-rich kitchen waste.

Carbon-to-Nitrogen Ratios

Composting Material	Carbon-to-Nitrogen Ratio (Estimated)
Aged chicken manure	7:1
Alfalfa hay	12:1
Food wastes	15:1
Lawn clippings	19:1
Fruit wastes	35:1
Rotted manure	15:1
Leaves	60:1
Straw	80:1
Sawdust	500:1
Wood	700:1

Sources: Composting and Mulching, University of Georgia Cooperative Extension; Home Composting Made Easy: homecompostingmadeeasy.com; University of Missouri Extension Service.

loving) bacteria. As these microbes multiply and consume organic material, they generate enough heat to raise the internal temperature of a compost pile to more than 160°F (71°C)—hot enough to kill almost any weed seeds and plant pathogens. But hot composting involves a little more work and preparation.

The trick to making a microbial oven out of your compost heap is to provide the right ingredients in proper proportion. This entails keeping piles of raw materials until you have enough of each to start cooking.

The ingredients you'll be collecting and adding to the compost pile can be broken into two categories: carbon and nitrogen.

Carbon. Think of carbon as anything that is dry and brown—fallen leaves, old hay, straw, wood shavings, and sawdust.

Nitrogen. Anything that is green or living counts as a source of nitrogen: lawn clippings, kitchen scraps (but no meat or dairy), and green crop residues—especially from alfalfa, peas,

Using several compost bins instead of one large pile allows you to transfer partially decomposed material from one bin to the next, mixing the material as you go.

beans, and other legumes that all have high amounts of nitrogen. Despite its usual color, fresh manure is also considered a green ingredient for compost because it's generally a good source of nitrogen (depending on the type of animal that made it). (See page 79 for more information on fertilizing with animal manure.)

The best ratio to fuel a hot compost pile is about 30 parts carbon to 1 part nitrogen. But you can't just measure and add exactly 30 times as much high-nitrogen material as high-carbon matter.

Easy Does It!

You'll find conflicting information regarding the best mix of green and brown ingredients to achieve the coveted 30:1 ratio. Many compost experts recommend one-third brown to two-thirds green; others recommend two-thirds brown to one-third green. The ratio by volume can vary depending on whether an ingredient such as fallen brown leaves are shredded (more condensed) or left whole (more volume). Likewise, if high-nitrogen green items are easily packed (lawn clippings or poultry, rabbit, and goat manure), you may need less green material.

Most of the high-nitrogen material you'll be adding to the pile already contains varying amounts of carbon. Livestock manure, for example, is usually mixed with shavings or straw, and the manure itself is a mix of carbon and nitrogen. In addition, high-carbon materials such as leaves and straw also tend to be fluffy, while high-nitrogen materials like chicken manure are generally more compact, so exact measurements are difficult. So, without getting too technical, if you mix about one-half brown items with one-half green items, you should come close enough to the right blend for hot compost.

By looking at all of your raw materials, you should be able to determine whether you have a carbon-heavy pile or a nitrogen-heavy pile, and then add ingredients that bring it closer to balance. Alfalfa meal can add nitrogen to the pile, while sawdust or wood chips will add carbon.

Most materials contain both carbon and nitrogen. The table opposite provides estimated carbon-to-nitrogen ratios for some common composting materials.

The Style of Your Pile

For small-scale home composting, you can use small plastic bins that keep the compost tidy and clean. You'll also find tumblers and rotating drums that speed the process along. However, most market gardeners and small-scale farmers will generate and use much more compost than store-bought bins will hold.

A small-scale grower with a market garden and some chickens can get by with a row of three or four compost bins, turning the material inside each bin several times until it achieves the desired crumbly brown consistency.

An advantage to having several bins instead of one large pile is that you can transfer partially decomposed material from one bin into the next, mixing it as you go. You can also keep one bin for finished compost while newer material cooks in the adjacent bins.

One of the easiest and cheapest ways to rein in a compost pile is by making a loop of wire fencing. Cut a section of 4-foot (120-cm) high fencing. Stand it on its edge and shape it into

Easy Does It!

If you're a neatnic and want to keep small bits of compost from falling through the spaces in your wooden compost bin, staple hardware cloth or heavy plastic netting (poultry mesh or deer fencing) on the inside.

a loop. Secure the ends with nylon rope or wire to hold it in place. You may need several metal fence stakes to keep the loop intact.

Wire fencing doesn't look as neat as wooden bins, but the material inside will receive air from its entire circumference. And when you want to get at the finished compost, undo the loop and pull away the fencing. For best results, make the loop at least 4 feet (120 cm) in diameter, which will allow enough material for a microbial population boom. Also, because the fencing is open to the air, keep the pile watered during dry weather. Use heavy-gauge metal fencing, not chicken wire, which can rust and break apart after several years.

Another option is a compost bin made of wooden planks (see page 76), which looks tidy and holds your material in a convenient cube. Use 4-foot (120-cm) planks to create the cube, leaving a space between each plank so that air can flow into the pile from outside. Cedar is a good choice of wood because it resists rotting, but it's expensive. Do not use pressure-treated lumber, which contains chemicals that can leach into the soil.

The best location for a compost pile is where you will use it. If it's in a far corner of your property, away from your home, gardens, and sources of manure, dragging materials out to the bin will be a chore. However, if it's located in reasonable proximity to where you do your weeding, mulching, and mucking, you'll be more likely to take advantage of the pile.

Choose a shady location, if possible, especially one under an oak or other deciduous hardwood tree. The tree's leaves will provide shade in the hottest months of the summer, which will keep the pile from drying out.

Layer by Layer

To simplify the creation of compost, I keep three bins: one with raw materials, one for new compost, and one for finished compost. When my "raw materials" bin is filled—usually with garden waste, leaves, and straw—I start a hot pile by bringing a wheelbarrow of chicken manure (mixed with shavings) down to the compost area and mixing it in layers with the raw materials. Everyone has his

Wooden compost bins help to keep your material neat and tidy. This bin is made by cutting notches into planks and interlocking the pieces at the corners.

own method, but for good results, you need to follow a basic recipe.

Most hot composters build their piles in layers. Before you add a single ingredient, start your pile by turning over the soil that will sit underneath your pile. Loosening the top 6 in. (15 cm) ensures good contact between soil organisms and compost materials. Next, add about 6 in. (15 cm) of rough material such as cornstalks or sunflower stalks. The gaps created by these coarse materials will allow air to more easily enter the bottom of the pile.

Now that you've created the foundation, add a layer of high-carbon material (6 in. [15 cm] if it's shredded or 12 in. [30 cm] if it's loose) followed by a roughly 6-in. (15-cm) layer of manure, grass clippings, or other green material. Short on green matter? Go ahead and sprinkle on a layer of alfalfa meal, feather meal, or other high-nitrogen organic material. (Don't use chemical fertilizers, which can harm soil microbes and are not an approved organic amendment.)

After the green stuff, add a few shovelsful of finished compost or ordinary garden soil. The soil will provide a starter population for the microbes that will multiply and begin to cook the ingredients.

Once you've added your loam or compost, turn a hose on the pile. Try to cover the material evenly until it's moist—like a damp sponge—but not soggy and clumped together. Compost piles that are too wet can inhibit airflow and create conditions that favor undesirable anaerobic (dank-loving) bacteria instead of the sweet-smelling and beneficial aerobic microbes.

After the water, add another layer of brown material, green stuff, then the compost or garden loam, and then another dose of water. In other words, rinse and repeat! Keep layering until the compost bin is filled.

Easy Does It!

Many organic farm and garden suppliers sell compost starter mixes for this purpose. The mixes, sometimes called compost activators, are essentially concentrations of soil microorganisms mixed with a food source to get them started. I've used several different types of activators, all with terrific results, but I've also had great results from adding microbe–rich soil or compost. If you're curious, you could experiment with one pile that uses compost starter and another pile that is activated with your own finished compost or garden loam.

Temperature and Organic Compost

If you want to market your food or compost as organic, your compost needs to adhere to the same organic guidelines as the rest of your business. First, your enclosed compost pile must maintain temperatures between 131°F (55°C) and 170°F (77°C) for at least three days. In addition, the raw materials used in the pile must have a carbon-to-nitrogen ratio of between 25:1 and 40:1.

If you don't adhere to these requirements, you'll need to apply your compost to food crops as you would apply manure. Organic-growing guidelines specify that a grower cannot use raw manure on food crops within 120 days of harvest if the food itself comes into contact

The Stages of Hot Composting

There are four hot composting stages: mesophilic (getting warm), thermophilic (getting hot), cooling, and curing.

Mesophilic Stage

Moderate-temperature microbes begin breaking down material, which causes the pile to heat up to about 104°F (40°C). As heat rises above this level, heat-loving microbes take over.

Thermophilic Stage

Heat-loving bacteria thrive, and their activity raises the compost temperatures to as high as 170°F (77°C). Ideal composting occurs at slightly lower temperatures—150°F (66°C)—so most compost makers turn the pile to keep it at about this temperature. Sustained high temperature will kill weed seeds and the organisms that cause plant diseases.

Cooling/Curing Stage

After the pile is turned several times, the raw materials will have been consumed, and the temperature will decline. Mesophilic (moderate-temperature) microbes return and continue to break down. Earthworms and other soil organisms also return during the cooling/curing phase, which can last several months.

with the soil. For grain crops and fruit trees, you can't apply raw manure within 90 days of harvest. (See "Animal Manure" opposite for more information.)

Personally, I use only organic materials in my compost, but it wouldn't qualify as certified organic because I'm not consistent about temperatures and moisture requirements throughout my pile. However, I usually make my compost in the fall and apply it to crops the following season, well before the required time for harvesting. Or I apply newly made compost as a top dressing for some raised beds in the fall and let it sit through winter.

To prevent the pile from drying out or becoming too soggy in rainy weather, spread a tarp over the top and let the microbes get to work. In a few days, you can pull back the tarp and stick a garden fork into the center of the pile to lift out a scoop. If the pile is cooking properly, you should notice a puff of steam (especially if the weather is chilly). You should also feel heat rising from the pile, and the smell should be earthy and sweet—almost like an oaky red wine. If the pile smells like rotten eggs or a swampy pond, you've probably added too much water—turn the material over to introduce air, encourage the growth of beneficial aerobic bacteria, and reinvigorate the pile.

A compost thermometer will tell you the temperature of the material inside your pile. Move it around to make sure the entire pile is cooking.

If you're serious about making hot compost, buy a compost thermometer. They sell for as little as $25 and look something like the thermometers you'd stick in a turkey or roast while it's baking in the oven—except the compost thermometer's stem is about 2 ft. (60 cm) long, allowing it to reach deep into the center of a pile.

A thermometer will take away the guesswork of determining whether your pile has become hot enough to kill weed seeds and pathogens. You can also move it from one spot to another in the pile to determine whether the entire pile is cooking evenly or there are just a few hot spots. In large composting operations, thermometers are a safety measure, warning you if the internal temperature is a potentially scalding 170°F (77°C) before you stick your hand into the pile. Although compost that reaches above 160°F (71°C) is dangerous to handle, it gets the job done when it comes to killing off the diseases and weed seeds.

After the compost pile has cooked for several days to a week, turn the material over. Use a fork to mix it up, pulling the compost from the bottom and sides and piling it on top and in the pile's center. If you have an adjacent compost bin that's empty, toss the compost from one bin into the next, mixing it up as you go. The compost should heat up and cool down again over the next week. At the end of this second cooking session, the material should begin to look partially decayed. Turn it at least one more time and let it heat up and cool.

When your compost is ready, use a spading fork to dig in and have a look. You may still find some recognizable stems, but you can give it a rest if the material looks as if it has substantially decomposed. Worms will quickly migrate into the pile and break it down further. And, now that the pile has heated up several times, fungi will take over the decomposition and break down coarser high-cellulose material until the compost is crumbly and dark. Allow at least thirty days for the compost to cool and for beneficial microbes to repopulate the pile.

Animal Manure

Animal manure was a part of nature's equation long before the first farmers began working the soil. Cows eat grass. Cows produce manure. Manure fertilizes the soil. Grass grows greener. Cows eat grass, and the cycle begins anew.

Organic growers who keep livestock are wise to take advantage of the equation on their own farms—recycling farm waste and cutting the bill on high-cost fertilizers. And growers who aren't fortunate enough to keep animals can seek a nearby supply of horse, cow, or poultry manure.

The benefits to an organic program are numerous: Manure adds organic material to the soil; it boosts phosphorus, nitrogen, and other nutrients while increasing the population of microbes; and, above all, it's inexpensive. But there are reasons for caution. In addition to its beneficial ingredients, manure (especially from horses) sometimes carries undigested weed seeds, which means that manure that hasn't been hot-composted can turn your well-kept fields into a germinating ground for thousands of unwanted weeds. Manure can also impart a bad flavor to certain crops—as manure decomposes, it releases compounds that squashes, potatoes, cucumbers, and a number of cabbage crops can absorb. It can also contain salmonella and E. coli bacteria, which sometimes survive inconsistent hot composting.

Some types of manure aren't recommended for vegetable farming at all because they can carry parasites that are harmful to humans. Manure taken from a non-organic dairy or livestock operation may carry hormones and antibiotics that were used to treat the animals. In addition, manure can sometimes hold heavy metals, pesticides, or herbicides from the animal's diet. Avoid dog, cat, and pig manures, which can pass parasites on to humans—even after the manure is composted.

You can eliminate some of these unwanted ingredients through hot composting, but certain harmful-to-humans bacteria such as salmonella and E. coli can sometimes survive in compost. That is why farmers who use manure should be cautious about the source of their manure and the way they treat and apply it.

Manure and Nutrients

Nutrients in livestock manure can vary greatly from animal to animal—and even among the same species—depending on what type of feed they consume. Manures are also often mixed with varying amounts of wood shavings or straw bedding material that affect the concentration of the key nutrients nitrogen (chemical symbol N), phosphorus (P), and potassium (K). The following are estimated N–P–K concentrations in common animal manures. Where the information is available, it includes analysis of fresh and dried manure.

Animal	N–P–K Concentrations (%)	Comments
Alpaca	1.5-0.2-1.1	High in nitrogen; usually sold in small quantity for home gardeners
Beef cattle: fresh	0.6-0.4-0.5	Low in nitrogen; may contain salts and weed seeds
Beef cattle: dried	1.2-2.0-2.1	High in phosphorus and potash; may contain salts and weed seeds
Chicken: fresh	0.9-0.5-0.5	High in nitrogen, can burn plants; manure from layers can be high in pH if feed contained ground oyster shells
Chicken: dried	1.6-1.8-2.0	High in all major nutrients
Dairy cow	0.57-0.23-0.62	Good fertilizer; some weed seeds; slow to decay; preferred over beef cattle
Goat	1.44-0.5-1.21	High in nitrogen, potash, and organic matter; low in weed seeds
Horse: fresh	0.6-0.3-0.5	May contain weed seeds; will burn plants if used fresh (hot)
Llama	1.7-0.7-0.66	High in nitrogen
Rabbit	0.6-0.3-0.5	High in nitrogen
Sheep	0.7-0.3-0.9	High in nitrogen and potash; few weed seeds; high in organic matter
Swine: fresh	0.6-0.3-0.4	Not for use on food crops; may contain parasites and diseases that are harmful to humans
Swine: dried	2.2-2.1-1.0	Not for vegetable farming
Turkey: fresh	1.3-0.7-0.5	High in nitrogen

SOURCES: Data from North Carolina State University; USDA's Natural Resources Conservation Service; National Sustainable Agriculture Information Service; and Elizabeth P. Stell, *Secrets to Great Soil* (Storey Publishing, 1998).

Certified organic farmers are required to follow even more stringent guidelines. Raw manure cannot be used on food crops within 120 days of harvest if the food itself comes into contact with the soil (this goes for carrots, lettuces, greens, strawberries, and most vegetables). For grain crops and fruit trees, raw manure cannot be applied within ninety days of harvest.

Adding manure can also create an imbalance of nutrients. Depending on the type of animal that produced it, significant amounts of nitrogen and phosphorus may occur in both raw and composted manure. While plants, the soil food web, and leaching quickly deplete the nitrogen, phosphorus can persist for a much longer time. And repeated additions of manure without soil testing can result in an overload of phosphorus.

Even manure that has been composted can create a nutrient imbalance when used indiscriminately. Tom Morris, a former farm manager with the Rodale Institute in Emmaus, Pennsylvania, and later an associate professor of soil sciences at the University of Connecticut, says some organic growers who use manure-based compost to fertilize their fields are creating phosphorous overloads. He stresses that any grower who is using regular applications of compost—especially compost derived from large amounts of manure—needs to test the soil every year and take note of the phosphorous content.

One way to avoid some of these problems is to apply manure in the fall and mix it into the soil using a rotary tiller or tractor. Planting cover crops immediately before or after the manure has been applied will further improve soil because the cover crops will store the nutrients and then release them into the ground the following spring. (See page 96 for more information on cover crops.) Conscientious growers should not apply manure late in the fall without integrating it into the soil; manure left above frozen ground can run off with melting snow and rain, potentially contaminating nearby water supplies.

Vermicomposting

Worm composting, or vermicomposting, is a way to harness the power of hungry earthworms to produce small amounts of nutrient-rich worm waste, called castings. Worm castings are considered the perfect organic soil amendment—rich in plant-ready nutrients and alive with beneficial microorganisms. Think of it as compost that has been re-composted and inoculated with microbes by passing through the guts of a thousand wriggling worms.

You can create this kind of compost indoors in the winter and outdoors in a shady spot during summer

Worms in a compost bin will speed up the breakdown of organic material and add microbe–rich castings to the mix.

Make Your Own Vermicomposter

What You'll Need

Electric drill with ¼-inch (6 mm) drill bit
1 plastic storage bin with cover (24 in. [60 cm] x 18 in. [45 cm] and at least
   12 in. [30 cm] deep)
Floating row cover or fiberglass screening (in the same size as your bin)
Bedding material (such as a mix of coir and compost)
2 lb. (1 kg) red wiggler worms (*Eisenia fetida*)

1. Drill about a dozen holes, evenly spaced, across the bottom of the bin.

2. Cover the holes with row cover or screening to keep the worms from slithering out: Cut the material to fit snugly across the bottom, and make sure there are no wrinkles to create a crawl space for the worms.

3. Fill the bin with about 6 in. (15 cm) of bedding material; a mix of coir (coconut fiber) and compost is ideal. Keep the material moist; the lid will help.

4. Add 1 lb. (0.5 kg) of red wiggler worms for every two people in your household. Red wiggler worms, which are the ideal worms for creating compost, can be purchased for about $25 per lb. from a mail-order catalog or vermicompost supplier (see "Resources" on page 356).

5. Begin feeding your worms immediately. They'll eat almost any vegetable waste your family produces—carrot and potato peels, apple cores, rotting lettuce leaves, and grass clippings (from organic lawns only), for example. Be careful with coffee grounds and citrus peel, as too much can acidify the worms' habitat. Do not add meat scraps, eggs, dairy, or fish that can create foul odors. Also don't overfeed the worms, or the uneaten material can grow mold and create a nasty odor.

months, allowing you to keep up the practice all year long. Suppliers sell dozens of different vermicomposting bins, some with stacking trays to make separating the worms from their castings easier (see "Resources" on page 356 for more information). I use the Can-O-Worms bin, which I picked up at an organic farming convention, and it has worked extremely well. But if you don't want to plunk down a hundred bucks for a fancy-pants worm house, take heart: It's easy to make your own vermicomposter using a plastic storage bin and a few household tools.

4 Tips for Successful Vermicomposting

Heed these tips and your worms will be happy and healthy.

1. In winter, keep the worm bin inside a basement. Once warm weather arrives, you can place the bin outside in a shady spot—the closer it is to the house, the more likely you will be to use it.

2. Worms do not like light or excessively hot or cold temperatures, so avoid opening the bin in direct sunlight.

3. Depending on how many worms you have, they should have created ample castings in three to six months. Unless they're exposed to bright light, worms will generally migrate to a level above their castings, so they can be scooped off the top of the mix and placed in another

container while you dig out the castings. After you separate the worms, add more bedding material and start another batch of vermicompost.

4. If your worm population doubles or triples, just add another bin and harvest twice as many castings next time.

Buying Compost

If your space is limited and your needs are great (which is often the case with small-scale farmers), you can purchase compost from nearby farms or a compost supplier. But not all compost is alike, and buyers should beware. Always check the ingredients of compost before you buy it to make sure the materials are safe for food crops and are produced organically.

Some composting operations accept grass clippings from landscapers who use chemical pesticides and weed killers. Others may use manure from farms that feed their livestock excessive hormones and antibiotics. And the machinery used to transport and turn compost can leak oil, fuel, or hydraulic fluid, contaminating the product. Even autumn leaves can be contaminated with auto exhaust if they're raked from a roadside with heavy traffic, says Karl Hammer, founder of the Vermont Compost Company in Montpelier, Vermont. Hammer recommends buying from a certified supplier or spending time getting to know exactly where all the ingredients are coming from.

Using Compost

So now that you've made compost, what do you do with it? Well for starters, you can mix it with peat moss and vermiculite

Manure, Compost, and Herbicides

Manure, compost, and hay aren't as safe and natural as they used to be. A relatively new class of herbicides can survive composting and remain active in the soil for years—even after being eaten and passed by grazing animals. That means the chemicals might be lurking in manure and waiting to harm your plants when you apply it. If your beans and peas falter and your tomatoes show signs of curled leaves and twisted stems, the culprit could be persistent herbicides. Sensitive crops include beans, lettuce, eggplant, peas, peppers, potatoes, tomatoes, and other broad-leaf vegetables.

Before buying manure or compost from someone who keeps livestock, ask if the animals were grazed on land that was treated with herbicides. The newer persistent herbicides are classified as "pyridine-carboxylic acids."

Also inquire if the animals ate hay or feed from a field treated with chemical weed killers. Farmers might not even know the answer because tracking hay to its source isn't always easy. Also, even if the herbicides are deemed safe for animals and don't kill grass plants, hay growers and suppliers might not consider them harmful. Herbicide labels advise farmers not to use manure for vegetable crops if the animals foraged on treated land, but not all warnings are heeded. The safest source of manure is from your own organically fed livestock or the animals of someone you know who uses organic feed and hay.

Manure isn't the only source of persistent herbicides. Golf courses and commercial landscapes may have been treated with these weed killers, affecting the grass clippings. Don't compost grass clippings from commercial sites, and don't use municipal compost on vegetable gardens unless you know the material is free from chemicals. If you mulch your beds with straw or grass clippings that have been treated, the chemicals can leach into the soil and cause damage.

Washington State University offers a way to check compost at home before you add it to the soil. The method involves planting pea seeds into three pots of clean potting mix and into three pots of mix containing 1 part potting mix and 2 parts of the compost in question. You label the pea plants and care for them identically, then grow for three weeks and observe for telltale signs of the herbicide: cupped leaves and twisted stems. For instructions on the test, go to puyallup.wsu.edu/soilmgmt/Pubs/CloBioassay.pdf

Brew Your Own Compost Tea

How do you make compost even better? By steeping it into microbe-rich compost tea—an organic elixir with the power to make plants stronger, healthier, and more resistant to disease.

But making compost tea isn't as simple as soaking compost in water. To encourage microbes, you need oxygen, so the water needs an air pump.

When properly made (with active compost and good aeration), compost tea produces billions of beneficial aerobic microbes that help plants resist pathogens. How? You can spray the billions of beneficial bacteria created in an aerobic solution onto leaves, where they will remain and feed off sugary solutions exuded by plants. Once the good bacteria are in place, they crowd out other, potentially harmful pathogens, preventing them from attacking your plants. And, while compost isn't usually a source of major nutrients, the micronutrients delivered in compost tea can help produce crops with higher nutritional value. You can also mix compost with fish emulsion or another foliar feed. Together the mixture can provide nutrients when plants aren't taking up enough material from their roots alone.

Pouring compost tea into soil at the base of plants adds bacteria, protozoa, and nematodes to the root zone. In a healthy soil, the microbes continue to multiply and boost biological activity. Their presence also helps protect plants from harmful fungi and bacteria.

Organic supply companies sell numerous kits for making compost tea, from small bucket-size brewers that sell for under $100 and make a few gallons at a time to commercial units that brew hundreds of gallons and cost thousands of dollars. You can also make your own compost tea brewer using household supplies and a few items from a department store or aquarium supply shop.

What You'll Need

1 large air stone (a 6-in. [15-cm] disc is ideal; a 6-in. [15-cm] wand will suffice). (You can also use three smaller stones and a splitter valve.)

Two 5-gal. (19-L) utility pails that haven't been used to store detergents or harsh chemicals (food-grade buckets are ideal)

1 heavy-duty aquarium air pump

6 ft. (2 m) of aquarium hose

5 gal. (19 L) of non-chlorinated water

1 large mesh bag to use as a strainer (Note: You can buy compost bags/filters from a supplier or use paint-filter bags from a hardware store. Or improvise with several items: Fine-mesh netting will work; nylon panty hose is also commonly used, but some brewers say panty hose, multiple layers of cheesecloth, or extremely fine mesh can filter out some of the larger beneficial microbes.)

1. Place the air stone or stones at the bottom of one pail and attach it to the pump. If your water contains chlorine, fill the bucket with water and run the pump for at least an hour, which should allow the chlorine to evaporate. (Note: This is a vital step; chlorinated water will kill the microbes that you're trying to cultivate.)

2. When your water is ready, put the air stone or stones at the bottom of the second pail. Fill the bucket half-full with compost but don't pack it down.

3. Add the non-chlorinated water from your other bucket until it's about 3 in. (8 cm) from the rim (you don't want to spill it).

4. The ideal temperature for microbe growth is 70°F (21°C), so if it's cold outside, move your arrangement to a warm room indoors. Plug in the pump and watch the bubbles. The air should circulate around and through the compost.

5. Once the bubbles are working, some brewers mix in food to hasten the growth of microbes, But there are cautions against adding molasses or sugars, which feed both good and bad bacteria. (See "8 Tea-Making Tips.")

6. Allow the air pump to run for twelve to twenty-four hours, stirring frequently with a stick or dowel. (Stirring the mix releases more microbes from the compost, allowing them to circulate and multiply.)

7. After twelve to twenty-four hours, the bacteria will be ready to use. Shut off the pump and pull out the air stone.

8. Pour the tea from one bucket into the other, using the mesh bag or mesh material as a filter.

9. If you're using the tea as a soil drench, pour it on your crops immediately. If you want to use it as a foliar spray, strain it again to remove any compost particles, which can clog a spray nozzle. The best flow will be achieved with a compost tea nozzle, which allows occasional particles to pass through. You can also pour the tea into a watering can and sprinkle it over your plants' foliage instead of spraying.

9 Tea-Making Tips

1. Do not store your tea for later use. Once the air pump has stopped, the aerobic microbes in the tea will survive for only a few hours without air. Ideally, you should use all your compost tea within four hours of brewing it.

2. It's best to apply compost tea in the evening or early morning, so time the brewing to coincide with those hours.

3. When applying compost tea with a sprayer, try to cover the entire surface area of each plant. And it goes without saying that the sprayer shouldn't have been used for any chemical pesticides, herbicides, or other materials that could harm either your plants or the microbes in your tea.

4. For extra nutrition in a foliar feed, add fish emulsion or seaweed extract moments before you apply it.

5. If the finished compost tea smells sour or like rotten eggs, you probably have too many anaerobic bacteria in the mixture. Discard the tea, which may contain harmful bacteria, and brew another batch, making sure there are enough bubbles to keep good bacteria thriving.

6. There are numerous recipes and variations for making compost tea, including teas that promote fungal growth over bacteria. Some brewers place their tea inside a mesh teabag and suspend the bag in the pail of water. Some use compost tea activators for increasing the microbe count. (See the "Resources" section on page 356 for more information.)

7. The National Organic Program restricts the use of some compost teas on food crops. Tea that is made with unapproved compost or with a sugary feeding material such as molasses may contain populations of potentially dangerous E. coli bacteria. Compost tea that does not meet NOP specifications can only be applied at least 120 days before harvest for crops that contact the soil and 90 days for grain, corn, and crops that do not contact the soil.

8. There are other restrictions for compost teas and manure teas. Before using compost tea, consult the NOP at ams.usda.gov/AMSv1.0/nosb

9. Compost that is used for tea must be made according to USDA/National Organic Program specifications, especially if molasses or another microbe food is added. Tea made from uncomposted manure must be applied under the same restrictions as fresh manure.

The Promise of Biochar

One of the newest ideas in soil improvement is actually more than 2,000 years old. Biochar—which is made today by heating wood or grasses in a closed container with reduced oxygen—was practiced by farmers in the tropical Amazon rain forest of Brazil, where rain-washed, oxidized soils are traditionally lacking nutrients. The native Indians created charcoal and added it to the earth, resulting in deep, rich topsoil that remains fertile today. Scientists interested in duplicating the technique are studying Brazil's terra preta (Portuguese for "black earth").

Making biochar isn't as simple as burning wood and throwing the unburned material into the soil. For high-quality biochar, wood is usually cooked inside a closed but not airtight metal cylinder that is heated to a high temperature with fire outside the container. It's then crushed into a soil amendment, mixed with compost and water to inoculate it with microbes, and added to soils.

Biochar is touted for its ability to add stable carbon to the soil, removing it from the carbon cycle and thereby reducing greenhouse gases. And its porous structure creates ideal soil space for microbes to flourish.

Not everyone is on board. Some critics say biochar benefits aren't fully known, and it's more expensive to make than compost.

You can find more about biochar and how it's made at biochar-international.org

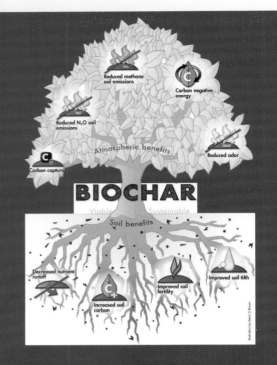

or perlite and use it in seed-starting mixes. (Just keep in mind that you should trust only hot-composted, pathogen-free compost for the important task of starting seeds.) You can also spread it over beds or fields and turn into the soil using a rotary tiller or tractor with a disk harrow attachment that breaks up the ground and mixes organic material into the topsoil.

Compost may also be used as a side-dressing: Sprinkle the material around existing plants and scratch it into the top 2 in. (5 cm) of soil. Many farmers top-dress their fields with several inches of compost each year, adding nutrients and improving texture. When fields are top-dressed in the fall, the natural cycle of freezing and thawing helps to break down the material and incorporate it into the soil. A top-dressing of compost in the spring can act as mulch, conserving soil moisture and preventing weeds from germinating.

Step 4: Mark Your Progress and Adjust Accordingly

Compost, cover crops, and manure are all beneficial to organic growing, but they will only get you so far. Even when combined with each other, they won't return some of the valuable soil elements depleted by the ongoing process of farming the land. That's why it is crucial to reintroduce essential minerals to the soil and test your soil regularly to prevent the land from becoming depleted.

The actions of planting and harvesting crops aren't sustainable because the crops remove essential nutrients from the soil. And when the crops are harvested and sold, the nutrients are effectively removed from the natural cycle of growth and decay. To return the minerals, you'll need to add organic fertilizers to the soil. But organic soil inputs tend to be more expensive than their chemical counterparts, so it makes both economic and scientific sense to test your soil before adding purchased fertilizers. Do-it-yourself kits can give you an indication of your soil's acidity (pH level) and its N-P-K (nitrogen, phosphorus, and potassium) content. But plants need much more than those three major nutrients.

There are seventeen essential soil nutrients for optimum plant growth (but three of them, carbon, oxygen, and hydrogen, are available from water and air). While even novice gardeners know about nitrogen, phosphorus, and potassium, "the big three," there are secondary nutrients and micronutrients that also benefit plants. A lack of these elements can hinder crops from reaching their full genetic potential. If even one ingredient is missing or deficient, it can become "the limiting factor" that holds your crops back.

In many cases, the lack of nutrients isn't noticeable to the naked eye. . . but insects can tell. For instance, the absence of calcium can lead to weaker cell-wall construction. Bugs have a sixth sense in identifying this weakness and will flock to feed on these softer-than-normal plants. So where you think you might have an insect problem because pests are invading your vegetable patch, you might actually have a soil deficiency. Understanding these key nutrients, testing for and noticing the signs of deficiency, and amending the soil accordingly are all helpful in cultivating soil that will become the foundation for healthy harvests.

Testing Your Soil

Major universities and county agricultural extension services offer soil analyses for a modest price. These tests usually include a check of the soil's pH, a search for potentially harmful levels of lead, a reading of the soil's potential for *cation exchange capacity* (CEC), and an inventory of major soil nutrients needed for crop production. Many university tests also include recommendations for fertilizers and organic amendments to bring deficient soil back into the range for healthy agriculture.

Private laboratories will also test your soil but often for a slightly higher price. Some will offer in-depth analysis of the organic content, microbial activity, trace minerals, ion activity, and other features. Labs can also vary in how they analyze soil for nutrients.

Check with other growers in your area for a recommended lab. Facilities that are

A reading of the soil's pH, its acidity or alkalinity, is part of most standard soil tests.

How to Take a Soil Sample

Dumping fertilizers on your fields without doing a soil test first is a little bit like taking cold medicine to treat a sprained ankle. It's misguided, wasteful, and ineffective.

To get the most benefit from fertilizers, you need to add the right amounts of the nutrients that are missing from your soil. And the best way to determine what's missing from your soil is to conduct a soil test.

A good soil analysis will tell you what fertilizers your soil needs, along with your soil's pH and its ability to hold nutrients before they leach away. A test will also indicate whether the soil has any contaminants, such as lead, that make it unsuitable for growing food crops.

Check with universities or private soil-testing labs in your area. Once you've found a suitable testing place, you'll need to follow that lab's guidelines for sampling and preparing your soil for a test. Instructions are widely available online, but most labs have a few basic recommendations for taking a soil sample:

- Late summer and fall are usually good times to take samples for a soil test because labs are much less busy than they are during the spring and early summer. In addition, soil is drier at this time of year, making it easier to collect and prepare a sample. And when you get your recommendations, you can apply limestone before winter, ensuring that the soil is ready for spring planting.

- Take at least five small samples from random spots around each planting field and mix them together to get a general reading for the entire field. If you know an area was recently fertilized, take separate samples from that area.

- To take your sample, scrape away the thin surface layer of loose organic material (it's usually less than ¼ in. [6 mm] deep). Then dig a clean hole about 6 to 8 in. (15 to 20 cm) deep, with a smooth wall on the inside of the hole. Use a straight-edged spade to take an even vertical slice of the soil that includes equal amounts from the top to the bottom of the hole. Do the same for each sample you take around the field.

- Mix all the samples in a plastic bucket. (Don't use galvanized metal pails, which can alter the sample.) When finished, you should have about 2 qt. (2 L) of soil.

- Sift the soil through a coarse screen to remove stones that are pea-size or larger. Also remove any roots or large chunks of wood chips and organic matter. (Alternatively, you could spread the soil in a plastic planting tray and pick out stones and other material by hand.)

- Allow the soil to dry in a shaded spot, and then pack it into an appropriate container (sometimes a box provided by the soil test lab, but often just a plastic bag with a zipper seal).

- If you're taking more than one sample, number or label each and keep a copy for yourself so you can match the results to the right plot of soil.

When you receive your soil report, it should indicate the amount of each nutrient that you need for optimum plant health. Save past soil tests so that you can compare the results with recent reports and mark progress in your soil-building efforts.

Organic farmers vary in their philosophy about how frequently to repeat soil tests. Some swear by taking tests every year or two, while others put off testing for several years once they know their soil is lead free, rich in nutrients, and high in organic material. If you're working to correct a problem, you'll want to get your soil tested more frequently.

geographically close to your own property may have a good perspective on local soil characteristics and can make more educated recommendations.

Whether you choose a private or university lab, once you have a record of what is lacking in your soil, you can follow a formula to improve it. The remedy may require adding organic matter, raising the pH by adding limestone, or boosting the presence of macro- and micronutrients in the soil. Stick with the same trusted laboratory for follow-up testing. That way, the methods remain the same, and the readings will be consistent with previous tests.

The Big Three

Nitrogen, phosphorus, and potassium are the most commonly known fertilizers and the nutrients that plants use the most as they manufacture food. The front of every bag at your local garden center acknowledges the ratios of these elements.

Nitrogen (N)

Adequate nitrogen is key to the growth of green, leafy material. Plants that produce lots of foliage are traditionally high nitrogen feeders. Lettuce and greens require nitrogen, as do

The right soil nutrients, along with ample organic matter and steady water will help produce healthy peppers, tomatoes, and other vegetables.

Nitrogen, phosphorus, and potassium are the "big three" of soil nutrients, often called by their symbols N–P–K.

all varieties of corn, which produce huge grassy stalks before bearing ears for harvest. Nitrogen is also a necessary building block of proteins and is an essential ingredient in chlorophyll, which allows plants to manufacture their own food from sunlight. Too much nitrogen—especially synthetic nitrogen that is quickly absorbed—causes rapid but weak leafy growth that invites diseases and insect pests.

Legumes (beans), peas, and cover crops such as alfalfa, vetches, and clover develop a partnership with specific soil bacteria that allows the plants to take in nitrogen and sequester it in the soil. If you pull up a pea or clover plant, you might see little white bumps stuck to the finer roots. These are the nodules where the nitrogen-fixing bacteria collect.

Nitrogen is often short-lived in the soil, either used up by plants or leached out of the soil with melting snow and rainwater. When plants are short of nitrogen, they can move it from one spot within the plant to another—usually, the lower leaves give it to new top growth, sometimes turning bottom leaves yellow. Organic nitrogen sources include soybean meal, cottonseed meal, alfalfa hay, feather meal, fish meal, or fish emulsion liquid (all slow-release). Fertilizer is a potent source of nitrogen, but it burns roots if applied directly to plants.

Phosphorus (P)

Phosphorus is most often associated with healthy blooming; it's always a high number in synthetic fertilizers made for the floral industry because it encourages fast growth and multiple blossoms. But phosphorus is essential in crop production, too, helping plants to develop strong roots and set fruit. And, like nitrogen, it's an essential ingredient in photosynthesis.

Plants that suffer from phosphorous deficiency may be stunted and display yellowed leaves that are occasionally tinged with purple. You can often add phosphorus through manure and compost, but it's also available as an organic fertilizer from ground-up rock phosphate, called *colloidal phosphate,* and bonemeal.

Converting Soil Test Results

Some soil test labs report the soil needs in parts per million (ppm) instead of lb. of fertilizer per acre. To convert ppm to lb. per acre, understand that the soil area where the fertilizer will be added is the plowing layer, which is about 6 to 8 in. (15 to 20 cm) deep. Soil in the plow area typically weighs about 2 million lb. (900,000 kg) per acre. So to convert ppm to lb. per acre, simply multiply by 2. If the recommended amount of a nutrient is 20 ppm, and your test results read only 10 ppm, you need to add 10 more ppm. Multiply 10 by 2 to get 20 lb. (9 kg) per acre. That's what you'll need to add. (Note that amendments are almost never pure. If your particular amendment is only 25 percent pure and 75 percent inert or other ingredients, you'll need to multiply the number again by 4. That would require 80 lb. [36 kg] of material per acre.)

Fertilizer labels indicate the percentage of actual nitrogen, phosphorus, and potassium in a bag. If a 40-lb. (18-kg) bag of blood meal is labeled 25-0-0, you know that 75 percent of the material is not nitrogen, meaning there's actually 10 lb. (4.5 kg) of nitrogen in the bag.

Potassium (K)

Overall strength and vigor, ability to fight diseases, and quality of fruiting are all attributed to adequate amounts of potassium in the soil. Potassium is also another essential ingredient in photosynthesis and is used up fairly quickly by plants.

Leaves on plants deficient in potassium will sometimes turn yellow and brown around the edges. The entire plant can seem to wilt, even when soil is sufficiently moist. For amendments, sources of potassium include wood ashes, manure, and compost, as well as rock potash and greensand. Greensand, which is mined in New Jersey, Texas, and Arkansas, contains varied amounts of iron depending on its source.

Adequate calcium in the soil helps plants build stronger cell walls that can resist insect damage.

Secondary Nutrients
Calcium, magnesium, and sulfur are secondary nutrients, not because they're less important than other nutrients but because your plants don't need as much of these elements.

Calcium (Ca)
You already know about bugs and their attraction to calcium-deficient plants. In addition to fighting insects with stronger cell walls, calcium contributes to the overall strength and vigor of crops. It also assists in retaining and transporting other nutrients throughout the plant.

Calcium can move through a plant only when dissolved in water and can therefore become unavailable to growing fruit during times of drought. Sometimes a lack of calcium can cause blossom-end rot in peppers and tomatoes, which can be a problem if plants don't get enough water. This can even occur when there is substantial calcium in the soil: Too much magnesium in the soil can block the uptake of calcium, which complicates something as simple as adding lime. Both dolomitic lime and high-calcium limestone will lower the acidity (raise the pH) of acidic soils, but dolomitic limestone also contains magnesium. Other sources of calcium are gypsum and superphosphate.

Micronutrients may help your strawberries grow stronger. Organic growers need a soil test to document the need for many micronutrients. Check with your organic certifier for more information.

Magnesium (Mg)

This nutrient activates plant enzymes, which are essential for controlling growth. It also plays a role in photosynthesis. Too much potassium can block magnesium uptake, as can soil that is too acidic. Plants lacking magnesium might have leaves that display yellowing between the veins (as opposed to around the edges). To remedy a deficiency, mix 1 teaspoon (5 ml) of Epsom salts in 1 gal. (3.8 L) of water, and apply the mixture as a foliar spray. Organic sources of magnesium include dolomitic limestone, greensand, and fish meal fertilizers.

Sulfur (S)

Sulfur helps root growth and development of seeds, assists in producing chlorophyll, and improves a plant's resistance to cold. Plants that are deficient in sulfur can appear stunted and show damage to leaves, which curl at the tips and appear dried and brown. Farmers and gardeners also use sulfur to make alkaline soils more acidic (especially for such acid-loving plants as blueberries). Adding gypsum can increase sulfur content in soil as well.

Rock Dust for Micronutrients

Organic growers who rely on manures and compost for much of their plant nutrition may find their soils low in some micronutrients. There are also growers who assert that plants need dozens more micronutrients than the seventeen listed as necessary. One popular way to add a wide variety of minerals, without overdosing on a few, is by incorporating rock dust into your soil or compost.

Rock dust is crushed rock—sometimes basalt or a mix of rocks containing multiple different minerals. It's a key part of the regimen for growers who aim to "remineralize" their fields for better plant and food nutrition.

When mixed with compost, soil microbes slowly break down the material in rock dust and make it available to plants. Suppliers sell rock dust by the bag for gardeners and small-scale farmers. Growers with multiple acres can order a truckload of rock dust or gravel dust from a quarry. Natural silt from the bottom of dredged riverbeds is also a source of minerals.

More information is available at the website: remineralize.org/

Micronutrients

You'll find multiple sources for micronutrients, and the absence of any one micronutrient can cause problems. But because micronutrients are used in such small quantities by plants, looking for individual sources without proof of deficiency isn't practical. A soil test will help you determine what's lacking and how to remedy the problem. If the report doesn't include specific recommendations, you can ask for them—they're worth any small additional expense.

Boron (B)

Beets need boron, which is helpful in sugar and carbohydrate production. It's also a must for better fruiting and development of seeds. Plants that suffer from boron deficiency develop slowly and can show yellowing at the bud tip, which can eventually dry out and die. Fruits may rot from the centers and ripen unevenly. Usually most acute during times of summer

Electricity in the Soil

All the action—as far as the transfer of nutrients from soil to plants is concerned—takes place within the root zone. That's where the microbes consume complex organic compounds and break the material down to basic elements. These dissolved nutrients carry electrical charges and are adsorbed onto the surface area of negatively charged soil particles.

Dissolved nutrients that carry negative charges are referred to as anions (pronounced ANN-eye-onz), and positively charged ions are called cations (pronounced KAT-eye-onz). These opposite charges attract each other. Clay particles and humus have a lot of surface area that is negatively charged, and therefore they attract and hold positively charged cations. That means these soil particles have a high cation exchange capacity (CEC). Light-colored sandy soils with low amounts of humus will show the lowest CEC. Their ability to hold nutrients is not very good, as demonstrated by the nutrients that quickly leach out of sandy soils when it rains.

Some important plant nutrients, including ammonium, calcium, magnesium, potassium, and zinc, carry positive electrical charges when they are in their dissolved form. When soil-testing labs report the soil's CEC, they are describing its ability to collect and retain positively charged elements like these. The higher the CEC, the more nutrients there are for plants. Adding organic matter to poor soils—especially to sandy soils—will help improve the soil's CEC.

drought, when roots aren't taking in much water, boron problems are also more pronounced in sandy and acidic soils. You'll find boron in organic matter and in Borax. To fix a deficiency, add 1 teaspoon (5 ml) of household Borax to 1 gal. (3.8 L) of water, and apply the mixture to the soil. One gal. (3.8 L) treats about 30 row ft. (9 m).

Chloride (Cl)

Chloride stimulates root growth, is essential for photosynthesis and plant metabolism, and aids in water circulation throughout the plant.

Copper (Cu)

Copper is beneficial for root growth and metabolism. Sandy soils with poor organic content can be prone to copper deficiencies. In addition, too much phosphorus and iron in soil can inhibit copper's availability to plants.

Iron (Fe)

In addition to assisting in plant metabolism, iron helps plants develop chlorophyll and is key for nitrogen fixation among legumes. Alkaline soils sometimes suffer from an iron deficiency. Signs of this include yellowing leaves with veins that are still green. Apply chelated iron to correct an immediate problem. Long term, add compost regularly and keep soil pH neutral (7.0) or lower.

Manganese (Mn)

Manganese helps plants metabolize carbohydrates and produce chlorophyll. Deficient plants can show yellowing and rolled leaves. Add organic matter (compost) and reduce soil pH to 6.5 or lower.

Molybdenum (Mo)

Molybdenum helps plants process nitrogen and assists in growth. Deficiency isn't usually a problem, except when growing cabbage-family vegetables in acidic soils. Remedy the deficit by improving pH with an application of wood ashes or limestone in the fall.

Nickel (Ni)

Nickel plays a role in plant metabolism and helps legumes fix nitrogen. When there's a deficiency, leaves can show yellowing and limited growth, similar to a lack of nitrogen. Plants don't need much nickel, and there's usually enough in the soil, but a high pH (alkalinity) can make nickel unavailable. Don't add too much lime.

Zinc (Zn)

Zinc helps regulate plant growth and carbohydrate production. Deficient plants can show yellow leaves that can drop as they age. To prevent zinc deficiencies, regularly apply compost made with aged manure.

How Much Is Too Much?

Be careful when adding phosphorus and other single amendments to your soil, since too much of one ingredient can tie up other nutrients. Be sure to follow directions provided with any purchased fertilizers or mineral additives. Also understand the recommendations made by soil-testing laboratories before attempting to follow the recommendations.

When reading a bag of fertilizer, check out the N-P-K ratio. A 50-lb. (22.5-kg) bag of balanced fertilizer that contains roughly equal parts of all three macronutrients might be labeled 10-10-10, which means it contains 10 percent each of nitrogen, phosphorus, potassium, and 70 percent other materials. Most organic fertilizers will have a lower numerical N-P-K ratio, such as 3-4-1. But organic fertilizers remain in the soil longer than their synthetic counterparts and must be broken down by soil organisms before they can be taken in by plant roots. The lower numbers and natural makeup of organic fertilizers make them less likely to cause damage than harsh chemical fertilizers, but you should take care nonetheless. Blood meal can burn roots when applied directly to plants, even though it's organic.

Step 5: Protect Your Investment

Once you've gotten your soil in prime condition—spending time, money, and considerable head scratching in the process—you'll want to protect it from any damage. That includes the use of pesticides (even organically approved ones), herbicides, fertilizers, and fungicides. Practicing good crop rotation will help prevent diseases and preserve soil nutrition naturally. But perhaps the best way to protect your soil is to plant cover crops that ensure that precious topsoil won't be washed away in a rainstorm or blown away by the wind. Integrating cover crops into your regular planting and rotation regimen will result in better soil and healthier vegetables.

Ask yourself how many times you see bare soil in nature. Unless you're in a desert or at the beach, chances are the answer is never. Your garden or farmland should mimic nature and be covered with some type of vegetation (dead or alive) throughout the growing season. In winter, it should be blanketed either with leaves, compost, or a dead or dormant cover crop.

The Benefits of Cover Crops

Cover crops and green manures shade the ground (and its organisms) from the hot summer sun, hold topsoil in place after crops have been harvested, and send out deep roots that break up compacted soil. In addition to preventing erosion in winter, cover crops can work all season long to recharge the soil between plantings or to provide a blanket of living mulch that fights weeds and adds nitrogen to the soil. They also provide habitat for pollinators as well as predatory ladybugs, lacewings, and other beneficial insects that keep pests at bay. Finally, rain will soak slowly into planted fields rather than running off fallow land, so soil is better able to handle surface water.

Some Cover Crop Strategies

One way I use cover crops in my small-scale operation is to sow peas as soon as the soil can be worked in early spring. I scatter the inoculated pea seeds about 1 in. (25 mm) apart across the entire bed and cover them with ½ in. (12.5 mm) of soil. In about four weeks, I begin harvesting pea tendrils for salads by cutting off the top 4 in. (10 cm) of the vines.

After a couple of cuttings from the same planting (around mid-May), I cut the peas down at ground level and turn them in to the soil. In two weeks, after soil microbes have done their magic, I plant a warm-weather crop in the same bed: tomatoes, eggplants, peppers, melons, cucumbers, or winter squash.

Here are some other uses for cover crops:

- Planting buckwheat in early summer after harvesting spring lettuce recharges the soil. In six weeks, the buckwheat will be ready to flower, and you can cut it down and till or disk it under before planting a fall crop of broccoli, cauliflower, or Asian greens.

- Pumpkin vines planted through a mulch of cut and flattened wheat, oat, or rye straw will set fruit that stays mud free, making the pumpkins cleaner and easier to harvest in the fall.

- Sorghum grass planted after crops are finished will catch any leftover nitrogen in the soil and keep it from leaching away. The nitrogen returns to the soil when you till the crops under. The tall-growing sorghum grass will also block out any weeds and will add more carbon to the soil when tilled in.

- Dutch white clover planted between rows or raised beds will preserve moisture and suppress weeds, working as living mulch. When clover gets tall, mow it down and compost the nitrogen-rich clippings.

- Buckwheat can be sown in an area that is thick with quack grass, bindweed, or another perennial weed. It grows rapidly in warm weather and shades out the sun, preventing weeds from growing taller. Once the buckwheat flowers, cut it down, till it in, and sow a second crop. The tilling will also break up any weak perennial weeds, which ordinarily would cause them to spread but not when you plant a second crop of buckwheat. The next crop will grow quickly in warm weather, just as the first crop, and it should exhaust any reserve energy in the roots of perennial weeds, effectively getting rid of them for good. I first learned about this natural weed control from Dick Raymond's terrific book *The Joy of Gardening* back in the 1980s.

By Any Other Name

Cover crops are used in so many ways that they carry different names to suit their purposes.

Catch Crops

When grown to draw up and store excess nitrogen in the soil, cover crops are catch crops. In conventional agriculture, they prevent excess chemical fertilizers from leaching into groundwater; in organic practice, they hold the nitrogen and recycle it, releasing nutrients when the plant material is returned to the soil food web.

Green Manures

When used primarily to fortify the soil, cover crops are referred to as green manures. Their function is to produce as much bulk or biomass as possible during the growing season so that, when they're tilled into the soil, they add high-bulk organic material to feed soil organisms. In the process, they sequester carbon from the atmosphere; when carbon in the crops is converted to humus, it reduces the accumulation of greenhouse gases. Cover-crop mixes that include legumes also add nitrogen to the soil, reducing the need for purchased fertilizers.

Mulches

When low-growing cover crops are sown around established plants, like clover between rows, they are called living mulches. They cool the soil, suppress weeds and—in the case of clover—add nitrogen to the soil.

Perennial Cover Crops

Clover, alfalfa, and a few other perennial cover crops can provide extensive root systems that break up compacted soil and encourage the growth of soil microorganisms.

After cover crops have done their primary job and protected the soil from washing away, you can cut them down and till them into the soil. Do this a few weeks before planting time so the microbes will have time to break down the material into usable nutrients. An alternative to this is to simply cut down the cover crops and plant through the dead material, which suppresses weeds and conserves moisture as natural mulch. The mulch will slowly break down throughout the season, adding organic content to the soil.

The Cover-Crop Recipe

Farmers in different geographical regions use cover crops differently, tailoring mixes that will provide the most benefit for their particular soil needs.

Clover, alfalfa, vetches, soybeans, and other legumes have the ability to fix nitrogen. Specifically, hairy vetch (a legume) can provide more than 100 lb. (45 kg) of nitrogen per acre when allowed to reach full size.

Winter rye grass is a good cover crop to plant late in the season. Grains have deep roots that break up compacted subsoil.

The best organic growers feed their soil microbes a two-course meal consisting of green and coarse cover crops.

Green Cover Crops. With soft, lush growth, these crops are like sugary fast-food appetizers, stimulating a population of microbes that grows quickly, consumes the matter, and then disappears. The action brings soil to life and provides food for higher members of the soil food web, but the results are not long lasting.

Coarse Cover Crops. Although crops such as dried straw, rye-grass stalks, and plant stems take longer to break down, they create stable organic matter (humus). Humus can persist in the soil for hundreds of years, collecting nutrients and improving the ability of plants to absorb them.

Grasses and Legumes

Materials that are high in carbon can actually cause soil organisms to consume nitrogen in the soil as they break down the rough material. For that reason, many growers plant mixes of high-carbon grains and grasses together with legumes. A grass and legume mix is also favored because the grass has deep roots that penetrate the subsoil and break it up, while the legumes add nitrogen. Peas and climbing vines can cling to the grasses as they grow.

Once you've built healthy soil, with stable organic matter (humus) and all the proper nutrients for optimal plant growth, there should be less reliance on purchasing inputs (fertilizers) for crops. If you test the soil and add only what's needed, and you practice methods to protect the soil, your crops should perform at their best.

Most organic growers don't simply use fertilizers as food for their plants. Organic growers feed the soil, which allows their crops to take care of themselves. Under the right conditions—good soil, adequate sunshine, and regular water, your crops will achieve their genetic potential, providing optimum harvest with a minimum of insect damage and disease.

Field peas are legumes, which add nitrogen to the soil. They will die over the winter and contribute to the soil's organic matter.

Nutrient-Rich Farming

There's a growing movement on the edges of the organic movement made up of farmers aiming to produce healthier, more nutritious food by improving soil conditions. Bio-nutrient and nutrient-dense growers note that our commercially available foods have become not only bland in taste but also much lower in nutrient content over the years. Why? Because the emphasis has shifted away from quality in favor of quantity. New varieties are developed for their ability to produce greater numbers of larger fruit, which taxes the soil and provides fewer minerals for each single tomato, apple, strawberry, or pear.

Traditional growers have also effectively mined many of the nutrients from the soil by planting crops without returning the necessary minerals. In addition to the sixteen key nutrients for plant growth, nutrient-dense growers assert there may be dozens more that our bodies require for good health and that plants, along with their partner microbes, require to attain their full genetic potential.

Nutrient-dense growers add ground rock dust and other mineral sources to their soils to achieve a better blend of micronutrients in the root zone. They also inoculate seeds with microbes and apply drenches with food substances that coat minerals and stimulate microbes to convert material into plant-ready nutrients. One of the ways they measure their progress is with a refractometer, often used by beekeepers and vintners to measure the sugar content in honey and wine grapes. Refractometers look like a tiny telescope, and they read the sugar and mineral content of liquids by measuring the degree of refraction or bending that occurs when light passes through a liquid. Squeezing a few drops of crushed fruit or plant sap onto a refractometer will tell the user how dense the dissolved sugars and minerals are inside the plant.

For more information, visit the website of the Bionutrient Food Association: bionutrient.org/. Also look for more information in the "Resources" section on page 356.

We Plow the Field . . .

*...and sow the seeds
and set in transplants
and pull the weeds
and watch for insects
and fight diseases
and water and feed,
and prune and pick...
and then we plow the fields
and plant again.*

Although small-scale organic farming is about as far as you can get from factory farming, the land operates something like a factory when it's properly managed. The fields are the assembly line; the sunshine, water, and soil are the raw materials; and the plants, insects, and soil organisms are your workers. On schedule with the seasons, your plants blossom, lure pollinators, and ultimately bear fruit.

The Living Factory

Your job as manager of the organic farm is to run the living factory, ensuring that work continues through droughts, diseases, insect attacks, weed pressure, and any other forces that threaten the production line. And like every other factory, timing is paramount. If it takes eighteen weeks for heirloom tomato seeds to produce sweet, juicy beefsteaks, you need to start their production eighteen weeks before the orders are due. Each vegetable, fruit, or berry must be planted, transplanted, watered, and harvested—all in time for delivery.

The vagaries of nature will no doubt have an impact on your factory and its workers. But coping with the uncertainties will ultimately determine your success. Go with the flow, adjust and reset your goals, strive to meet them, and continue as the season progresses. Regardless of the hardships, you must plan your farm and farm by your plan.

A Time for Everything

I arranged this chapter in a month-by-month format because that's the way a season progresses—with new chores such as those listed below added each month:

- Ordering seeds and organizing your indoor planting materials

- Starting the first seeds indoors

- Prepping the beds and hardening seedlings for their outdoor home

- Direct sowing outdoors and transplanting seedlings

- Fertilizing, watering, trellising, weeding, and watching for pests

- Harvesting early crops and protecting vegetables from deer, woodchucks, and other critters

- Planting short-term cover crops and sowing second and successive crops

- Stretching the season; sowing garlic and crops that will winter over; and cleaning up, composting, and cover cropping

- And, finally, reflecting on the season and taking notes that will help you plan the next year

The time line I've used is tailored to growers in agriculture Zone 6 (see the zone maps on page 384 and the inside back cover), but it provides a logical progression of the chores for growers in most of the northern United States and southern Canada. If you're growing in Zone 7, which is warmer, start the calendar one to two weeks earlier. If you're growing in Zone 5, which is cooler, start the calendar one to two weeks later. Make adjustments as you record your observations and experiences.

The goal is to create your own schedule and assign these basic tasks to the time that makes most sense for your part of the country. Southern growers will experience an entirely different season for many cool-weather crops, which are best planted in late fall and winter and then allowed to grow and mature before the hot weather arrives. All growers should keep a record of their own schedule for future reference, noting the early and late frosts, dry spells, heavy rains, the effect of microclimates, and how you dealt with such problems as insects, weeds, diseases, and animal incursions. Take note of which varieties performed best and which new items fell short of your goals.

Online Tools

A good place to start is a website with an online calendar that provides a planting schedule based on your zip code.

All Things Plants. A social website dedicated to gardening, All Things Plants offers spring and fall calendars with a range of planting times for starting seeds indoors, transplanting outdoors, and direct seeding outdoors. Try it out at allthingsplants.com/apps/calendar

Make Your Own Schedule

Using Google Documents or Microsoft Excel, create your own calendar, designating a cell for each week of the growing season on the top and listing individual vegetables in a vertical column on the left. Mark your first and last frost date, and then calculate and enter key dates for each vegetable: when to plant seeds indoors, when to harden transplants and set them into the garden, and when to sow directly in the soil.

Recommendations on the back of seed packets will tell you how long each individual crop needs indoors before it's ready to transplant. Also pay attention to germination temperatures. Some crops like corn won't germinate if the soil temperature is too cold. Also know that some vegetables such as cucumbers and squash need only a few weeks before, while peppers, eggplants, celery, and others need longer. (For information on specific crops, see chapter 5.)

Burpee. Burpee seeds offers another calendar that includes a region's first and last frost dates, together with a grid showing vegetables and a range of planting dates (indoor seed starting, outdoor direct-seeding, and such for vegetables and flowers). The web address is: burpee.com/gygg/growingCalendarNoZipCode.jsp

For my postal zip code in Zone 6a in southeastern Massachusetts, the site gives me a slightly more conservative schedule than I actually use because I stretch the seasons with floating row covers. Note below the differences shown in the Burpee calendar for a variety of other regions:

· Bismarck, North Dakota (zip code 58504; Zone 4): shows a last frost in mid-May and a first frost in mid-September, with corresponding planting suggestions.

· Boise, Idaho (zip code 83634; Zone 6): shows a similar planting schedule to mine with a last frost date in mid-April and first frost in mid-October.

· Waco, Texas (zip code 76655; Zone 8): offers much different recommendations, including sowing cabbages indoors in January. Last frost of spring in mid-March and first frost of winter in mid-November.

JANUARY: Warm Thoughts and Icy Days

The winter winds whip snow across planting beds, and the soil beneath is cold and hard. But already the days are getting longer. Spring will soon be here, and there is plenty to do before then. January is the time to inventory seeds and seed-starting supplies. Once the inventory is complete, it's time to order new seeds.

The midwinter lull also allows time to sit at the computer emailing customers and planning for the coming season. When and where will the farmers' markets be? If you have a Community-Supported Agriculture (CSA) operation, how many subscribers will return for another year? What about restaurant sales—should you hike your prices to keep pace with the rising cost of supplies? All of these details are best settled in the month of January, before the growing season begins in earnest.

Testing Saved Seeds

Seeds have varying longevity—some, such as onions, last only a year, while many others can last longer if they're stored well—that means cool and dry in airtight containers away from direct sunlight. If you have a large stock of good seeds from last year, you may want to test them.

Count out ten seeds (twenty if you have enough to spare) and place them on a moistened paper towel. Drape another moistened paper towel over top, and roll up the layered towels into a tube. Put the rolled-up paper towels inside an unsealed plastic bag and place it in a warm spot—on top of the refrigerator is good. If the seed packet tells you germination occurs in one week, check the seeds a day or two before the end of that time period, and check them again every day for three or four days past the estimated germination time. If nine out

When buying seeds, look for organic choices first. They'll have the familiar green and white USDA Organic logo.

Organic Hobby Farming

Easy Does It!

Gather together a few basic supplies and use the handy suggestions below to make your seed-starting routine go more smoothly.

Reading Glasses. I don't generally use glasses for reading, but I do keep a pair of drugstore reading glasses handy for starting seeds. They're especially helpful when dealing with Brassica seeds in small containers. Even harder to see are tiny black basil seeds or black-seeded lettuce seeds on dark potting mix; using glasses will make it easier.

Chopsticks. One of my favorite seed-sowing tools is a bamboo chopstick sharpened to a point in a pencil sharpener. Touch the tip to a moist sponge and then to a tiny seed. The seed will cling to the point of the chopstick, allowing you to press it neatly into a hole in the starting mix.

Scoops. A hand trowel or a big serving spoon and a large cup to use as a scoop will help you spread soil mix and smooth it into seed flats, plug trays, or cell packs.

Watering Jugs. I save a lot of 1-gal. (3.8-L) plastic jugs, washing them thoroughly to keep on hand for watering or mixing up liquid organic fertilizers. Four of them fit nicely into one of those plastic milk crates. You can use them to mix a half-strength fish emulsion and seaweed liquid fertilizer or just keep them filled with water. Pour into a watering can for dispensing.

Spray Bottles. I always moisten my seed-starting mix before planting, but if the mix ever looks a little too dry when I'm ready to sprinkle in seeds, I use a spray bottle to mist the top. Spray bottles are also a good way to water newly seeded trays without washing away the topsoil and uncovering seeds.

Seed Dispensers. You'll find numerous contraptions to help you plant tiny seeds. Or you can try tap-tap-tapping on the seed envelope and watching as a line of tiny Brassica seeds rolls along a crease in the envelope flap.

One variation (Seedmaster) looks like a plastic trowel with a V-shaped blade. Inserts with different-sized seed holes fit into slots midway up the blade. Seeds roll through the insert and down the V-shaped blade to where you want them.

Several models of plunger seed dispensers are available. They look basically like a large syringe, with a clear cylinder that holds the seeds and a spring-loaded shaft on a plunger. Push the plunger, and the seeds roll into a groove in the shaft, dropping out one by one.

A suction-operated seeder has a soft ball for a handle, which is attached to a long, hollow shaft with removable nozzles for different sized seeds. Squeezing the ball and releasing it creates a suction that picks up one seed. Squeeze again and the seed drops where you want it.

Dial dispensers have a petri-dish—sized body with funnel-like dispenser. The dish has a rotating top with different-sized holes that allow seeds to roll down the dispenser and out.

It's easy to lose sight of tiny black basil seeds, which is why I always keep a pair of drugstore reading glasses handy.

of the ten seeds sprouted, you have good germination—90 percent. If fewer than half of the seeds germinated, throw them all out and buy new seeds. You might keep the seeds even if only slightly more than half sprouted, but sow them at double the rate to account for the possibility that some won't germinate.

Good seeds are a vital part of your operation. If you have any doubts about the quality of your seeds, don't scrimp. Buy new ones that are packaged for the current year. As I mentioned, timing is important, so it's not worth sowing seeds and waiting a week or two only to find out that germination was bad. The inconvenience of having to sow another crop is frustrating, and the lost time that you spent waiting for defective seeds to germinate could be a costly disruption to your planting schedule.

Seed-Starting Supplies

The first real chore of the year is taking inventory of your seed-starting supplies. If you're in the habit of saving equipment from year to year, you can dramatically reduce the annual expense of starting seeds. But you don't want to be caught short on supplies, so take time in January to count your planting trays, cell packs, peat pots, plug trays, and so on. If you're just getting started, consider the options (which are outlined in the following pages) and stock up. Whatever choice you make for sowing seeds, just be sure you have enough on hand to begin the season.

Cell Packs

Inexpensive plastic cell packs are an easy way for small-scale growers to get started. Cell packs are the common four-, six-, and nine-packs that you bring home from the garden center if you buy seedlings. The individual cells are typically square shaped and narrower at the bottom than the top, making it easy to pop young plants out of the container—soil mix and all.

You'll often find cell packs in plastic carrying trays. The standard size for a tray is about 10 in. (25 cm) x 20½ in. (52 cm) (inside dimensions), and they're referred to as 1020 trays, or just ten-twenties. Some cell packs also come with a clear plastic lid that fits over the entire tray and acts as a miniature greenhouse. Cell packs vary in size and are usually described by the number of cells that fit inside a tray—forty-eight, seventy-two, or ninety-six. The larger the number of cells in a tray, the smaller each cell. (Note: It's a good idea to cut the cell packs into manageable groupings of four, six, or nine cells before you fill them and start seedlings. The smaller packs will be easier to handle once the flimsy cells become unwieldy with soil.)

Flats are also available in a common 17 in. (43 cm) x 17 in. (43 cm) size, with matching rigid carrying trays.

You can also plant seeds directly in a 1020 tray filled with seed-starting medium, then transplant them to individual pots before the roots get too crowded.

Cell packs are made to be disposable, but you can recycle the plastic packs for several seasons before they ultimately crack or become crumpled beyond repair. For longer life, don't leave empty trays out in the sun. The best treatment is to wash cell packs immediately after use. Sterilize cell packs by soaking in a 1:10 solution of bleach and water for 10 minutes. (The bleach will kill any diseases that might harm the next year's seedlings.) Dry them for a day, and then put them away until you need them again.

Plug Trays

Plug trays, like cell packs, have tapered cells you can fill with seed-starting mix and place into a carrying tray. However, plug trays are much more durable—made to last for multiple seasons. I have some that are ten years old. They're also more rigid, and the cells are often more conical, usually tapering sharply to a large drainage hole at the bottom. Some have channels that steer roots downward to the bottom hole. As roots protrude from the bottom and become exposed to air, they naturally dry out, causing numerous small roots to branch out. The result is a tight plug with thick feeder roots that will rapidly grow deep into the soil once transplanted.

Peat pots are an alternative to using plastic pots and cell packs. Plant them pot and all in the field.

Plug trays come in a wide variety of sizes for the flower and greenhouse trade, including some that hold just a few grams of growing mix per cell. Because organic growers don't use fast-acting fertilizers, larger plugs are more desirable for us. You'll have to experiment a little to figure out which size works best for your needs. (Note: Unlike cell packs, plug trays are sturdy enough to tote around even filled with soil and seedlings.)

Plug trays are made to last more than a few years. Peat pellets (shown in inset) expand when watered to create a pot contained in a light mesh.

Soil Blocks

Soil blocks are like pots without the containers—essentially cubes of packed growing material pressed and molded with a cube-making tool. These are becoming more popular with organic growers because they cut down on the use of plastic. Soil blocks are a do-it-yourself option, so once you've purchased a soil blocker to make the planting cubes, your only additional cost each season is the growing mix.

There are small hand-held models or larger stand-up–sized blockers. And you can buy molds to make cubes in multiple sizes. The smallest molds will make twenty ¾-in. (19-mm) mini-cubes for tiny seeds. Larger-sized molds will produce 2- and 4-in. (5- and 10-cm) cubes for big seeds or replanting smaller seedlings. You can even buy inserts that create an indentation on the top of each cube, allowing you to pop a mini-cube (complete with young seedling) directly into a larger cube for second-stage seedling growth.

Soil blockers take a little practice, and the blocking mix has to be just the right consistency (like mushy oatmeal) for the cubes to retain their shape. The tiny blocks also have a tendency to dry out quickly, so you have to water them regularly. But soil blocks have benefits beyond conserving materials: Plants grown in soil blocks will extend their roots in all directions until the root tips reach the air, so the plants are air-pruned on five sides of the cube instead of just at the bottom.

Individual Pots

Like cell packs and trays, you can save plastic pots for several years, provided you limit their exposure to sunlight. Individual pots are a good choice for onions, leeks, parsley, and basil that can stand up to the abuse of a little pulling and tugging to separate intertwined roots.

Sowing seeds at ¼-in. (6-mm) intervals inside a 4-in. (10-cm) pot allows ample room for onions and leeks to grow rapidly through the first six weeks of their lives. When it's time to transplant, soak the soil thoroughly, remove the plants—soil and all—and then pull them apart, untangling the roots as you go. Individual pots are also good for seeds that perform best in flats (shallow trays without divisions). Celery, parsley, and some other seeds germinate best when grown close together. After they sprout, you can thin them to avoid overcrowding. And if you want to start larger transplants outdoors, use larger pots for small tomato, pepper, and eggplant seedlings grown in cell packs or plug trays.

Peat Pots

Many organic growers don't like buying and using plastic. If you count yourself among them, try pots made from compressed peat moss, cow manure, or coconut husks (coir). All are biodegradable, allowing growers to plant the pots in the field along with their contents. Peat pots are convenient, and they eliminate clutter; when you transplant the seedling pot and all, you don't need to collect and wash all the containers. But they're also more expensive than cell packs, and you have to invest in new pots every season.

If you choose to use peat pots, take care to bury the rims of the pots beneath the soil level when you plant them in the field. Exposed peat rims can wick moisture away from the seedlings' roots.

Homemade Containers

You'll find many ways to cut costs when sowing seeds, but saving random-sized plastic containers, cereal boxes, soda bottles, and egg cartons would not be my choice. When you sow seedlings every two weeks to keep a succession of vegetables growing, you can't afford to use odd-sized materials. Having the same-sized planting containers with adequate drainage holes and knowing that they will fit comfortably into functional trays is worth the money you'll spend to get started.

I made this planting board for outdoor seeding after seeing one at Many Hands Organic Farm. On the bottom are pushpins set in a grid; on the top is a handle. Push the pins into moist soil to leave a grid of evenly spaced depressions. Then sow one seed in each hole for even spacing. I also have another board with spaces set at 2-in. (5-cm) intervals.

Easy Does It!

If you're inclined to make some of your own materials, newspaper pots are a good option. Wrap a 5-in. (12.5-cm) -wide strip of newspaper several times around a small-diameter can (a small tomato paste can is ideal). Slide the paper tube a couple of inches above the open rim of the can and fold the ends in loosely. Then slide the paper tube off, turn it around, and slide it back onto the can so the bottom of the can faces the folded-in ends. Press the can firmly against a hard surface to crush the folded-in ends and create the bottom. Then slide off the paper pot and fill it with potting mix. Newspaper pots will break down in the same way that peat pots dissolve in the soil. They're biodegradable, quick to make, easy to use, and cost just a few pennies for dozens. Avoid using newsprint with colored ink.

One exception may be rigid yogurt cups if they're all uniform in size (maybe you have a favorite brand and can accumulate five to ten cups per week). If they all fit nicely into a tray, there's no harm in saving them. Just remember to drill holes in the bottom of each cup for drainage.

Let There Be Light

Once you know you're on target with your seed-starting supplies, it's time to look at your lighting. The biggest problem novice growers face when starting their own seedlings is that the plants get spindly—and the main cause of this problem is inadequate light. If you have a greenhouse with high-quality grow lights, you can just set the lights on a timer and flood the young seedlings with twelve to sixteen hours as soon as they germinate. If your bulbs haven't been replaced in a long time and they're approaching the end of their viability, buy new bulbs before it's time to plant to ensure that things go smoothly.

In mid-February my leeks and onions are ready to go under the light table. Other seedlings will be sown in March and April.

Absent a lighted greenhouse, small-scale growers can make their own light tables using a tiered table with fluorescent light fixtures over each level. Start by building a sturdy frame using 2 x 3 (38 mm x 64 mm) lumber, leaving 14 in. (36 cm) of vertical space for each level. Use plywood for shelving—not particleboard, which can swell and begin to deteriorate if it gets wet. For efficient spacing, your light table should measure 2 ft. (60 cm) deep x 4 ft. (120 cm) wide, which should hold four 1020 trays on each level. Attach fluorescent light fixtures above each tier, using chains or hooks so the lights are vertically adjustable. As the seedlings grow taller, you can raise the lights to keep them just a few inches above the foliage.

The least expensive fixtures are standard T12 shop lights, 4 ft. (120 cm) long, with two bulbs per fixture. Each bulb uses 40 watts, or 160 watts per tier. T8 lights and T5 lights are more expensive, but they give off more light per watt.

If you don't want to fuss with lights that adjust their height, bring the plants closer to the light source. Place very young seedlings on top of an upside-down 1020 tray, bringing them closer to the light. Once the plants grow taller, remove the trays to give the plants more vertical space. Use new bulbs with full-spectrum light and a high output of lumens.

Easy Does It!

Use a seed-starting formula and not general-purpose growing mixes for starting seeds. Potting mixes may be coarser and heavier, with inadequate drainage and airflow. Numerous organic mixes are available from suppliers across the United States and Canada. (If you join a regional organic farming cooperative, you should be able to order seed mixes and other supplies at reduced prices.)

The key ingredients in organic seed-starting mixes are usually peat moss, perlite, and/or vermiculite, along with disease-free compost. Perlite is a volcanic rock that is heated and popped like popcorn. The lightweight white pellets improve aeration and drainage in soil mixes, allowing optimum conditions for fast root growth. Perlite also holds water on its surface area, helping the mix retain moisture. Vermiculite is a mineral that expands when heated. Like perlite, it's used to lighten or fluff up seed-starting mixes.

Don't buy commercial mixes that contain fertilizers unless they are labeled organic. Some popular brands of commercial mixes at hardware and garden supply stores contain wetting agents and chemical fertilizers that are not approved for organic production. Even if you're not certified organic, the fertilizer in seed-starting mixes isn't really necessary. Seedlings don't need any soil nutrients to assist in germination or in early-stage growth because everything they require is contained within the seed kernel. Only once the plants show their first true leaves will they benefit from a weekly feeding of half-strength liquid organic fertilizer. (Fish and seaweed emulsion is a good choice, since it contains limited nitrogen and high-quality phosphorus and micronutrients to encourage strong growth.)

Another medium that is gaining in popularity among growers is coir, also called coco peat. It's made from the milled husks of coconuts and is typically sold in compressed bricks that expand when you add water. I mix a small amount of coir into my purchased organic seed-starting mix because I like the way its stringy texture absorbs water. Peat-based mixes often shed water when the peat is dry, but adding a small amount of coir to the mix (just one or two bricks in a 30-lb. [13.5-kg] bucket of mix) will help the medium absorb water.

I'm testing growing Walla–Walla onions in spring. Many growers sow these big, sweet onions in late summer and allow them to overwinter for rapid growth the following season.

Planting Mix

I'm a do-it-yourself kind of grower, but I still buy seed-starting mix from an organic supplier. I use the starting medium on all of my small-seeded plants—cole crops (such as broccoli and cauliflower), lettuce, and basil. I don't want to risk poor germination in a mix that's too heavy. Besides, I generally start these crops in small containers, so I don't use that much mix.

I also create my own potting mix, which I use for transplanting seedlings once they are large enough to handle. The potting mix that I make is also acceptable for starting larger seeds, such as squash, cucumbers, melons, and pole beans.

Many growers blend their own mixes from compost, high-quality garden soil, builders sand, and peat moss. If you take this route, make sure the mix is not too heavy or seedlings may not drain well enough.

Odds and Ends

Make sure your grow lights are all working and that your electrical timer is functional. Also check that you have enough liquid fertilizer for when the seedlings develop true leaves. Finally, don't forget to gather the odds and ends that are helpful for starting seeds.

FEBRUARY: Sowing the First Seeds

My seed starting begins with onion family crops in the middle of this chilly month. Those who grow celery or artichokes from seed can start plants at the same time because they need a long head start to the growing season.

I start onions and leeks in mid-February because I like to get a jump on the season with long-day onions. Long-day onions, which are the best choice for growers in the northern United States and Canada, get a signal to start setting their bulbs when the days become longer than the nights. If plants don't reach a decent size before the days get long, the onion bulbs may not grow very large. Southern growers should plant short-day onions. (For more information on growing onions, see page 208.)

February is also the time to check on the condition of your outdoor growing area. Are beds covered? Is garlic sprouting from an October planting? If the area needs mulch, put it down now to keep the repeated cycle of freezing and thawing from pushing bulbs out of the ground. Also take inventory of the tools and equipment you'll need for the growing area. From fencing, tomato stakes, and cages to floating row covers, mulches, and pest controls—if you need any supplies, now is the time to order them.

Onion seedlings started in February will be ready to transplant outdoors in April under the protective canopy of a floating row cover. Green onions will be ready to pick in June.

Speaking of ordering, do you plan on starting a chicken flock or adding chicks to an existing one? If so, get the order in now. Chicks delivered on the first of April will be ready for the outdoors in early May, when the weather has warmed. In this way, the youngsters will still have plenty of time to mature to lay eggs in September before the cold weather hits. (For more information on keeping chickens, see chapter 7.)

Late winter and early spring are also good times for pruning fruit trees and berries. (See chapter 6 for additional information on pruning.)

MARCH: A No-Nonsense Pace

Before March is over, I start some seeds outdoors. I direct-seed spinach, Swiss chard, salad turnips, peas, and radishes during the last week of March for a late-spring harvest. Raised beds with low tunnels and floating row covers help keep the soil warm and well drained for the young plants. (For more information, see page 121.)

Planning Is Key

Growing for a market and following a detailed plan isn't as simple as sowing your favorite crops on a sunny Saturday morning in May and then hoping they'll be ready for your midsummer barbecues. Meeting the demands of customers means you may have to start broccoli, cauliflower, and Chinese cabbage under lights in early March; sow spinach and mesclun under floating row covers in early April; and sow flats of lettuce every two weeks throughout the spring, summer, and early fall.

If you've never followed such a structured planting routine, the pace can be a little daunting. But once you've formulated a plan, it's no more difficult than each individual task. What's that old saying—inch by inch, it's a cinch? Experiment and discover what works for

Soil Temperature Is Key

Different vegetables have different soil temperatures for best germination. Seeds that are sown in cold soil may rot and succumb to fungal diseases. To know precisely what the temperature of your soil is, measure with a soil thermometer. These devices look like a dial–style roasting thermometer; some have digital readings. To use, insert the probe into the soil to get an accurate reading of the soil temperature before planting your seeds. The table below provides a listing of soil germination temperatures for some common vegetables.

Soil Temperatures for Seed Germination

Crop	Temperature Range	Optimum Temperature	Number of Days
Beans	60°F–85°F (16°C–29°C)	80°F (27°C)	7
Beets	50°F–85°F (10°C–29°C)	85°F (29°C)	5
Cabbage	45°F–95°F (7°C–35°C)	85°F (29°C)	4
Corn	60°F–95°F (16°C–35°C)	95°F (35°C)	3
Cucumbers	75°F–95°F (24°C–35°C)	95°F (35°C)	3
Eggplant	40°F–80°F (4°C–27°C)	95°F (35°C)	5
Lettuce	40°F–80°F (4°C–27°C)	75°F (24°C)	2
Peppers	65°F–95°F (18°C–35°C)	85°F (29°C)	8
Radishes	45°F–90°F (7°C–32°C)	85°F (29°C)	3
Squash	70°F–95°F (21°C–35°C)	95°F (35°C)	9
Tomatoes	60°F–85°F (16°C–29°C)	85°F (29°C)	5

Sources: Data from the University of Georgia; Ft. Valley State College; the U.S. Department of Agriculture.

you. Log information into a spreadsheet or use an ordinary datebook planner, noting the date for starting seeds, moving them from the light table to the cold frame for hardening, and then putting the transplants into the raised beds. You can also program the dates into your computer calendar and set email alerts.

Doing the Math

March is the time to break out some of your math skills. Consider all of the vegetables you'll need for the coming year and then figure out when you'd like to harvest them (within reason and accounting for their natural season). For example, maybe you want to harvest a particular vegetable for a scheduled farmers' market: Count backward from the harvest, subtracting the number of days it takes for your plants to reach maturity. That information is usually printed on seed packages, but check whether the number of days listed starts at the time the seeds are sown or at the time seedlings are transplanted outdoors. If it's the latter, add four to six weeks to the schedule.

Here's another example: Let's say you want to harvest tomatoes in mid-July, and the seed packet for the heirloom variety that you're growing—Stupice—tells you the harvest will be ready fifty-six days after transplanting. Count back fifty-six days from July 15. To meet that goal of a mid-July harvest under perfect growing conditions, you'll need to set out the transplants by May 20. First, check if the weather is warm enough at that date. If it's not, you'll have to provide protection from the cold or change your goals. If it works, add one week for hardening off the seedlings in a cold frame and five more weeks for growing the seedlings indoors. That brings you back to the first week in April. If you don't start your tomato seedlings that far in advance, you won't meet the deadline. It's that simple.

You'll need to do the math for every crop you intend to grow—from simple radishes that are direct-seeded in April and harvested three weeks later to winter squashes that you plant as soon as the hot weather hits and harvest in early autumn. For some growers in northern climates, the season won't allow for every variety of winter squash or melon or other long-maturing vegetable. If a vegetable takes 120 days to harvest and you can't set it out until after all danger of frost has passed, you'd better have a season that's at least 120 days long. That's one more reason to do the math before you plant.

If you're growing crops for a CSA, you'll have to think of ten to twelve crops that will be ready for immediate harvest when you decide to begin your subscription. For example if your season begins in the first week of June, the following crops should be ready to harvest: lettuce, spinach, pea tendrils, salad turnips, radishes, rhubarb, scallions, green garlic, mesclun, snow peas, herbs, and Asian greens. To ensure that all of those crops are ready for harvest on the first day that customers come to pick up their shares, you have to plant them early enough to allow them to mature. If you don't think you'll be able to deliver on enough crops by early June, you may want to change the start time for subscriptions.

Many Times to Sow

To make a complicated task simpler, break it into smaller parts. First consider that you'll have to adhere to natural times and dates, depending on the hardiness zone in which you are growing crops (see the zone maps on page 384 and inside back cover).

In Zone 6, where I grow my crops, five key dates determine most of my seed-sowing schedule: March 1, April 1, April 15, May 1, and May 15. These dates are based upon the average last frost of the season in my area, which the U.S. Departure of Agriculture lists as late April. Vegetable plants that can tolerate frost are usually

An inexpensive soil thermometer will take the guesswork out of sowing direct seeds outdoors and improving germination. Seeds planted too early can rot in the cold, wet ground.

safe to plant outdoors a few weeks before that—on April 15. Because cool-weather crops such as cabbage and broccoli would take too long to grow if sowed directly in the soil, I have to count back six weeks from April 15 to arrive at March 1—the date when I sow them indoors on the light table. In five weeks (April 1), they will be ready to harden off in a cold frame. And in another two weeks (back to April 15), I can move them to their permanent place in the field. (Of course, I watch the weather reports, sometimes biting my nails, and I use floating row covers if a frost is expected.)

Once the cool-weather crops are hardening outdoors, the warm-weather crops (tomatoes, peppers, and eggplants) get their turn under the lights. They'll be ready for the cold frame in four to five weeks and ready for the field another one to two weeks after that (May 15), pending the weather forecast. Just as I determined my seed-starting dates by counting backward from my last frost, you can set your own seed-starting dates.

Among the more schedule-specific crops are a number of perpetual and successive crops. Perpetual crops can be harvested all season long. These include kale, Swiss chard, and pole beans, which are all must-grow crops for organic farmers who want a season-long harvest. Successive crops are generally short-lived vegetables that you can plant every two to three weeks to provide a continuous harvest all season long. Some of them don't handle heat well and may need to be shaded from the sun or skipped during the hottest weeks of the growing season: radishes, Asian greens, mustard, and lettuce mixes are generally best planted in cooler weather. You can sow carrots and beets throughout the season. Bush beans love the

If you want a beautiful bed of lettuce in May or June, start the seedlings indoors in early April. Mix reds and greens for variety.

heat and are best planted during warmer months. Look for this growing season information as you order seeds for individual varieties.

A few crops fall somewhere between perpetual and successive. Zucchini and summer squashes are a prime example—they begin pumping out fruit as early as seven weeks after transplanting and continue for another month or two. But as the summer progresses, the vines start to lose vigor, and production slows down. The solution is long-range successive plantings. Sow your first summer squashes indoors and transplant into the garden after all danger of frost has passed. About five weeks later, as soon as the plants start to set blossoms, direct-sow more summer squashes in a different bed. The second crop will come into production about the time the first crop loses its gusto. Many growers in New England manage three separate crops of summer squashes in the short growing season.

In March it's time to pull the mulch back from the garlic patch. These Chesnok Reds were planted in late October.

Seed sowing is not an exact science, and each grower develops his or her own rhythm. But if you keep records from year to year and take note of your successes and setbacks, you will develop a workable system for your own geographical area.

APRIL: Extend the Season and Begin

April is the perfect time to get a jump on your crops in the great outdoors. Not only can you set the natural stage for your veggies, but you can also create raised beds to help them overcome the boggy effects of the season.

Raise Them Up

Without raised beds, my main planting area would be soggy for the better part of April. But it's not. At ground level, the soil is wet and unworkable. In some years it's flooded with puddles. But above the muck, in raised beds, my plants are thriving, pushing roots through the well-drained soil, and soaking in the sun beneath translucent canopies of white floating row covers.

Several years ago, after waiting until late May to begin my gardening, I saw an ad for used cinder blocks and decided it was a good time to experiment with raised beds. Four of my raised beds stand two cinder blocks high (about 16 in. [40 cm] tall) and 3½ blocks wide (about 40 in. [101.5 cm] on the inside and 50 in. [127 cm] on the outside), and they vary in length from 15 ft. (4.5 m) to 25 ft. (7.5 m). There's no mortar between the blocks, but the holes are filled with loam, and I plant aromatic herbs, strawberries, or marigolds inside the blocks. I loved these raised beds so much that I bought new blocks the following autumn and made more.

The height and width of cinder blocks are ideal for sitting down on the edges and reaching across to sow seeds, pull weeds, or harvest mesclun. And the soil depth is outstanding. Considering that I had more than a foot of good topsoil in the garden before building raised beds, the additional 16 in. (40.5 cm) of compost and sandy loam inside each bed make the overall topsoil depth more than 28 in. (71 cm). The soil drains early, warms up quickly, and provides moisture and nutrients more than 2 ft. (60 cm) below the surface. Best of all, the beds allow me to sow my crops in mid-March and throughout the soggy month of April.

You don't need cinder blocks to create permanent raised beds. Almost any kind of sturdy building material will work. I've also used 2-in. (5-cm) -thick, rough-hewn planks from a local sawmill. The unfinished wood is only a few dollars for a 10-ft. (3-m) -long x 6-in. (15-cm) -wide section, making it easy to build 20-ft. (6-m) beds that are 12 in. (30 cm) high.

Even if you don't want to make permanent raised beds, raking soil into mounds for temporary raised beds each season will give you many of the benefits, without the rigid structure. Just be aware that the beds will slowly flatten out over the season and will need to be raked back into place every year. If your soil is wet, that can delay the start of your planting season.

Many large-scale organic farmers who use tractors over acres of fields make raised beds with an attachment that discs the soil and shapes it into raised beds, complete with flat tops and rows for planting. Some attachments even lay down a sheet of black plastic mulch.

Raised beds allow soil to drain and warm up earlier in the season. These beds also have chicken wire fencing to ward off rabbits and groundhogs.

The Wonders of Raised Beds

Raised beds are certainly not a practical option for large-scale farmers with acres of fields to plant. But if your growing space is limited, raised beds can help you get significantly more vegetable yield from your limited land. The following are some of their other benefits:

- Raised beds preserve soil structure because they don't need to be tilled every year, preventing the breakup of pores and pockets created by earthworms and soil microorganisms.

- Because you walk on the paths between beds, the soil inside doesn't get compacted.

- Soil has great tilth—loose and friable—making it easy to pull weeds that never really get established. Most weeds prefer compacted soil.

- Loose, deep topsoil is ideal for growing beets, turnips, and other root crops—carrots will grow long and straight.

- Many insect pests fly just a foot or two aboveground searching for young plants to feed on. When the host plants are growing above the bugs' flight patterns, finding their meals may be harder.

- Rabbits and other small mammals can't get at your crops as easily. (Although I'm sorry to say that woodchucks can still scale the beds and cause serious damage.)

One more advantage to permanent raised beds is that you can easily add structures to stretch the season and protect your crops from pests. (For more information, see page 120.)

Setting the Stage for the Growing Season

In addition to creating raised beds, you can take a few small steps at this time of year to start the growing season right:

- Top-dress your beds with compost—then make more compost. (Make sure your compost is finished, well seasoned, and contains no fresh manure when you apply it in the spring.) As the temperature rises, soil microbes get busy; put them to work by making a fresh batch. You can also turn over the batch of leaves and garden residue that you placed in the bins back in the fall (see chapter 3 for more information on composting).

- Remove straw and mulch from the garlic plants, and sow lettuce seedlings in between the garlic.

- Begin sowing successive crops of lettuce. Heads that take forty-five days from transplanting will be ready by late May if you get them into the garden by mid-April.

- Build tomato cages, bean trellises, and other structures now—before you plant the crops.

- Sow successive crops such as mesclun, radishes, and spinach as well as peas, pea greens, and onion sets for scallions. Just press them into the ground in any spot where there's a little extra space, and you'll be harvesting them for salads in just a few weeks. (Note: For scallions, plant the bulbs an inch deeper than you'd plant onions, which will result in a longer section of blanched stem.)

- Check your hoses and sprinklers. If you use drip irrigation, check your fittings and give the system a test run.

MAY: Plant and Protect

In much of the country, backyard gardeners are just waking up to the fact that planting time is nigh. Meanwhile, if you've stretched the season and started seedlings indoors, you'll be ready to harvest your first crops before others even begin to rake out their beds.

Pest Prevention

Because you've already planted a good number of crops, May is the time to begin keeping an eye out for insect pests. As soon as the sun warms the earth, adult insects emerge from pupal dormancy and look for food and a mate—not necessarily in that order. But you can catch the adults before they lay eggs and start another generation. The first damage many growers

see around this time are tiny holes chewed in the leaves of mustards, arugula, radishes, and Asian greens—it's the telltale mark of flea beetles on Brassica foliage. But you can control flea beetles and other flying insects by one very simple physical barrier: a floating row cover.

Row covers are my primary plant protection when the weather forecaster warns of an unexpected frost. They hold in heat, warm the soil, and block desiccating winds. They're my best defense against insect pests.

Lettuce sown in March is ready to transplant to the garden in April. Here I am planting some seedlings between the rows of garlic.

Begin looking for the first signs of pests when the weather warms up in May. These squash bug eggs on the underside of a zucchini leaf will be more common in June and July in my New England garden. If you find them, scrape them off and crush them before they hatch.

Floating row covers come in varying sizes and weights, my favorite being an all-purpose weight that is sold in bulk rolls for growers with more than a tiny patch of turf to protect. The material blocks heavy winds, protects against frosts, and impedes insect pests, but it doesn't substantially heat up the soil during summer months. (If heat is a problem, choose a lighter weight.) A 250-ft. (76-m) length of 83-in. (2-m) -wide cover costs about $60 and is more than enough to protect all of my beds for the entire season. This almost-7-ft. (213-cm) width fits nicely over a 4-ft. (120-cm) bed.

It's Transplant Time

May is peak time for transplanting warm-weather vegetables. The big three of the heat-loving Solanums are peppers, eggplants, and tomatoes. Because these vegetables all love heat, I like to warm up the soil first with a sheet of mulching fabric.

Plant tomatoes deep in the soil to encourage strong roots. A full-size tomato plant has so much foliage that the network of roots needs to be long and strong to draw in enough water to support the leaves. Peppers and eggplants, however, must stay at the same depth they were planted as seedlings. Portions of their stems buried underground will not sprout new roots, as would those of tomato plants. Instead, the stems could simply rot, especially if the soil is excessively moist.

Other Tasks

May is the time to pick those Asian greens, arugula, pea tendrils, and salad turnips you planted earlier. Then you can sow one more crop of these cool-weather standards before the hot weather sets in. Here are a few other tasks suited to the warm spring weather:

Make Your Own Floating Row Covers

Because floating row covers protect crops from almost every insect pest, they can be used all season long. They work on any crop that doesn't require insect pollinators, and even on those that need pollinators as long as you remove the covering when plants start flowering so the bees can do their magic.

Here's how to make a protective low tunnel for floating row covers that will shield your raised beds and protect crops from insects and the cold.

What You'll Need

Hacksaw

One 100-ft. (30-m) roll of ¾-in. (19-mm) PVC tubing

Two 10-ft. (3-m) lengths of 1-in. (25-mm) PVC electrical conduit

Row cover material

Rope (optional)

Screw eyes (optional)

Clothespins (optional)

1. Use a small hacksaw to cut the tubing into twelve 8-ft. 4-in. (2.4-m 10-cm) sections. (These pieces will form the arches to support the row covers.)

2. Cut the conduit into eight 15-in. (38-cm) pieces.

3. Cut the conduit sections with a saw at a 45-degree angle on one end and a 90-degree angle next. This will create one pointed end on each piece.

4. Pound the pointed ends into the ground at 3-ft. (1-m) intervals along the outside of each raised bed, allowing a few inches to remain aboveground. If your beds are ringed with cinder blocks, pound the stakes into the soil in each hole.

5. Insert an 8-ft. 4-in. (2.4-m 10-cm) section of tubing into the 1-in. (25-mm) rigid sections that are already anchored into the ground.

6. Pull or push the tubing to adjust the height to where you need it.

7. Cover the beds and arches with row cover material, which you can do in several ways. If the raised beds are wooden, insert screw eyes at the corners and run a rope around the perimeter; attach cloth with clothespins. Or use U-shaped pins pushed through the cloth and into the soil.

8. Peel the cover back when you need to inspect the crops or sample the harvest.

- Direct-seed bush beans, pole beans, corn, pumpkins, winter squashes, cucumbers, and melons after all danger of frost has passed, and the soil is sufficiently warm.

- Plant basil seedlings wherever you have space, and immediately start a new crop of basil in seed flats. Successive sowings of basil are essential to keep the basil from growing bitter, which occurs when plants begin to flower and happens no matter how vigorously you pinch back the flower stalks.

- Keep sowing carrots, beets, lettuce, and greens. However, unless you have a spot with partial shade, stop sowing spinach and radishes until cooler weather returns.

- Direct-sow nasturtiums for their edible flowers to include in salad mixes. Soaking the seeds overnight improves germination and early growth. Although young seedlings often start slowly, they quickly stretch out once the weather turns hot. Don't give nasturtiums high-nitrogen fertilizers or the plants will be all foliage with few blossoms.

Full-sized garlic won't be ready until July, but you can begin harvesting green garlic in late May and early June.

- Harvest green garlics. I plant my garlic cloves about 3 in. (8 cm) apart so I can pull the thinnings for green garlic, which are essentially garlic scallions.

- Watch for damage from animal pests, which become active in spring; woodchucks can be especially destructive when they find their way to the broccoli patch.

If you plant early, you can pick early. May offers the first chance at crisp, sweet radishes, salad mix, and Asian greens.

Big-Time Animal Pests

Insects and diseases damage crops by small degrees. Bigger problems happen when larger animals invade your growing space.

Deer

When large-scale damage happens—an entire row of plants disappears in one night, for example—the problem is likely deer. Deer prefer broadleaf weeds and the leaves and twigs of shrubs that grow along woodland edging. When they find a convenient source of food in a vegetable garden, they can be extremely destructive. Bush beans, pole beans, pumpkins, sweet potatoes, squash, lettuce, and peas can be on their menu. Telltale signs are the hoof marks they leave behind.

Repellents include garlic-scented sprays, bags of human hair, and other products, but the best option is fencing placed around the perimeter of your growing area. Growers have two options for fencing: non-electric and electric.

Non-Electric Fencing. Deer are great leapers, so non-electric fencing will have to be 8 ft. (2.4 m) tall to stop persistent deer. But deer can also crawl, so fencing needs to be no farther than 6 in. (15 cm) off the ground. Also make sure there's spacing for deer to see the fence. (Don't place it directly against a wooded area, or the deer might run into it and knock it down.)

If building an 8-ft. (2.4-m) fence sounds like too much effort, try a lower fence first. Deer that are only casual visitors might be spooked by anything in their way. Even a few strands of monofilament fishing line strung knee high and nose high may be enough to turn them in another direction.

If you don't have the budget or inclination to set up a permanent barrier, plastic deer fencing is an inexpensive option. The flexible netting can be rolled up and stored away when not needed. If you do use netting or fencing, make sure it completely surrounds the area you want to protect. Deer that get inside but can't get out may do more damage than if the area were left unfenced.

Electric Fencing. Some growers use highly visible ribbons of electric fencing called Hot Tape. When deer first encounter the tape, they will remember, but only if it gives them a charge. To be most effective, the fence should be baited to produce the maximum effect from a first encounter. A fact sheet from the University of Massachusetts advises using two strands of Hot Tape baited with a metal tab smeared with peanut butter. When deer take the bait, they get zapped and will remember the experience. (Space the baited tabs about 30 ft. [9 m] apart and keep the fence baited for several weeks.)

In rural areas, dogs can be an effective deer control, but most communities don't allow dogs to roam free, where they will frighten deer. A dog with an electric collar that is outdoors during daylight hours can keep deer away by day.

Voles or Meadow Mice

Small things can do big damage. Voles are only a few inches long, but their vegetarian diet and potential for high population can create serious problems. They feed on beets, cabbages, lettuce, spinach, broccoli, kale, Brussels sprouts, sweet potatoes, and root crops such as carrots and turnips. They can tunnel under snow in the winter and gnaw at fruit trees, damaging the bark and inviting disease. One fat vole that makes its way into your cold frame can wipe out multiple flats of young plants, especially cabbage crops.

You can trap voles using ordinary mousetraps baited with apples, strawberries, or peanut butter. You can also use unbaited traps as long as they

are set in the pathways that voles travel. Set the trap in front of a vole's burrow, perpendicular to the path. When the vole crosses, the trap should snap. When you do catch a vole, wear disposable gloves to remove it from the trap since the animals often carry diseases.

Keeping an outdoor cat (a mouser) can help control vole populations. Other predators are weasels, foxes, coyotes, and owls. Because voles reproduce so quickly, predators are sometimes not enough to keep the population in check.

Cultural controls, including mowing the lawn low and limiting mulch around the base of fruit trees, may discourage voles.

There are also numerous baits and poisons for vole control, but none is advised for organic growers.

Woodchucks (aka Groundhogs)

Gardeners wish it were easy enough to scare a groundhog away with a shadow. The truth is, these squirrel relatives can become persistent pests, raiding your crops every day.

Woodchucks are especially drawn to broccoli, kale, and cabbage family crops, but they will eat almost anything—beans, carrot tops, peas, and lettuce. Corn and hot peppers might be the only plants they don't destroy.

Trapping woodchucks is not difficult, but the solution may be only short-term (read on). If you want to trap the animals, use a humane, cage-style trap so you can release possums, skunks, or other animals that you might catch by accident. If you catch a woodchuck, be prepared to exercise a less-than-humane treatment. Woodchucks can carry rabies, and most states prohibit you from relocating and releasing them. (Needless to say, you should handle the animals carefully.)

Traps can be baited with cabbage, kale leaves, or strawberries. Some experts say the best way to trap them is to leave the trap unbaited but set up a temporary fence that funnels the woodchuck into the front door of the trap. However, because woodchucks are territorial, trapping them is usually only a temporary solution. After the critters are gone, another one (or pair) may take up residence.

Fencing is the best option, but it's not easy. Because woodchucks can climb and dig burrows, the fencing has to extend underground at an angle away from your crops. Then if the varmint tries to burrow underneath, it will bump into the underground fencing.

A University of New Hampshire bulletin advises that fencing should be 3 ft. (1 m) high because woodchucks can climb. Bend the fence at the top (at a 45-degree angle away from the garden), which forces the animal to crawl upside 0down to scale it. Another option is to leave the top of the fencing loose: When the woodchuck tries to scale the top, it will bend, causing it to fall backward. Fencing also has the benefit of protecting your beds against hungry rabbits.

If you want to go all the way, electric fencing is effective, with one wire at 4 in. (10 cm) aboveground and a second wire 4 to 5 in. (10 to 12.5 cm) above that. Battery-powered fencing is available at many garden centers.

There are other tricks you can try:

- Place foil pie plates around vegetables and set up shiny pinwheels to scare the woodchucks away (limited success).

- Use a scarecrow or owl figurine (limited success).

- Set up clean jars of water around the garden; woodchucks may see their reflections and run (limited to fair success). Maybe they see their shadow.

JUNE: The Heat Is On!

With summer's bounty set to peak in a few short weeks, it's time to take care of *everything*—but pay special attention to the heirloom tomatoes.

Perfect Tomatoes

Keep indeterminate tomato vines under control by pinching off suckers. Find the first blossom cluster and prune off the succulent branches that grow from leaf nodes beneath it (below the hand in photo).

Late-May transplants will be at least knee-high in June and starting to branch out. As seedlings gain size and vigor, growth can sprout in rampant fashion from leaf nodes. You might need to remove these branches (called suckers), depending on the type of vines you're growing.

If vines are *determinate*—meaning they stop growing as soon as the first wave of fruit appears—the suckers should remain in place for maximum harvest. If the vines are *indeterminate*—meaning they keep growing and setting new fruit until the vines are killed by either disease or the end of the growing season—you need to remove many of the suckers.

A tiny sucker on a vigorous indeterminate vine can develop into a full-size branch in weeks. Left unattended, suckers can create a tomato vine that's all branches, foliage, and flowers, with little energy to set full-size fruit. To solve that problem, pinch off all suckers from the bottom leaves until the last sucker below the first set of tomato blossoms. Then allow only two leaders to grow. Remove suckers before they are 4 in. (10 cm) long.

Tomato Cages

Caging tomatoes is a good option as long as you don't use flimsy store-bought items. The best tomato cages are made from concrete mesh, the heavy-gauge steel fencing used to reinforce poured concrete sidewalks. It's available at most construction supply stores in 150-ft. (46-m) rolls, which will give you twenty-three tomato cages for about $120.

The cages take time to make, but once finished, these cages will last a lifetime (despite a little surface rust). The finished cages are 5 ft. (1.5 m) tall with 6-in. (15-cm) squares, allowing plenty of space to sneak your hand inside to pick tomatoes.

These heavy-duty tomato cages are especially good for vigorous cherry tomatoes, which can't really be tamed with stakes or store-bought tomato cages. Use them for Juliet and Black Cherry—two extremely fast-growing cherry tomato varieties with potentially huge harvests. And if you don't need twenty-three tomato cages, cut a few cages in half and use them to rein in zucchini squashes, cucumbers, or other vine crops.

Raise Them Up

To keep your tomato vines neat and productive, you'll need to stake or cage them. Tomato vines left to sprawl invite damage from rodents, slugs, and ground insects. Prevent this potential damage by training your tomatoes to grow inside a rigid cage or against sturdy stakes.

Basket Weave Staking

The easiest way to provide support for a long row of field-grown organic tomatoes is by basket weaving, also called stake and weave. Here are four easy steps to show you how to do it:

1. Pound sturdy stakes between every two tomatoes—one stake, then two vines, then one stake, then another two vines, and so on. Place two stakes or a heavier stake at either end.

2. When plants are 2 to 3 ft. (60 to 90 cm) tall, tie twine to one of the double stakes, then walk the twine through the plants as if you are skiing a slalom run. When you get to a stake, wrap the twine around it two times to hold it in place.

3. When you reach the end stake, wrap the twine around it a few times, turn around, and do the slalom again in the opposite direction, creating a second support about 8 in. (20.5 cm) above the first. The twine will create a basket weave that keeps tomato vines from sagging under the weight of the fruit. Start the first line about 12 to 18 in. (30 to 45 cm) aboveground with the next one about 8 in. (20.5 cm) higher. To save time in a long field, use a spool of twine on a dowel so that it unravels as you walk.

4. When you reach the top of the vine, cut the twine and tie it to one of the end stakes.

5. As the tomato vine grows taller, repeat the process.

Dealing with Nuisances

Whether you grow tomatoes with stakes, basket weave support, or cages, you'll need to watch for the start of fungal diseases. In addition to pruning off suckers, I like to remove the lower leaves of the tomato vine before the plant bears fruit and those leaves begin to yellow. Pruning the understory foliage increases air circulation and prevents leaves from contacting the soil, where they can pick up fungal diseases. In addition to pruning, ensure that your plants are getting the proper soil nutrients—especially the right amount of phosphorus, potassium, and calcium—which help plants grow stronger cell structure and fight fungal diseases.

Keep an eye on all plants to spot early signs of disease and insect damage. I like to spend ten to fifteen

Pruning off the bottom leaves of tomato plants in June can improve airflow and prevent fungal diseases in July and August.

Organic Bug Killers

Just putting the word "organic" in front of a pesticide doesn't make it safe. It means only that the bug-killing compounds are derived from plants instead of being man-made. When concentrated, the plant-made toxins can become potent weapons against vegetable pests. Some can also do harm to birds, bees, earthworms, and natural predators. And if applied carelessly, they can harm the farmers who use them. Be sure to read labels, follow directions, and heed warnings.

One reason why organic practices are easier on the environment is the philosophy of dealing with insects. Controlling a specific pest begins with identifying the insect and understanding its life cycle. Organic farmers use the least disruptive practices first—handpicking pests, destroying their eggs, removing their breeding materials, and using physical barriers such as floating row covers. The next step is introducing natural predators like ladybugs or parasitic wasps and, finally, spraying. Even spraying can be low impact, with bacterial pesticides like Bt, which is toxic only to a specific pest when it eats the material. Broad-spectrum organic insecticides, which kill most insects they contact, are the last resort.

What follows below are descriptions of some organic materials—including pesticides—you might reach for if insects gain the upper hand.

Azadirachtin (Neem). This botanical pesticide is made from oil in the seeds of the tropical Neem tree and works as a contact pesticide, an "ingesticide" and a fungal control. Neem kills numerous beetles when applied when they are young (Colorado potato beetles, Mexican bean beetles, and flea beetles) by disrupting the insects' molting cycle.

Azera. A combination of Neem and Pyganic (azadirachtin and pyrethrin), it kills a wide range of insects, including cabbage worms, Japanese beetles, aphids, cucumber beetles, and squash bugs (when sprayed on young nymphs). The mixture of two botanical insecticides results in a broader range of pest killing. It also kills beneficial insects.

Bt. The bacterial pesticide Bacillus thuringiensis is a reliable control for soft-bodied caterpillars, especially those that attack broccoli, kale, and other cabbage-family plants. When young caterpillars eat foliage sprayed with Bt, the ingredients attack their digestive systems, and the insects die. Bt sprayed directly on insects does not kill them. Be sure to cover all the foliage when spraying Bt—the tops and bottoms of leaves and inside the new growth where young caterpillars feed. Different strains of Bt work on different pests, including Colorado potato beetles. Read labels and check with your certifier.

Clay and Minerals. Fine-powdered kaolin clay (sold as Surround) is mixed with water and sprayed on vegetable crops. When it dries, the foliage is coated with a fine powder that disguises the plant or at least makes the food unappetizing to insect pests. Kaolin on cucumber vines repels cucumber beetles; on apples it deters fruit pests. A drawback is that it leaves a powdery residue on your harvest.

Diatomaceous Earth. It looks like fine white powder, but it's actually the fossilized shells of microscopic diatoms. DE, as it's called, has sharp edges that pierce an insect's skin or shell causing it to dehydrate. Sprinkle it around seedlings to deter cutworms, spread it over crops with a hand-held duster, feed it to animals to prevent worms, and sprinkle it in chickens' dusting holes to kill parasites on their feathers. DE does not pose a threat to birds or mammals but will kill beneficial insects. And it loses its effectiveness when it gets wet.

Floating Row Covers. Using lightweight spun poly fabric is a highly effective but low-impact way to keep insects away from your vegetables. Drape the cover loosely over crops and secure it to the ground with pins or weights (rocks). Row covers are available in multiple weights. They can also be stretched over loops of stiff metal wire or semi-rigid PVC tubing to create a breathable and water-permeable low tunnel. A tunnel has the advantage of stretching over taller crops rather than simply draping the material on plants.

The best way to buy row covers is in bulk rolls. Row covers are used by farmers big and small but are ideal for market gardeners and small-scale growers who can gather up the material, store it in a shed, and use it again the following year. Joining a local organic farming organization can bring the benefit of co-op prices on row covers and other supplies. Or check with an organic supply catalog.

   Pyrethrum. A contact-kill insecticide made from pyrethrins extracted from chrysanthemum flowers, it's commonly sold under the name PyGanic. Pyrethrum kills a wide range of pests, including whiteflies, thrips, aphids, and some beetles (potato, cucumber, Mexican bean, and flea beetles). It will also kill beneficial insects, so use it as a last resort. Pyrethrum degrades quickly in the sun, and should be sprayed or dusted directly on insects. Don't mix soap with Pyrethrum; it will degrade faster. There are man-made versions (pyrethroids such as Permethrin) that are not approved for organic production.

   Soap Sprays. Soap and water insecticides will kill such soft-bodied insects as aphids, whiteflies, and mites. You can mix your own concoctions with dish soap and water, or add vegetable oil for smothering action, too. Several applications may be necessary to stop a heavy aphid invasion. Soap sprays should not be used on sunny days since they can damage foliage when the sun is bright.

   Spinosad. This pesticide works both as a contact killer (sprayed directly on pests) and as a residual control for insects that eat the material on foliage. It's derived from a type of soil bacteria. Spinosad is toxic to numerous beneficial insects, including bees, so it should be used with care.

Organic Materials Review Institute

Any plant fertilizer, soil additive, bug spray, animal feed, or product that will be used on a certified organic farm in the United States or Canada must be approved by the Organic Materials Review Institute (OMRI), which is a nonprofit independent organization that's been around since 1997. OMRI reviews products to ensure they comply with USDA organic standards.

   Whether you're a certified organic or just an environmentally conscious grower who follows organic practices, you'll find the OMRI website a useful resource. It has a searchable listing of products with National Organic Program's restrictions on how they can be used by certified farmers. It also has a list of resources on organic farming.

   For a fee, you can become a subscriber and receive newsletters, product updates, access to archives, and email support for questions. Go to omri.org

Homemade Garlic-Pepper Spray

Gardeners have long used garlic, onion, and pepper sprays as natural repellents and pesticides for flea beetles, aphids, whiteflies, and soft-bodied insects. This basic recipe can be modified: Make the spray with garlic alone for slugs; pepper alone for aphids; or the full combination for an all-purpose bug killer.

What You'll Need

1 large garlic bulb (8 peeled cloves)
1 small onion, peeled
1 tablespoon (15 ml) ground cayenne pepper (or 3 fresh cayenne peppers)
1-quart (1-L) canning jar
2 cups (275 ml) hot water
Cheesecloth
½ teaspoon (2.5 ml) liquid castile soap
Rubber gloves and safety glasses

Directions

1. Dice and grind the garlic and onion.

2. Add garlic and onions with the cayenne pepper.

3. Add mixture to jar and fill with the hot water. Screw on lid and allow to steep overnight at room temperature.

4. Strain through cheesecloth to remove any particles that might clog your spray nozzle.

5. Add soap and mix well.

How to Use

For a weaker solution to use as a repellent, mix ¼ to ½ cup (59 to 118 ml) with 1 gal. (3.8 L) of water and coat tops and undersides of leaves. Use full strength to kill aphids and other insects. Refrigerate the unused mixture; it will last up to about six weeks.

   CAUTION: This mixture can be irritating to eyes and skin. Wear safety glasses and gloves when mixing and applying. Also, before applying to your entire crop, test your plants' sensitivity by spraying one branch or one small patch. Wait a day to make sure spray doesn't damage plants. Don't apply products containing soap on sunny days; you could harm foliage. Certified organic farmers should check ingredients with their certifier before using.

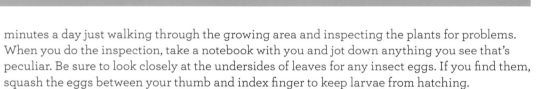

Make Your Own Cages

What You'll Need

One 150-ft. (46-m) roll of concrete mesh

Bolt cutters (not wire cutters)

Heavy-duty leather gloves

1. Cut one 6½-ft. (2-m) section from the roll with the bolt cutters. Snip the wire as closely as possible to the edge of each square (that will leave you with twelve full squares and a row of open 6-in. [15-cm] wires at one end).

2. Pull the two sides of the section together to create a loop.

3. Using leather gloves, bend the free wires through each opposing square to secure the cage. (Be careful, because the mesh is springy and the loose ends can be sharp.)

4. Once you have your secured loop of mesh, cut off the horizontal wires at one end to create a ring of spikes. (That's now the bottom of your tomato cage.)

5. Lower the cage over your tomato seedling and push the spikes into the ground.

minutes a day just walking through the growing area and inspecting the plants for problems. When you do the inspection, take a notebook with you and jot down anything you see that's peculiar. Be sure to look closely at the undersides of leaves for any insect eggs. If you find them, squash the eggs between your thumb and index finger to keep larvae from hatching.

Sometimes plants don't move as quickly as they should. If peppers are slow to get started, give them a drink of organic fertilizer made from fish and seaweed emulsion. It contains a good dose of natural phosphorus, which is sometimes difficult for transplants to take in when roots are still undeveloped. Any cabbage family crops that aren't growing under floating row covers may need to be sprayed with an organic pesticide, especially if you see white butterflies fluttering over the foliage. Cabbage butterflies don't harm the plants, but their larvae—green, soft-bodied caterpillars called cabbage worms—will cause significant damage. The larvae chew holes in leaves and create an ugly mess with their frass (the polite word for poop). Spraying with the organic bacterial pesticide Bt (Bacillus thuringiensis) should keep them in check. Bt is only effective if consumed by young caterpillars, so spray it on the tops and bottoms of the leaves, as well as the new growth where the little critters are feeding.

Another organic pesticide is Spinosad, which is made from soil-dwelling microorganisms. The material works to some degree as a contact pesticide but is more effective once ingested by the pests. Use it on cabbages, broccoli, cauliflower, collard greens, kale, Asian greens, and other cabbage-family crops. Some growers like to alternate applications, spraying Bt first and then spraying Spinosad ten days later.

Get a Head Start on Weeds

Speaking of pests, stay ahead of weeds. Scratch, hoe, cultivate, or hand-pull whatever you can. In addition, keep the grass mowed around the perimeter of your growing area to prevent weeds from setting seeds that could cause problems later. The remains of crops that you've harvested need to be pulled and sent to either the composter or the chickens. If you have time, sow a fast-growing cover crop to suppress weeds before planting another crop. If not, top-dress the area with aged compost and get the new veggies going.

It's Onion Time

Finally, June is the time to pull young onions for thinnings. And I don't mean little scallions; I mean the big boys—Walla-Wallas, Ailsa Craigs, and other giant, sweet-tasting salad onions. I always overplant onions and pull the largest ones early, while the tops are still green. Peel off the dried layer to reveal the onion's shiny inner skin, and the vegetable looks like a marvelous piece of artwork.

Early onions—picked while the tops are still green—look like rare jewels at the farmers' market. Peel the outermost layer of skin off the onion to expose the bright, shiny bulb.

Beat the Bugs of Summer Naturally

Summertime is when bugs are at their peak. But when you want to beat them naturally, you have to hit them at the weakest spot in their life cycle, which sometimes occurs before you see them in the bug days of summer.

There are hundreds of different insect pests, and none of them has a clue that the beautiful crops you plant each year are meant for customers and not for them. Left unchecked, they can lay eggs, damage your crops, and repeat the process with another generation.

Check with fellow farmers, your local cooperative extension service, or the website of a nearby land-grant university for information on the pests that are specific to your region. Meanwhile, here's my Dirty Dozen of the bug world. Chances are they may be gnawing at you—and your plants.

Asparagus Beetles

Description: Tiny, fast-flying beetles, (slightly larger than a cooked grain of rice) attack plants in late spring. There are actually two different pests, asparagus beetles (orange with black dots) and spotted asparagus beetles (orange and black, with white spots on their backs). Adults appear in spring and lay eggs on new growth. The eggs hatch and larvae begin devouring stalks and new foliage.

Control: Look for the brightly colored adults. Hold a can of soapy water underneath them and knock them into the can (not always easy since they can be quite fast). If you don't mind looking silly, catch them in a child's butterfly net and squish them. Use a soft broom to knock the larvae off the foliage (they aren't good at climbing back up the stalks). A natural predator, the Eulophid wasp, attacks the eggs.

Beet Leaf Miners

Description: Adults look like tiny houseflies. They lay tiny, almost invisible, white eggs on the undersides of leaves. Larvae hatch and burrow into the leaves, feeding on the inner cells and leaving ugly brown trails. Different miners attack spinach, beets, chard, and the common weed lamb's quarters. After feeding, larvae drop to the ground, burrow in, and pupate, emerging in about two weeks to do more damage with second and third generations in the same season.

Control: Stop them by rotating crops and covering plants with a floating row cover. Blue sticky traps can catch adult flies before they lay eggs. As a last resort, spray the organic pesticide Spinosad or Neem when you first see eggs. (Look for pinhead-sized white eggs on the undersides of leaves when the first true leaves form.)

Cabbage Worms

Description: There are several different cabbage worms that attack cabbage, broccoli, kale, Asian greens, and other Brassicas—cabbage butterflies, European cabbage worms, and cabbage loopers.

Control: Cover young plants with floating row covers. (I also keep a small butterfly net in the garden to catch the white butterflies that produce tiny green worms.) Spraying plants with the bacterial pesticide Bt (Bacillus thuringiensis var. kurstaki) kills caterpillars without harming beneficial insects. To be effective, Bt must be sprayed on the leaves where young caterpillars are feeding. Spinosad also works, but it's harmful to bees and beneficial insects.

Colorado Potato Beetles

Description: These clumsy looking brown and white striped beetles chew leaves and lay fluorescent orange eggs on foliage. They attack potatoes, eggplants, and sometimes tomatoes.

Control: Crush eggs and handpick adult beetles. Once pinkish, sluglike larvae appear, it may be too late for hand picking. Spray foliage with the biological insecticide Bt var. san diego, which specifically targets potato beetles. Oddly, the adult beetles, which are capable of flying, only have enough strength to crawl to young potato plants in the spring. Large-scale growers have had success in trapping the beetles with trenches around potato fields. The trenches are lined with black plastic, which is too slick a surface for crawling insects, so the beetles get stuck and die on the hot plastic.

Corn Earworms/European Corn Borers

Description: Corn earworms and European corn borers are different pests, but I'm listing them together as highly destructive corn pests. Earworms can have more than six generations in the South.

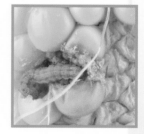

Control: Block worms from getting to your corn with a clothespin over each ear at the point where silks enter. Parasitic Trichogramma wasps, ladybugs, and green lacewing larvae can be purchased to control earworms. Ladybugs and lacewings eat corn borer eggs, while parasitic wasps kill larvae. To stop both earworms and borers, control crop residue. Crush or shred corn stalks and then hot-compost. Till the soil 2 in. (5 cm) deep in early spring; this will expose the pupating earworm larva to the cold. Bt var. kurstaki works on corn borers but not as well on earworms.

Beat the Bugs of Summer Naturally

Cucumber Beetles

Description: These are serious enemies of cucumbers, squash, pumpkins, and melons. Adults attack young seedlings, chewing cotyledons and laying eggs in the soil at the base of plants. Larvae eat the roots, pupate, and emerge as a second generation. The beetles also spread bacterial wilt disease, which kills plants quickly. Adults feed on the dying plants and spread the disease to others.

Control: Stop beetles with floating row covers over young plants. Catch and kill the tiny beetles (spotted beetles are yellow with black spots; striped cucumber beetles have black and yellow stripes). Yellow sticky traps can catch adults. PyGanic spray can also be effective against cucumber beetles, but it has to contact them directly. It also kills beneficial insects, so watch drift and take care not to use it when bees are active. Blue hubbard squash planted at the edges of the field will lure pests away from cucumbers, melons, and summer squash. If plants become infested, spray them instead of spraying all your other crops.

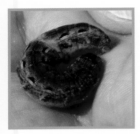

Cutworms

Description: It's heartbreaking to see a row of newly planted seedlings mowed down at ground level by these pests. Soil-dwelling cutworms are the larvae of several different night-flying moths. They crawl along the soil and chew down seedlings, then come back to eat the fallen plants after they've wilted.

Control: Stop them with cutworm collars—strips of newspaper wrapped around each stem so that an inch is buried and an inch protrudes from the soil. Sprinkling a ring of diatomaceous earth around each seedling helps deter them, too. Predatory nematodes are extremely effective natural controls when released into the soil.

Flea Beetles

Description: A persistent pest of radishes, mustard greens, arugula, and a few other Brassicas, flea beetles chew tiny holes in leaves, leaving them unattractive. The pests themselves can be difficult to see because they're so small (flea-sized). They also move like jumping fleas when you get close enough to see them.

Control: Floating row covers will keep flea beetles from reaching crops if the material is secured tightly to the ground. Arugula can be used as a decoy or trap crop to keep flea beetles off of radishes. A last option is spraying with Spinosad, which may kill other beneficial insects.

Mexican Bean Beetles

Description: These ladybug relatives are definitely not beneficial. They are about the same size as their orange–and–black cousins but have a tan shell with tiny black dots. Adults lay minuscule eggs on the bottoms of leaves, and fuzzy yellow larvae strip leaves into skeletons.

Control: Handpick adults and spray with Neem or Spinosad, Destroy crop residue to stop future problems. The parasitic Pediobius wasp can be purchased to kill larvae.

Squash Beetles

Description: If you have Mexican bean beetles, you might think by looking at squash beetles that the Mexican beetles had just grown larger and started eating your squash. The insects look very similar but have different favorite foods. Squash beetles chew the rims of summer squash leaves and lay eggs underneath. Fuzzy yellow larvae chew all parts of the leaves.

Control: Catch adults and drop them into soapy water. (Why soapy water? The soap breaks the surface tension and makes the bugs sink.) The best time to handpick adult beetles is during the midday heat.

Squash Bugs

Description: Ugly, foul-smelling, dark-brown squash bugs (Can you tell that I hate these things?) chew foliage and suck plant juices, eventually weakening summer and winter squash vines. You can sometimes see them crawling under the shade of large leaves that have fallen to the ground.

Control: Lift the leaves, grab adult beetles, and drop them into a can of soapy water. Create your own traps by leaving a square of corrugated cardboard on the ground between squash plants overnight. In the morning, lift the cardboard, grab the bugs and drop them into the soapy water. Don't squish them with your fingers or you'll suffer the discomfort of stinky hands no matter how many times you wash. Look for clusters of cinnamon-colored eggs on the undersides of leaves; scrape them off with a butter knife and wipe the knife in your can of soapy water.

Squash Vine Borers

Description: Another pest, the squash vine borer can be especially destructive with larvae that tunnel into squash stems and eat the vines from the insides out. Look for vines that suddenly droop and show a pile of sawdust around the stems.

Control: Later plantings grown under covers will avoid squash vine borers. By the time the vines bloom and you remove the row covers, the adult pests (clear-winged red and black moths) will be out of season.

JULY: Harvest, Clean Up, and Plant Again

This is the start of the peak growing season, with lettuces, carrots, beets, beans, cucumbers, greens, and summer squashes producing new crops every day. The great tomato parade begins in July, too—usually starting with a few cherry tomatoes and early varieties such as Stupice. Before long, the main crop of beefsteaks will be reaching its peak.

July is also the month when cool-weather crops produce the last of their harvest. Pull spent broccoli, cauliflower, and kohlrabi and discard them with one of two destinations in mind: again, the chickens or the compost pile.

If the end of one season and the peak of another aren't enough to keep you busy, July is also the time to start sowing crops for the fall. That's right—a second crop of cabbage, broccoli, cauliflower, peas, lettuce, collards, Asian greens, and cucumbers awaits those who take advantage of the opportunity.

Beat the Heat

Just as seedlings need protection from the cold in winter, they also need a little TLC to prevent them from burning up under the scalding midsummer sun. You can start seeds in flats indoors

If space is limited, borrow techniques from urban farmers. Strawberries, herbs, and ornamental lettuces can serve a dual purpose as decorations and dinner. Use a compost-rich planting medium and water twice a day during the heat of summer.

Planting in containers provides great opportunities to combine different plants, color schemes, and interesting containers. I grow Lacinato kale with Purple Ruffles basil in pots *(above)*. Even small-growing tomato vines are possible in patio containers *(at right)*.

under lights or underneath the gentle protection of shade cloth outdoors. I prefer starting my seedlings in flats on my patio deck, where they receive full sunshine but are also close to the hose. That guarantees they will receive a drink of water in the morning and in the evening.

The summer heat will quickly take its toll on container plants, too. I typically grow a handful of plants in large containers around my pool and patio. Instead of impatiens, canna lilies, and petunias, I fill my pots with Bright Lights Swiss chard, red-leaf lettuce, Lacinato and Rainbow Lacinato kale, Purple Ruffles basil, and other vegetables that look as good as they taste. One of my favorite arrangements is a single Lacinato kale plant in the center of a large pot surrounded by compact Purple Ruffles basil. The dimpled blue-green kale leaves look magnificent growing like palm fronds over the deep purple basil plants.

The challenge with potted plants that are left to endure the midsummer sun is that they can dehydrate quickly, especially if the foliage is larger than the container it's growing in. Make sure the pots are filled to the brim with a good compost-enhanced growing mix, and keep the water coming twice a day during hot spells.

Ready to Harvest

The garlic you planted last fall should be ready to pull by early July. Look for the tops to start drooping, but don't wait until they turn brown or else the cloves might begin growing new roots. After harvesting, brush off the loose dirt (but don't wash them under water). Place the cleaned-up garlic on a rack, stalks and all, and dry it out for a couple of days—just keep it out of the sun and rain during this time. Later, trim the tops (or braid them if you grow soft-neck garlic), and store the garlic in a cool, dry place until you need it.

Hang garlic in a dry spot that's out of the sun. Harvesttime should be a reminder that you need to place another order for fall planting, while suppliers still have a good selection.

Picking the garlic should remind you that it's also time to order more for fall planting. Some growers save the best and biggest cloves for replanting, and then sell and/or eat the rest. If you want to do that, you'll save money and ensure that next season's garlic crop will be as big or bigger than this year's. If you plan to order new garlic—or save some garlic and order other varieties—do it now. Farmers often sell out of the best garlic varieties before the plants are even harvested. Get your order in early to ensure the best selection.

Shallots will also be ready for harvest this month. Give them the same treatment you did the garlic, but understand that most shallots are even better at storage. Finally, keep picking and parceling out your big sweet onions. Even if they're still not ready for the full harvest, you can include a giant onion in each CSA share or bring a bunch of them to the farmers' market.

Tomatoes should be arriving fast and furiously by the end of the month. Keep picking them as they ripen—especially the cherry tomatoes. And keep the plants evenly watered to prevent cat facing and splitting fruit. Split tomatoes usually occur when a tomato vine is starved for water and

Easy Does It!

Don't pass out all your garlic in CSA shares during midsummer, when all of your other crops are pumping out food like gangbusters. Garlic keeps well, so hold off until the crops start to thin out and then add some garlic cloves to keep the variety and weight of the shares up.

suddenly takes in a huge amount. The plant stores the liquid wherever it can—including in the fruit, which swells up faster than the skin can stretch.

Just as with the other months, keep sowing those successive plantings of lettuces and greens. I know it gets monotonous, but when the weather is hot, customers want to enjoy cool, light meals. Salads are in high demand, which means that beautiful heads of lettuce are, too. Look for heat-tolerant varieties to mature in July and August. One of my favorites is Devil's Ears lettuce, a large multicolored Bibb-style lettuce that has a sweet leaf and crunchy ribs like romaine.

Finally, take time to smell the flowers! Pick some of the edible nasturtium blossoms you planted in May; they add an eclectic and elegant touch when scattered across the top of fresh-cut lettuce mix.

Keep tomato vines evenly watered to prevent split and damaged fruit.

The Right Way to Water

The heat of July means it's a great time to check your watering techniques. Irrigate crops early in the morning before the sun gets hot and evaporates the incoming water. Drip irrigation

Peppery-sweet nasturtium blossoms are a perfect complement for cool leaf lettuce. Don't give nasturtiums too much nitrogen fertilizer or you'll end up with beautiful vines that never flower.

Drip irrigation tubing can work above or below. To conserve water and prevent fungal diseases, set irrigation drip tape under soil, mulch, or black plastic sheeting.

is the most effective delivery system, bringing water slowly to the root zone where it is absorbed by thirsty plant roots. It's also a way to conserve water in dry regions because the moisture drinks in before it evaporates.

Hand Watering. A hand-held hose is the next best way to water. But while that's fine for a small garden, hand watering is very time-consuming for a large plot of land. A compromise would be to hand-water sensitive crops such as tomatoes, taking care not to splash soil and water onto the leaves (that's how fungal diseases are spread), and use overhead irrigation (sprinklers) for the rest of the growing area.

Overhead Irrigation. While overhead irrigation is the least effective method, it's still better than just relying on rainfall. If you use sprinklers, set the spray to coarse droplets to prevent evaporation in the summer heat. In addition, water during the early morning hours. Sprinkling at midday causes water to evaporate; late-evening sprinklers soak the foliage, which often doesn't dry out before nightfall, creating perfect conditions for fungal diseases.

To determine how much water your plants are actually getting, irrigate the growing area and then check the soil immediately after you finish. Dig a hole in the soil near your plants (but not so close as to cut into the roots). Observe how far down into the soil your water actually reached. If it's all concentrated in the top 2 in. (5 cm), you need to increase watering, irrigating deeply but doing it less frequently. Also consider mulching plants with fabric, plastic, or weed-free straw to help shade the soil and preserve moisture. Mulching is another way to conserve precious moisture in dry climates and during summer droughts.

AUGUST: It's the Best Time of Year. . .
So What's Bugging You?

There's really no finer time of the year for a farmer. Tomatoes, peppers, and eggplants are all pumping out fruit; cucumbers and melons are ripening under the late-summer sun; and fruit is already swelling on the runaway vines of winter squashes.

This month, you'll need to harden off and transplant second-season crops into the main garden. Hardening off in summer means giving seedlings protection from direct sunlight. Bring them outdoors on a cloudy day or place them in a shaded area, then gradually increase their exposure to direct sun. Using shade cloth may be helpful.

It's Time to Take Pests Seriously

August is also peak time for insect activity. Bean beetles can run rampant over the French fillet beans, white caterpillars flutter like giant snowflakes above cabbage crops, and tomato worms the size of your thumb chew leaves, stems, and even fruit. Some organic growers choose to battle the summertime pests with beneficial insects—fighting bugs with bugs so to speak. But not all predatory insects are created equal. Here's a gallery of some of the more popular choices.

Lady Bird Beetles, most often called ladybugs, will devour aphids and other soft-bodied insects.

Ladybugs

In my pre-organic days, I was almost ready to spray our Stanley prune-plum tree, which had become covered with aphids. As I studied the leaves, I saw a ladybug land on the tree, and I decided to let nature take its course. In a few weeks, the tree was covered with ladybugs and their larvae, which look like tiny orange-and-black Gila monsters. They quickly devoured the entire aphid population and moved on.

This was one of those "organic moments." Today, almost nothing is more uplifting than seeing a few bright orange ladybugs patrolling the leaves of my vegetables. The insects are capable of eating up to fifty aphids per day. And while their appetite for aphids is what draws accolades from gardeners, ladybugs make meals of other soft-bodied pests as well, chowing down on mealybugs, leafhoppers, scales, and mites.

You can order ladybugs by the quart, gallon, or small bag from a number of sources. A quart of ladybugs contains more than 15,000 of the insects, probably sufficient for a large garden. A gallon will cover 1 to 3 acres (0.4 to 1.2 ha), with application rates varying by crop. Ladybugs arrive in a dormant stage. You can release them in the morning or evening shortly after they arrive, or you can store them in the refrigerator—just make sure they don't get wet, and the temperature doesn't drop below freezing.

Parasitic Nematodes

These microscopic roundworms meander through the soil attacking soft-bodied insects. They pierce the insect's skin, burrow inside, and release toxins that kill the hosts. Then they reproduce inside the dead insect, only to burst out to kill again. Predatory nematodes control many soil-dwelling insect pests and are effective against cutworms. You'll find different species of nematodes for different climates and pests.

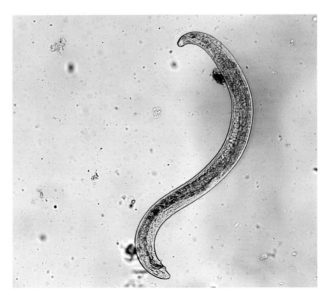

Green Lacewings

Green lacewings are delicate creatures with emerald bodies and gossamer wings, and they lay white oval-shaped eggs on the end of thin silken threads. Not so for the ugly-bug larvae, which devour aphids and other soft-bodied insects along with their eggs.

Predatory nematodes patrol the soil and kill grubs and cutworms, making them very effective in controlling pests naturally.

My wife, Val, holds a giant praying mantis that we found in our garden in late summer. These voracious predators eat good bugs as well as bad.

Praying Mantis

Looking like some science-fiction robot bug, the praying mantis feeds on almost anything that moves, including its own brothers and sisters. These bugs are indiscriminate predators—meaning they would just as soon devour your honeybees as they would eat your garden pests. For that reason, some organic growers don't recommend using them for insect control. But they are beneficial and consume lots of bugs. Mantis egg cases should be tied to a twig or a bunch of grass a few inches above the ground to prevent ants or other insects from damaging the eggs.

Other August Chores

Bugs aside, there are plenty more chores for you to do in August. (What? You thought this was going to be easy?)

First, remove and discard spent crops. Compost them if the plants are free of diseases and insect pests. Otherwise, toss them in the trash. Leaving the plants in the field will provide a safe haven for pests to overwinter.

Next on the agenda is to keep sowing lettuces. With the cooler weather coming, the heads will fatten up nicely without going to seed. Try Bibb and romaine lettuces now that the heat is waning and the heads are less likely to twist in the sun. And don't forget the salad mixes. Clear a patch now and sow mesclun or microgreens for a September harvest. If the weather is hot, use shade cloth to protect the seedlings.

Depending on the end of your season, early August may be time to plant your last crop of summer squash. Squash borers will not bother seeds sown at this time of year. In fifty days, when picking time is nigh, you'll be harvesting.

Attracting Beneficial Bugs

If you allow some of your herbs—dill, cilantro, and fennel—to flower and go to seed in late summer, they will provide food for local beneficial insects that will help you battle pests. It's not just the purchased beneficial insects that will keep your pests in check; the right plantings can also attract and keep local beneficials as well.

Predatory insects also like a varied habitat, so a nearby meadow garden, tall grasses, and non-invasive weeds make a good habitat. Also keep a water source. Rocks in the birdbath will give them a place to land and drink.

Below is a listing of beneficial insects and information on how you can attract them to your garden.

Green Lacewings

Green lacewings and their larvae are fierce predators of aphids. Plant dill, cilantro, fennel, and daisies to attract. Also keep crops watered so lacewings can find a place to drink. Don't panic when there are a few aphids on a vegetable. They'll provide food that keeps lacewings and their larvae close to home.

Ground Beetles

There are multiple species of shiny, black beetles that scurry along the ground, and most of them feed on garden pests—cutworms, earwigs, slugs, grubs, and more. Leave an area of tall grass—also called a beetle bank—where beetles can hide during daylight hours.

Hoverflies

Adults hover like hummingbirds over flowers, but they don't eat pests. The larvae, however, take care of aphids by the bunch. Plant cosmos, alyssum, dill, and fennel to attract.

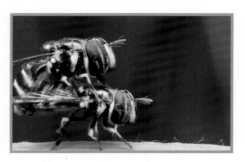

Ladybugs

Adults and larvae eat aphids and other soft-bodied insects. Attract them with dill, fennel, Queen Anne's lace, and other carrot-family plants. You will need to let the plants bloom since it is the flowers that are the attraction.

Paper Wasps

Wasps and hornets prey on cabbage worms, cabbage loopers, and other soft-bodied caterpillars. Grow mint, dill, goldenrod, and meadow flowers to keep them happy. Encourage nesting by making small wooden boxes (a few inches deep with open bottoms). The wasps will make nests underneath the boxes so don't put them where you're likely to disturb them.

Parasitic Wasps

You may never see them because they're small, but tiny parasitic wasps perform large deeds, laying eggs in the larvae and eggs of tomato hornworms, asparagus beetles, and other garden pests. Lure them with dill, fennel, mint, and other herbs, letting the plants go to flower.

Soldier Beetles

These predators attack both good bugs as well as the bad, but cucumber beetles are on their menu, so that alone makes them worthy of your respect. Attract them with goldenrod and mint family plants (catnip, catmint, and spearmint).

Tachinid Flies

There are multiple species that feed off common garden pests. Adults lay eggs on or near cutworms, cabbage worms, squash bugs, and other pests, and the larvae burrow in for the kill. Lure them and keep them around by planting dill, fennel, and parsley (if you let the parsley go to flower).

Cool fall weather is ideal for sowing microgreens. Some can be ready to harvest just a few weeks after sowing.

You'll also want to sow carrots in August. Yes, carrots. The sweetest, crunchiest carrots you'll ever taste are the ones pulled from near-frozen soil in November and December. But to harvest them at that time, you will have to plant them in mid-August. Rake the bed smooth and sow the carrot seeds according to the directions on the packet. You should be able to harvest these in October, but leave some in the ground awhile longer. They will grow sweeter when the temperatures dip.

Ready for more work? Transplant fall Brassicas into the field or garden. Broccoli, cauliflower, kohlrabi, and other cool-weather crops may need shade cloth for protection on very hot days, but they will size up quickly in late August and early September. Don't forget the Asian greens: Plant Pak Choi, Senposai, Hon Tsai Tai, Chinese cabbages, and others in August and September. You'll be surprised how nicely Chinese cabbages head up in the cooler temperatures of September and October without bolting to flower.

Also sow your last crop of basil. Spring-planted basil gets tough and bitter after flowers form, so you'll want to have sweet, tender leaves for all of the slicing tomatoes you pick in late August and September. Keep sowing other herbs as you need them. For example, you should sow cilantro every two weeks for a steady supply to accompany your tomatillos. In addition, sow late radishes, turnips, and parsley for fall harvests.

Keep everything watered and weeded throughout the long month. Weeds that go to seed will cause headaches when the young plants germinate next year. And finally, make time to harvest your peppers, tomatoes, and eggplants. Pick each when they are at their peak: Tomatoes should be fully ripened on the vine—whether that means red, green, orange, black, or striped. They also should be tender to the touch but not mushy. Eggplants should be a little soft but still springy enough to immediately return to their shape when you give them a little squeeze. Purple varieties are generally ripe when their skins take on a satiny sheen. Peppers will be thicker, sweeter, and more flavorful if you allow them to ripen to their true color. In many cases, that's red, but some bell peppers turn from the ordinary green to yellow, orange, purple, or chocolate if you allow them an extra two weeks on the vine.

Fast-growing microgreens can be sown in late August and early September. Kales, beets, chards, mustards, and Asian greens sown in a mix can be harvested in two weeks, as soon as the leaves have formed. Second cuttings and successive sowings can keep you supplied until winter.

SEPTEMBER: It's Back to School and Back to Work

Summer is still here, but autumn is at hand. And while your tomatoes may be looking a bit shabby, there's still enough sun to ripen fruit.

This is a great time to build hoop houses that will keep your plants from freezing during the winter. Keep seeding mesclun, lettuce, spinach, Asian greens, arugula, and radishes. Any fast-growing green that can be harvested within forty days is fine for sowing in September. Also pick crops that are producing—pole beans, peppers, eggplants, potatoes, and anything else that hasn't been consumed by insects or plant diseases. And finally, don't give up on watering. While September is a cool month, it can also be dry. Crops won't be able to give you their best unless they are watered.

OCTOBER: It's Garlic Season

Harvest all remaining tomatoes, peppers, eggplants, and cold-sensitive crops before the frost. If some tomatoes haven't quite ripened, that's OK. You can always make green tomato relish, a fitting end to the year. Once you've finished your harvest, get rid of any crop residue that looks as if it's harboring diseases. Then put all of the healthy leftovers to use in making compost, adding the ubiquitous fallen leaves for carbon and animal manure for nitrogen.

Next, make a note of all of your beds and take samples for a soil test. This is the best time of year to do testing because the labs are less crowded. Testing now will also allow you plenty of time to adjust the pH and nutrient levels before the next big planting season.

Before the season ends, there's one more crop to plant. Garlic sown in October and November will be ready to pick in June and July.

November is the time to look back on the previous year's harvest. This is my kitchen counter filled with produce for three CSA customers in the late summer. It takes work to pick and prepare the produce, too.

Now you can sow cover crops to protect your soil. Depending on the date of your first expected frost, you may not have time to plant a mix of oats and legumes, which die back after the frost. For a late cover crop, try winter rye; it sprouts quickly and grows a few inches before halting when the ground freezes, resuming its growth in the spring.

And while you're harvesting so many other crops, you do get to plant one more. October is garlic month for many growers. You should be receiving those orders you placed during the summer right about now (if you don't already have them). If you've saved your own garlic, sow the largest cloves to get a big crop next year. You can also create a patch of green garlic as you did in the spring. Plant the largest cloves with adequate spacing, then save the smaller cloves for your green garlic patch. Sow them closely and pull them as you need them.

If you have a fall CSA share, continue distributing winter squashes, Asian greens, and other Brassicas, along with root crops, potatoes, and onions. If your CSA has ended for the year, thank your members for their support and let them know some of your plans for next year. Reaching out now will keep them interested in signing up again in the coming year. You might even create a short survey to find out what members liked and didn't like about your program, which could help you improve your business and enable you to sell even more shares next year.

NOVEMBER:

It's Cleanup Time

November is a great time to tackle unfinished tasks: Clean and tune up your tools, mend fences, and move vegetables to storage in the cold frame. If you still have crops growing under hoop houses, keep them watered and free of insects. Cabbage aphids sometimes appear on broccoli, Brussels sprouts, and kale during late October and early November. Blast them away with hard spray from a hose if you find them.

When our cover crops are sown and the other beds are covered with compost for winter, I always leave a few beds open for a late harvest. These white carrots are crispy and sweet when pulled from the ground after a frost.

Your power equipment will need maintenance before you put it to bed for the winter. Drain gas tanks and run the fuel until the engine stops. Any tools that you might need to use occasionally will need fuel stabilizer to keep newer oxygenated gasoline from gumming up carburetors. Don't drain the oil in your machines unless you replace it at the same time—you could forget the oil is missing next spring and damage an engine.

In addition, clean and put away any hand tools. Use a steel brush to scrub dirt off the blades of shovels, rakes, hoes, and other metal tools. Sharpen blades if they need it and oil or repair wooden handles.

Now that you have frost to worry about, it's time to disconnect your drip irrigation system and let it drain to prevent icing, which will split hoses and fixtures. Also disconnect garden hoses, roll them up, and bring them inside.

Finally, don't be afraid to throw things away. November is a good time to take stock of what you need and what you don't, as well as what you need to replace. When you clean your seed-starting supplies, for example, toss out the cracked pots and leaky trays; you can order more in January.

DECEMBER:

Time for a Break!

Sure, you can always find something else to do—but take a rest. You've worked very hard all year long and deserve a break.

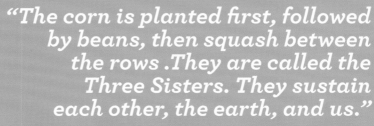

"The corn is planted first, followed by beans, then squash between the rows .They are called the Three Sisters. They sustain each other, the earth, and us."

—Elizabeth Haydon, *The Dragon's Lair*

To tweak that old real estate saw, there are three key principles for small-scale organic vegetable farming: diversity, diversity, diversity. Growing a wide variety of vegetables and herbs leads to an ever-changing harvest, providing a steady flow of crops that stretches from spring until fall, keeping growers and customers looking forward to something new every week.

Good Bug, Bad Bug

Diverse cropping is also an important component in the organic equation. Growing multiple vegetables and herbs attracts a varied insect population—one that polices itself and ensures the appropriate balance of predators to pests. Maintain a healthy mix of plants at every point of the season, and you'll attract the predatory "good bugs" that devour many of the "bad bugs."

If you add further to the diversity by planting flowers, berries, herbs, or a meadow garden, you'll create a habitat that encourages alpha-predators to move in and get comfortable. Spiders, wasps, tiger beetles, birds, toads, and other roving hunters can feast on a living smorgasbord all season long. The benefit to you: As the season progresses, these indiscriminate hunters will vary their diet from slugs to cutworms to earwigs to potato beetles to corn borers to squash bugs.

A healthy mix of plants also means that many different flowers will be blooming throughout the season, attracting and keeping a population of bees and other pollinators. Some flying insects that feed on nectar also prey on garden pests. And all the myriad insects attracted by the multiple plantings help keep any one specific bug—like the little green caterpillars on your Nero di Toscana kale—from getting out of control.

A third benefit of diversity is that it allows a grower to rotate crops. Crop rotation—another practice in organic growing—preserves soil nutrients, combats diseases, and thwarts pests that look for a nearby meal after overwintering in the soil. (See chapter 3 for more information.) But you can't practice good rotation if you grow only one or two vegetables. Farmers who plant a diverse assortment of herbs and vegetables can separate their crops into groups based on similar diseases, insect pests, and nutritional needs, and then rotate the groups accordingly.

Diversity also protects against financial losses from plant diseases, droughts, flooding, and uncontrollable natural disasters. If a ruinous blight spreads through your tomatoes, the varied plantings will ensure that there are still cucumbers, garlic, melons, and pole beans to provide

A diverse mix of vegetables, herbs, and flowers will encourage predators—including spiders *(left)*, tiger beetles *(center)*, and wasps *(right)*.

Pollinators come in all shapes and sizes. If your plantings are varied, you'll encourage Black Swallowtail butterflies *(left)*, Monarch butterflies *(center)*, and honeybees *(right)*.

an income. In a monoculture system, where entire fields are planted with just one or two crops, a single pest or disease can severely reduce or even eliminate an entire season's income.

Organic Choices

Selecting the right vegetables and herbs—and the right varieties of each crop you grow—is almost as important as overall diversity. Finding plants that perform best in your geographic area is part experience and part listening. Talk to other growers about their successes and setbacks. With a little personal advice, it can be easier to sort through the hundreds of varieties available from the hundreds of tempting seed catalogs.

There are tomato hybrids that resist common diseases, lettuce varieties that stand up to the summer heat, beans that germinate in cool soil, cucumbers with all female flowers, and others that emit less of the cucurbit odor that attracts insect pests. And there are eggplants, peppers, tomatoes, and kales with different growing habits, flavors, shapes, and colors. Disease-resistant hybrids may seem like the perfect choice for organic growers, but surprisingly, hybrids aren't always the first choice.

Why Choose Heirlooms?

Many organic growers choose open-pollinated, old-world varieties for their crops. The choice is part philosophy and part pragmatism. A few generations ago, open-pollinated varieties were the only show in town. Many open-pollinated flowers and vegetables are called by the non-scientific term heirlooms. They were the standard in farming and gardening. Seeds passed from grandfather to father to son, and those who were saving seeds sought out the best-quality fruits from their strongest, most prolific plants.

The intent was to duplicate the highest-quality crops in subsequent years. And the result was that each generation of the same vegetable became better, more flavorful, stronger,

Part of the attraction of heirloom tomatoes is the variety of color, shape, and size. Pictured are Pink Oxheart, Brandywine, Tigerella, Black Cherry, and burgundy–hued Black Krim.

and more adapted to the immediate climate and soil conditions than their parent plants. Through minuscule changes over centuries, open-pollinated plants developed the most desirable traits that their caretakers sought. In addition to growing stronger with subsequent generations, open-pollinated vegetables developed regional peculiarities—colors, flavors, aromas, and shapes that made them different from similar vegetables grown just a few hundred miles away.

Seed saving changed when small family-run farms declined and farms became suppliers for larger geographic areas. Plant hybridizers working for large-scale producers turned their focus toward improved efficiency—intentionally cross-pollinating vegetables to produce tomatoes that wouldn't bruise easily, peppers that stayed firm and fresh during an extended truck ride to market, and corn that was so predictable and uniform in its harvest time that a massive field could be machine-picked in a single day. Now, thanks in part to small-scale organic farmers and gardeners, open-pollinated varieties have made a comeback.

Lost in the desire for convenience were the subtleties of flavor, color, texture, and individuality that gave so many old-fashioned varieties their charm. Hybrid seeds have another disadvantage: While the plants grow strong and resist diseases, saving and replanting seeds won't result in a predictable second generation. Some hybrid seeds, though,

The GMO Threat to Organic Seeds

The term genetically modified food (also known as genetically engineered food) refers to crop plants that have been modified in the laboratory to enhance desired traits. Genetically engineered crops pose a growing threat to organic seeds.

Certified organic growers and seed producers, who are not allowed to use genetically engineered varieties under the National Organic Program, try to protect their crops from chemical pesticides and other forms of contamination. But pollen from a genetically engineered crop can drift onto organic fields and cross with organically grown plants. The threat is greatest in corn, and more recently in beets and Swiss chard. It's difficult and expensive to test for the presence of genetically engineered material in seeds.

The GMO threat, combined with the reality that many of the best parent strains for creating new hybrids have been patented by large agribusiness companies, means that seed choices are becoming more challenging for organic growers. The Organic Seed Alliance, a nonprofit organization that lobbies for organic growers and seed producers, has more information on its website: seedalliance.org

have been isolated and planted and replanted so many times that their genetic material has stabilized, and the parent plants will produce nearly identical offspring.

Heirlooms Rediscovered

Small-scale growers who specialize in heirloom varieties can gain a marketing advantage over larger commercial farmers by tapping into the nostalgia of bygone times, as well as promoting the superior flavor of old-world varieties. Top chefs and discriminating customers now pay premium prices for heirlooms. And many of the older varieties seem every bit as tough and disease resistant as the newer hybrids—especially the varieties that have adapted to local conditions.

Growers can benefit greatly by adding these old-world vegetables to their mix—especially if the vegetables are harvested at the peak of ripeness and sold at a roadside stand, farmers' market, or bundled into that day's CSA (Community-Supported Agriculture) share. After all, the food our agricultural ancestors so highly treasured was meant to be picked from the vine and eaten fresh, not placed in cartons, packed into refrigerated trucks, and hauled thousands of miles away to a supermarket's produce aisle. Vegetables were grown to be picked and eaten—without a lot of time in between.

The Joy of Heirlooms

The recent infatuation with heirloom varieties stretches beyond history. Many of them just plain taste better; others wow the eyes with their color and shape. Here are some of my favorites.

Rattlesnake pole beans

Beans

Rattlesnake pole beans are colorful, tasty, and prolific. Dark green pods are streaked with purple on vines that reach 9-10 ft. tall (2.7-3.0 m) (I ran out of trellis space by late July and picked until fall.) Ready to harvest in 65 days. Heirloom Seeds—heirloomseeds.com/beans.htm—and others.

Beets

The Chioggia Italian heirloom beet from around 1840 hits the bull's eye with concentric rings of red and white, like some peppermint candy. It doesn't taste like peppermint but does have good beet flavor and green stems with chard-pink stems. And it won't leave red marks on your cutting board. You pronounce it key-OH-jah. Ready to harvest in 55-60 days. Sustainable Seed Company—sustainableseedco.com—and others.

Chioggia Italian heirloom beets

Broccoli

Early Purple sprouting broccoli is an English heirloom that is planted in the fall to overwinter for a surprisingly sweet and prolific harvest in the spring. Vigorous plants produce lots of purple broccoli sprouts. Baker Creek Heirloom Seeds: rareseeds.com

Cabbage

Early Jersey Wakefield is an early producer of interesting pointy heads, ideal for growers with limited space. One head makes a great summer slaw. Ready to harvest in 75 days. Annie's Heirloom Seeds—anniesheirloomseeds.com—and others.

Collards

Variegated collards have tender leaves with unique, sometimes frilly leaves. Some plants turn yellowish white when the plant gets close to flowering. Ready to harvest in 80 days. Southern Exposure Seed Exchange—southernexposure.com—and others.

Variegated collards

Corn

Before Golden Bantam people would eat only white corn. Massachusetts farmer William Chambers changed that attitude when he introduced a mutant with bold corn flavor. Golden Bantam is an older variety that will get starchy in a few hours, so it should be eaten immediately after picking. Ready to harvest in 78 days. Heritage Harvest Seed: heritageharvestseed.com

Cucumbers

White Wonder has also been called Jack Frost, Albino, and Ivory King. Buff white fruit is 5-6 in. (13-15 cm) long and good for pickling or slicing. Ready to harvest in 60 days. Victory Seeds: victoryseeds.com

Lettuce

Les Oreilles du Diable or Devil's Ears lettuce is similar in shape and flavor to the Amish heirloom Red Deer Tongue lettuce. But the devil is in the details. Heads are large and slow to bolt, with contrasting burgundy and green leaves. This is one of my personal favorites. Ready to harvest in 50 days. Siskiyou Seeds, Williams, Oregon: siskiyouseeds.com/

Melons

The Charentais French heirloom produces 2-3 lb. (1-1.5 kg) of fruit with gray-blue-green skin and sweet, aromatic orange flesh. Ready to harvest in 85 days. Seed Savers Exchange: seedsavers.org

Devil's Ears lettuce

Peppers

Thomas Jefferson grew the Bull Nose pepper, but it originated in India in the mid-1700s. The plants are prolific and early. Harvest in 55 days for green; 80 days to red. Seed Savers Exchange: seedsavers.org

Radishes

Beauty Heart is a beautiful radish by any other name. (It's also called Red Meat of China and Watermelon Radish.) When planted in late summer for a late fall harvest, it can reach 3-4 in. (7.6-10 cm) across. The "beauty" happens when you slice into the pale green roots; you can't help but say ooh! when slicing into its deep magenta center. Harvest in 55 days. High Mowing Organic Seeds: highmowingseeds.com

Squash

Trombetta Italian heirloom is a climbing summer squash that won't be bothered by squash beetle, bugs, or borers as it pumps out mild-flavored pistachio-colored fruit. Squashes are best picked when 12 in. (30 cm) long but are still crisp and nutty-flavored at 18 in. (46 cm); they will reach more than 4 ft. (120 cm) long if you turn your back. Ready to harvest in 58 days. Renee's Garden: reneesgarden. com

Trombetta Italian heirloom squash

Tomatoes

What? Not Brandywine? Sorry, the Amish heirloom is a terrific tomato, but production can be iffy. Meanwhile, I like the color and flavor of Pineapple, a yellow-and-red marbled heirloom. Other favorites are: Black Krim (75 days), a large mahogany-hued tomato; Aunt Ruby's German Green (85 days), with surprisingly sweet fruit; and Amana Orange (85 days), an apricot-colored tomato that is sweet and juicy. Sources: Sustainable Seed, sustainableseedco.com; Baker Creek Heirloom Seeds, rareseeds.com.

Pineapple tomato

Top Vegetable Crops for Small-Scale Growers

There are hundreds of different vegetables with a dizzying array of colors, shapes, growing habits, and climate preferences. With so much variety, the economy-minded grower will experiment and find that some vegetables outperform others—in the garden, at market, and in the kitchen.

The following A-to-Z directory profiles many favorite crops of organic growers, chosen for their productivity, flavor, market appeal, and ability to provide a strong and steady harvest with limited space. Each entry includes basic instructions for sowing, transplanting, weeding, feeding, protection from pests and diseases, harvesting, and marketing. Also included are listings of some recommended varieties (both heirlooms and hybrids), along with the number of days to harvest, called out in parentheses when noteworthy, to indicate the expected harvest time from transplanting into the field or garden.

This directory is only a starting point. Wise growers will consult the pages and websites of seed suppliers, especially those with open-pollinated seeds. I never let a year go by without taking my chances on at least a half-dozen new varieties—either on the recommendation of a friend or just giving in to the urge when poring over the pages or websites of seed catalog favorites—Johnny's Selected Seeds: johnnyseeds.com; Southern Exposure Seed Exchange: southernexposure.com; Baker Creek Heirloom Seeds: rareseeds.com; the Seed Savers Exchange: seedsavers.org; Fedco Seeds: fedcoseeds.com, and High Mowing Organic Seeds: highmowingseeds.com

Asparagus *Asparagus officinalis*

Basics: Asparagus is an investment—a perennial vegetable that won't produce an appreciable harvest until its 3rd year but can last for 20 years if well tended. It provides an early-season harvest for most growers, with spears poking through the soil shortly after the latest frost date. Asparagus performs best with a lot of space. Plant at least 10–15 crowns per person.

Details: Fertile, well-drained soil; neutral to slightly acidic with a pH of 6.0–7.0. Hardy in Zones 3–8. Grow in full sun or partial shade.

Ease of Growing: Easy to moderate.

Varieties: All-male hybrids are generally more productive than older Mary Washington or Martha Washington, which produce female plants that waste the plant's energy on seeds. Jersey King, Jersey Prince, Jersey Knight, and Jersey Giant are productive all-male hybrids. Purple Passion is a vigorous heirloom variety (male and female plants), which produces thick, violet stalks; it can be planted closer in rows and still thrive.

Companions: Aromatic herbs are said to help repel asparagus beetles. Parsley, basil, and tomatoes are recommended neighbors but not onions. Asparagus takes up lots of space and is a perennial, mixed planting so companion planting is often unworkable.

Vegetable Families

Knowing the relatives of your vegetables can help you determine a plan for rotating. Under a crop-rotating plan, growers move their crops from one plot to another each season, taking advantage of the characteristics of each grouping. For instance, legumes leave nitrogen in the soil, so you might follow them with corn, which needs high amounts of nitrogen. Squash should not follow cucumbers or melons because those crops are in the same family, drawing many of the same nutrients from the soil and harboring the same diseases and insect pests. By planting root crops or a mix of lettuce and herbs or Brassicas instead, a pest's life cycle is interrupted.

Alliums (Garlic and Onion Family)

Garlic, onions, and other alliums are grown for their sweet and spicy bulbs or foliage. Onions, garlic, shallots, scallions, leeks, and chives are all important alliums—each with its own characteristics. One benefit to growing alliums is they repel many garden pests, including deer, mice, voles, and aphids. Use that to your advantage when companion planting. Most alliums are easy to grow and provide a strong yield in limited space.

Apiaceae (Carrot and Parsley Family)

Formerly known as the Umbelliferae family, this group was named for their umbrella-shaped flower heads. But most growers never see the flowers of members like parsnips and carrots because the plants are biennials that are harvested before their flowering year. Many culinary herbs such as dill, fennel, parsley, and cilantro are members of the group; some are annuals. Leave a few of these herbs to bloom because they all attract beneficial insects.

Brassicas (Cabbage Family)

Organic vegetable farmers grow so many Brassicas that it's hard to plan a rotation scheme. There are also multiple names for the group, including crucifers and cole crops. Some are planted twice or more in the same season, making it hard to find a spot where you haven't grown them for 4 years. Radishes, turnips, arugula, cabbage, cauliflower, mustard, collards, kale, and most Asian greens are Brassicas.

Chenopodiaceae (Beet Family)

The beet group includes Swiss chard and spinach, which share common traits, nutritional needs, and pests.

Compositae (Daisy Family)

How is it that lettuce and sunflowers are related? It's the shape of their blossoms, which are actually clusters or composites of multiple flowers.

Cucurbits (Cucumbers, Melons, and Squash Family)

Another big group, this includes cucumbers, melons, pumpkins, and summer and winter squashes, which all share common insect pests along with their space in the plot. Successive crops of zucchini and summer squash can keep the harvest coming from July to October.

Legumes (Bean and Pea Family)

Legumes include pole beans, bush beans, soybeans, cowpeas, peas, and snow peas—along with such cover crops as vetch and clover. They're easy to grow and provide a quick and heavy harvest. And they recharge the soil with nitrogen. Traditionally, heavy-feeding crops like corn are grown the year after legumes. Always use a legume inoculant to get the best benefits from nitrogen-fixing nodules on the roots.

Solanaceae (Potato Family)

This group includes tomatoes, potatoes, peppers, eggplants, and tomatillos. With the exception of potatoes, Solanaceae crops need warm soil and will be damaged by even a light frost.

Easy Does It!

How to Grow: Asparagus can be started from seed, but it's quicker and more productive to begin plants from 1-year-old crowns purchased from a garden center or mail-order supplier. Plant crowns in deep, loamy, and well-drained soil. Dig a trench about 8 in. (20.5 cm) deep and loosen soil at the bottom. Add phosphorus and minerals according to soil testing and work them into the soil. Also mix in compost, but do not make the soil compost-heavy, since the plants don't like to be waterlogged. Space the roots 18 in. (45 cm) apart in rows that are 3–4 ft. (90–120 cm) apart. Spread the roots over a small dome of soil inside the trench with the tiny asparagus bud at the top. Cover with about 2 in. (5 cm) of soil and keep the beds watered. As the plants grow, add soil until the trench is filled. For Jersey Giant, Jersey Knight, and Jersey King, dig a shallower trench—about 5–6 in. (12–15 cm) deep. For Purple Passion, plant crowns 12 in. (30 cm) apart. With all asparagus, it's important to care for the crop after harvest. Keep the patch weeded and watered throughout the first year so tall fernlike foliage can thrive. Healthy plants will store more food in the roots, allowing for a healthier crop next spring. Don't cut down foliage until it has died and turned brown in the fall.

Pests: Asparagus beetles. Two species of tiny beetles (a little bigger than a grain of rice) attack plants in late spring. Asparagus beetles are orange with black dots and lay eggs on spears. The eggs hatch and gray sluglike larvae begin devouring the new foliage. Spotted asparagus beetles are similar in size but are black and red with six large spots on their backs. Both are fast moving and hard to pick by hand. Try it anyway by knocking them into a cup of soapy water held beneath them. Also cut and use any spears that bend over at the top like a candy cane. (That's a sign of beetle damage.) When foliage appears, use a soft broom to brush larvae off the plants. (They aren't good at climbing back up the stalks.) As a last resort, use the organic pesticides neem and pyrethrum. (See page 126 for more information.) A natural predator, Eulophid wasps, attack eggs. Another pest, the asparagus miner (larvae of a small fly), can damage spears and spread disease. To reduce infestations of all insects and diseases, cut down and remove old foliage after it turns brown in the fall.

Diseases: Fusarium crown/root rot and rust, two fungal diseases. Order disease-free crowns from a reputable supplier. Remove any diseased plants from your bed, along with surrounding soil. Water the soil but not the foliage, and don't plant a new bed of asparagus where it has been grown within the past 5 years. Jersey Giant and Apollo are resistant varieties.

Challenges: Weeds can be a problem, especially when plants are young, and space between them is high. Don't let perennial weeds like quackgrass, bindweed, or wild violets get the edge. Pull weeds and mulch with leaves or straw; doing this will conserve water and prevent weed seeds from germinating.

Harvesting: Cut spears at ground level with a sharp knife. Don't harvest any spears for the first season, then pick for about a month during the second. A decent harvest should be available in year three, with a full harvest after that. Asparagus beds can last for 20 years if well maintained.

Marketing Tips: Purple Passion is an attractive color for markets.

Beans

Hot, cold, even raw, beans are prolific growers. Whether you grow pole varieties or bush, beans are easy to grow. Grow classic green and yellow varieties or try beans with purple and streaked coloration.

Bush Beans *Phaseolus vulgaris*

Basics: Bush beans are the speedsters of the legume world, providing a harvest just 45–50 days after sowing. But because bush beans provide all their harvest over a 10-day to 2-week window, you'll need to make successive sowings. Many growers rely on bush beans for a first bean crop of the year, and then move to pole beans, which can provide a steady harvest from midsummer until the first frost.

Details: Soil temperature must be above 65°F (18°C) for good germination, with a soil pH of 6.5–6.8. Seeds are viable for up to 4 years.

Ease of Growing: Easy to moderate.

Varieties: Sungold and Rocdor are good yellow-pod beans. Royal Burgundy is a purple-pod bean that tolerates cooler soil. Jade, Blue Lake Bush, and Derby are good choices for green beans. Romano Gold, Navarro, and Roma II yield strong harvests of flat pods. French filet bean choices include green Maxibel, yellow Soleil, and purple Velour beans.

Companions: Similar to pole beans. Beans like cucurbits and Solanaceae family plants. Avoid growing bush beans in the same bed where other beans have been grown in the past 3 years.

How to Grow: Sow directly in blocks or rows, spacing seeds about 2 in. (5 cm) apart. When seedlings emerge, thin to 4 in. (10 cm). Do not soak bean seeds before planting, but spray them lightly with water, sprinkle in some legume inoculant, and shake them up to coat the seeds before planting.

Pests: Similar to pole beans, although they seem to be a greater magnet for Mexican bean beetles.

Diseases: Similar to pole beans, but because they are not open to the air as much as pole bean foliage on a trellis, bush beans are more likely to suffer from anthracnose.

Challenges: Mexican bean beetles can be destructive, especially when they follow successive plantings, becoming a larger problem with each new crop.

Harvesting: Pick beans when young and tender for the best quality. To prevent the spread of fungal diseases, don't pick beans when the foliage is wet.

Marketing Tips: Mix yellow, green, and purple beans together for a different kind of three-bean salad.

Pole Beans *Phaseolus vulgaris*

Basics: For ease of growing, yield per square foot, and consistency of harvest, nothing beats pole beans. Grown on trellises, teepees, or mesh netting, they take up precious small ground space and will yield a harvest for 2 months when most bush beans falter after 2 weeks. That's important for an ongoing harvest to fill CSA bags.

Bush beans take less time to fruit than pole beans.

Details: Fertile, well-drained soil, with a pH of 6.5–6.8. Soil temperature: 60°F–70°F (16°C–21°C). Germination: 6–12 days. Seeds are viable for up to 3 years.

Ease of Growing: Easy to moderate.

Varieties: Emerite is a green French filet bean meant to be picked when pencil thin. But it doesn't get tough or stringy when left to grow a little larger. The vines are strong, harvest is bountiful, and the flavor and texture are outstanding (60 days). Fortex (60 days) is another top choice, with productive vines, a long growing season, and slim, stringless pods that reach 11 in. (28 cm). (Rattlesnake is an heirloom that bears nutty-flavored, long, green pods, streaked with purple; it's a great-tasting, attractive bean that is more than a novelty. Heat-loving asparagus beans, often called yard-long beans are a novelty, and Red Noodle variety is striking in shape and color; some don't like the strong flavor, even when the pods are picked young.

Bush Beans or Pole Beans?

Beans are indeed a magical fruit, with different growing habits. There are snap beans, lima beans, dry beans, and broad beans—all with similar but not identical growing requirements.

So should you grow bush beans or pole beans? The answer is both. Bush beans will give you a quick harvest from the time of sowing—about 50 days. But the harvest will last only a few weeks. Then you'll have to pull the crop and rework the soil for another vegetable. Pole beans will take longer to blossom and bear fruit, between 55–75 days, but once the harvest begins, it will last much longer. Some varieties keep producing throughout the summer and fall, ending with the first hard frost. Pole beans are also easier to pick, growing on a trellis or tripod that puts the harvest at eye level—with no stooping.

If you do decide to grow both bush and pole beans, pay attention to pests. Mexican bean beetles can start with your bush beans and jump to pole beans when the bush beans are pulled. Get rid of crop residue as soon as the harvest is finished and handpick any pests you see.

Among the best sellers and the most flavorful are French filet beans, which command top prices for their crisp, tender texture and nutty flavor. All beans are terrific to grow because the large seeds are easy to see and sow, the plants are vigorous, and the rewards are outstanding for a small area.

Companions: Cucumbers, eggplant, tomatoes, and chard. (Pole beans are one of the classic "Three Sisters" plants for companion planting; the other two are corn and winter squash. Pole beans crawl up the corn stalks as they grow, while winter squashes hug the ground and block out weeds.) Beans also provide nitrogen because they're legumes. Avoid planting beans near basil, alliums, beets, and Brassicas.

How to Grow: Sow pole bean seeds in single rows, not blocks at the base of a tall trellis (at least 6 ft./2 m). Plant seeds 1 in. (2.5 cm) deep and about 3 in. (7.6 cm) apart on both sides. After true leaves appear, thin to 6 in. (15 cm) apart and feed with fish emulsion, preferably a blend with higher nitrogen than phosphorus and potassium. After several sets of leaves appear, the plants will look like bush beans, until the delicate trailing vines start to snake from the tops. It's important to train these vines up the trellis, and not let them get tangled together and hug the ground. This is a little time-consuming but worth the effort. Use a length of twine to rein in the vines, bringing them in contact with the trellis. After that the growth will take care of itself, snaking around the supports, growing thicker, and climbing higher.

Pests: Ants, slugs, earwigs, and cutworms can devour young seedlings, but if the soil is warm, the plants should be able to keep ahead of the insects. The more persistent pests are Mexican bean beetles, ladybug relatives that are about the same size as their orange-and-black cousins but have a dull beige shell with tiny black dots. Adults lay minuscule eggs on the bottoms of leaves, and larvae hatch into yellow fuzz balls with black spines. The larvae will quickly devour leaves, leaving nothing but the ribs and veins behind. Control them by handpicking the adults before they lay eggs. Spray with Spinosad, which works on contact and by ingestion, but avoid spraying when bees and other beneficial insects are in the vicinity. A calm, dry evening is best. Prevent future generations of bean beetles by destroying crop residue. If the pests become a serious problem, forgo planting bush beans, which attract the early generation and get successive generations rolling throughout the season. Growers in the Mid-Atlantic region have reported successful control of bean beetles with the parasitic Pediobius wasp. The beneficial insects should be ordered to time their release with the first notice of larvae—while they are still tiny, just after eggs hatch.

Diseases: Stem rot, anthracnose, rust, and other fungal diseases can damage crops, especially in damp conditions. Practice good rotation and plant cover crops to encourage beneficial microbes that limit the development of pathogenic organisms. Bean yellow mosaic virus and bacterial blights can also become problems. Avoid picking beans when the foliage is wet; plant resistant varieties. Kentucky Wonder pole beans are among the varieties most susceptible to bean yellow mosaic virus.

Challenges: Planting in cool weather can result in bad germination and slow or spotty growth. Beans can also be stunted by a lack of nitrogen, even though they are legumes. Always use a bean inoculant, which adds to the beneficial soil bacteria that colonize roots and allow plants to fix nitrogen from the atmosphere. Adequate soil nitrogen is also helpful for developing bean seedlings before the bacteria become established.

Once pole beans are trained on a trellis, the vines will wind around the structure as they grow upward, bringing the beans to eye level. Avoid picking beans while foliage is wet to prevent the spread of diseases.

Bean beetles can destroy a crop if not caught and dealt with early. Planting too closely can result in crowding, which blocks weeds but can also invite pests and disease at the expense of a productive harvest.

Harvesting: Pick beans when the foliage is dry. For snappy filet beans, harvest pods when they are thinner than pencil width.

Marketing Tips: Filet beans are preferable to large pods for the market as well as the kitchen. Trellis-grown beans should be straight, uniform in width, and free of blemishes. Keep beans evenly watered to avoid "tadpole beans," short pods with fat bottoms and thin, undeveloped tops.

Beets *Beta vulgaris*

Basics: Beets are related to spinach, chard, and amaranth and are an easy-to-grow addition to small-scale organic CSAs. Sow early-maturing varieties in the spring for a quick harvest and larger-root varieties for later in the season. Or sow successive crops every 2–3 weeks for season-long harvest. Even a single harvest can be picked for a long time, thinning plants by pulling the largest roots, then letting smaller beets grow.

Details: Soil temperature: 65°F–85°F (18°C–29°C), with a pH of 6.5–7.5. Germination: 5–7 days. Seeds are viable for up to 4 years.

Ease of Growing: Easy.

Varieties: Early Wonder Tall Top (45 days) has handsome foliage and good root size. Red Ace (50 days) is a good all-around beet with uniformly round roots. Cylindra (55–60 days) produces long cylindrical roots for slicing. Italian heirloom Chioggia beets (55 days) have a bull's-eye pattern in red and white that's visible when the roots are sliced.

Companions: Cabbage, garlic, onions, and lettuce. Low-growing greens can shade young plants and cool the soil for beets planted in midsummer. Avoid planting near runner beans and pole beans.

How to Grow: Work compost and organic material into the soil before planting. In late spring, sow seeds directly, ½ in. (12.5 mm) deep in rows or blocks. Space seeds 2 in. (5 cm) apart and thin to 4 in. (10 cm) for full-sized roots. (Note: Beet and Swiss chard seeds are actually clusters of seeds that are sometimes hard to separate and may produce four or five plants where you only want one. To thin them, snip them off weaker seedlings with fine scissors at soil level. Don't pull the seedlings or you could damage the roots of nearby plants.) Keep beets watered all summer to prevent the soil from drying out. Don't use high-nitrogen fertilizer unless you are growing solely for their tops.

The Color Purple

What do Graffiti, Purple Haze, Royal Burgundy, and Indigo Rose have in common—besides their vibrant violet hue? They all have high concentrations of *anthocyanins*, the same antioxidants that give blueberries their deep, dark color. In addition to being healthy and beautiful, violet veggies are in high demand.

Purple is the new green at the produce counter. It's the color people want to put on their plate for novelty, curiosity, and for optimum health. So if you want to lure the curious and the health conscious—or maybe just entice the kids to eat more fresh veggies—try a few of these eye-popping varieties.

All Blue Potatoes

Purplish brown skins are almost dull when compared to the beauty of this spud's interior. The insides look like a purple tie-dyed shirt. All Blue is also a great-tasting potato, with a good texture for roasting, baking, or mashed. Plants are tinged with purple, too.

Fairy Tale Eggplant

Almost nothing you grow is as pretty as these streaked, purple-and-white miniature veggies. Fairy Tale hybrid is ready to pick 50 days from transplanting, making it one of the earliest eggplant varieties available. It's also tasty and firm, with thin skin and scant seeds.

Graffiti Cauliflower

The first time I saw this cauliflower, I thought it had been dyed. The purple is so brilliant it looks artificially bright, as if it needed a switch on the side to turn off the electricity. Graffiti is easy to grow—and it's ready to harvest 80 days from transplanting outdoors. It's self-blanching, meaning you don't have to tie up the leaves, but I'd still recommend it. The first year I grew it, a few dark spots appeared on the otherwise magnificent heads.

Indigo Rose Tomatoes

Stunning blackish purple tomatoes grow on compact vines, making Indigo Rose a good patio tomato plant—it's certainly a conversation starter. Fruits have orange interiors and are packed with flavonoids, a type of antioxidant believed to fight asthma, cancer, heart disease, and stroke. Indigo Rose bears fruit 80 days from transplanting.

Purple Haze Carrots

Deep violet skin and bright orange interiors will make salads colorful enough to take you beyond Jimi Hendrix, to about A.D. 900 in Afghanistan. That's where the first carrots were cultivated in shades of purple, white, and yellow. Despite its old-world color, Purple Haze is a hybrid, not an heirloom. It's an easy-to-grow Imperator-style carrot, with crisp, sweet roots that are ready to harvest in about 70 days. It's best for raw eating, since cooking dulls both the color and flavor.

Purple Passion Asparagus

Striking purple spears are sweet and productive, despite being mixed male and female plants. Used raw in salads, they add a colorful hue, but when cooked, they revert back to green. Grow them from 1-year-old roots. Purple Passion can be spaced a little closer than Jersey hybrid asparagus.

Royal Burgundy Bush Beans

Not only are they colorful, the deep purple pods make them easy to pick. Royal Burgundy beans also hold up well in cooler soil, meaning they can be planted a week or more before other varieties. Harvest in 55-60 days. If you prefer climbers, try Purple Podded pole beans, which are ready to pick in 60 days.

Pests: Beet leaf miners can cause serious problems to the foliage of beets, spinach, and chard. Adult pests look like ordinary houseflies that lay eggs on the undersides of leaves in late spring. Larvae hatch and burrow into the leaves, eating the material inside and leaving tunnels that cause brown paper-thin patches. After feeding, larvae drop to the ground and pupate. Several generations can appear each season, causing increasing damage to successive crops. Inspect the undersides of leaves for tiny (almost invisible) white, oval-shaped eggs. Use floating row covers stretched over hoops for the first crop of the season. Stopping them in the spring will prevent successive generations. Also get rid of lamb's quarters (an edible weed that serves as an alternate host to beet leaf miners). Use blue sticky traps to catch adult flies. As a last resort, spray foliage with Spinosad when adult flies are active.

Diseases: Root rot is the most damaging beet disease, caused by several types of soil fungi. Plants may die back shortly after germinating, with blackened roots. Adult beets can have damaged and deformed roots. Avoid root rot by practicing good hygiene. Rotate crops and avoid planting seeds in cool, wet soil. Use cover crops in between plantings, especially grain cover crops that are tilled back into the soil. The addition of organic material stimulates beneficial soil microbes that can fight root rot. Plant beets in well-drained soil with high organic content.

Challenges: Most problems occur from soil that is too cool, causing poor germination. Cool weather (in the 40°F/4°C range) can also stunt beets that have already germinated. Use a floating row cover to protect young plants from a late spring cold snap. Leaf-miner damage to beet tops can make them unappealing for market.

Harvesting: For best flavor, pick beets before the roots reach 3 in. (8 cm) across, unless the beets are large winter storage varieties.

Marketing Tips: Use the thinnings with small roots as beet greens. Mix round, cylindrical gold and white beets together in one bundle. For added interest, slice one Chioggia beet at the farmers' market to let customers see the red-and-white bull's-eye pattern.

Beets aren't all red and round. There are golden beets, white beets, and deep magenta beets with colorful interiors and leaves in shades of green or red.

Broccoli *Brassica oleracea* Italica Group

Basics: Broccoli can be finicky if not given the right growing conditions but with a little attention is a reliable performer. It's more productive than cauliflower because after the main broccoli head is harvested, numerous side shoots can be picked for several weeks. That kind of harvest may be time-consuming for large-scale growers, but market gardeners and small-scale farmers can usually get at least three meal-sized harvests from each plant. Broccoli can be planted in both the spring and fall.

Details: Broccoli loves the cool weather and will germinate in soil temperatures as low as 40°F (4°C), but it germinates best at soil temperatures of 65°F–75°F (18°C–24°C). That makes it an ideal candidate for starting indoors in plugs or cell packs and then transplanting outdoors. Germination: 5–10 days. Seeds are viable for up to 3 years.

Ease of Growing: Moderate to difficult but easier for autumn crops transplanted to the field in late August for an October harvest.

Varieties: Green Goliath (55 days) and Early Packman (55 days) are good early-season varieties. Green Magic (60 days) tolerates heat. Arcadia (70 days) is a good fall choice. Apollo (60–90 days) is a heavy-sprouting variety that produces numerous tender side shoots after the small main head is picked; it's a sweet choice for autumn harvests.

Companions: Alliums, Swiss chard, cucumbers, potatoes and tomatoes. Avoid beans and strawberries.

How to Grow: Start broccoli indoors about 8 weeks before transplant time. Sow in well-drained soil as soon as the ground can be worked in spring. Plants will tolerate frosts. Fertilize with fish emulsion or another balanced fertilizer every few weeks. Keep well watered and weeded.

Pests: Cabbage butterflies and other Brassica pests flock to broccoli. Protect seedlings with floating row covers. When plants become too tall for a cover, alternate spraying Spinosad with Bt to control caterpillars. Broccoli is also a favorite woodchuck food. If you have woodchucks, protect your garden with fencing. In the fall, cabbage aphids can appear; spray with a forceful hose blast. Keep soil minerals high for resistance to insects.

Diseases: Botrytis (gray mold) and mildews, both fungal diseases can damage crops. Avoid overwatering and high-nitrogen fertilizers, especially in cool weather. Soft rot is caused by bacteria that can damage heads. Avoid overhead watering, especially at night. Varieties with domed heads are less susceptible than flat-head broccolis. Rotate crops and keep pH near or above 7.0 to avoid clubroot.

Challenges: Hot weather, insects, and inattention can leave plants ragged looking, especially at or beyond harvesting the main head. Keep side shoots picked to avoid flowering. Scout regularly for insect damage.

Harvesting: Cut heads when flower buds are still compact. Early-morning picking helps prevent wilting. For more sprouting, especially in varieties grown for side shoots, cut the central head while it's still small.

Marketing Tips: Organic customers are generally tolerant of a little insect damage, but finding little green caterpillars in freshly cooked broccoli florets is a stomach turner. Keep crops healthy and watch for pests and their eggs. Wash broccoli thoroughly before bringing it to market and encourage customers to rinse it, too. Soaking heads for a few minutes in salted water will kill caterpillars that are too small to see. Rinse broccoli thoroughly after soaking to remove both the salt and any bugs.

Many broccolis will produce abundant side shoots after you harvest the main head. Keep the plant watered and pick frequently, before the new florets blossom.

Brussels Sprouts *Brassica oleracea* Gemmifera Group

Basics: Brussels sprouts take a long time to mature—nearly 100 days from transplanting for some varieties. But the vegetables taste better when picked after a frost, so the long season shouldn't be a deterrent. In fact, they perform best in climates where the sprouts (or baby cabbages as my daughters used to call them) can size up during cool weather.

Details: Rich, loamy soil with high organic content. Soil temperature: 75°F–85°F (24°C–29°C), with a pH of 6.0–7.0. Germination: 5–10 days. Seeds are viable for up to 5 years.

Ease of Growing: Medium to difficult.

Varieties: Falstaff (98 days) is an early-maturing open-pollinated variety with maroon and green sprouts that are tasty and eye-catching, but not reliably large. Green hybrid varieties are more consistent performers. Bubbles (82 days) is about the earliest variety you'll find, with tight, green sprouts. Royal Marvel (85 days) is early and productive. Churchill (90 days) is dependable and less fussy.

Companions: Potatoes, cauliflower, and aromatic herbs. Avoid strawberries. Interplant with small-growing lettuce, which will benefit from the shade of the taller Brussels sprouts during the summer months.

How to Grow: Amend soil with plenty of organic compost. Compost not only provides nutrients but also helps the soil hold water. Mulching with straw or leaves will also help retain moisture during the summer months. Because Brussels sprouts are best when picked after a frost, start seeds indoors about 5 months before your first frost. In 6 weeks,

Brussels sprouts will keep longer if you trim the leaves and harvest the entire stem with sprouts still attached.

harden seedlings, then transplant to the garden about 100 days before the first frost. If frost arrives around October 1 in your area, early May is the right seed-starting time with transplanting in June. That way plants will be ready to pick when the cold weather arrives.

Pests: Brussels sprouts are plagued by the same pests as other cabbage crops. Protect young plants with floating row covers until they grow too tall. Watch for cabbage worms and spray with Bt and/or Spinosad as a last resort. In late fall, cabbage aphids can become a problem, clinging in colonies to the tiny sprouts and rendering the crop unappetizing. Encourage ladybugs, lacewings, and other predators. Also watch all your cabbage crops for aphids as the weather turns cooler. Blast aphids off with a forceful hose or use an organic pepper and garlic spray.

Diseases: Like other cabbage-family vegetables, Brussels sprouts are susceptible to fungal diseases; clubroot, downy mildew, and alternaria leaf spot are sometimes problems. Fertile, biologically active organic soil can help plants resist insects and diseases. Rotate crops, dispose of diseased plants, and add lime (or wood ashes sparingly) to keep pH near neutral. Clubroot is more prevalent in acidic soils.

Challenges: Brussels sprouts have shallow roots and need steady moisture.

Harvesting: Before picking, you'll need to stimulate the plant into sprouting mode. Remove the bottom three or four leaves from the plant in midsummer. About one month before your harvest date, cut off top growth to stimulate development of tight, well-sized sprouts.

Marketing Tips: Instead of picking individual sprouts, cut the entire stem, then remove the leaves but keep the sprouts attached. (Brussels sprouts will also keep better in a root cellar if harvested this way.)

Cabbage *Brassica oleracea* Capitata Group

Basics: If you have the space, cabbage is a worthwhile crop, depending on the variety. The biggest drawback for small-scale growers is that it takes up valuable ground space with a low return on investment. To maximize value, grow red cabbages, savoy cabbages, or compact varieties. Cabbages are prone to most of the same Brassica pests and diseases, but they can be controlled with row covers and the organically approved sprays Spinosad and Bt.

Details: Likes fertile, well-drained soil with ample N-P-K. Germination: 5–8 days. Soil temperature: 70°F–80°F (21°C–27°C), with a pH of 6.5–7.5 (slightly alkaline soil inhibits clubroot). Seeds are viable for up to 4 years.

Ease of Growing: Moderate.

Varieties: Early Jersey Wakefield (62 days) is a fast-growing early cabbage with an attractive

Savoy cabbage has an appealing dimpled texture. When harvesting and trimming, keep a few outer leaves on each head for appearance.

conical shape that saves space. Golden Acre (65 days) is another compact early variety. Savoy King (125 days) and Savoy Perfection (90 days) are good savoy choices. Red Acre (67 days) is a compact early red cabbage.

Companions: Alliums, beans, carrot family plants, lettuce, and spinach. Avoid pole beans and tomatoes.

How to Grow: Start spring cabbage indoors about 6 weeks before the last frost date. Harden seedlings and transplant as soon as the soil can be worked (raised beds helps for earlier planting). Keep plants watered and weeded. Foliar feeding with fish emulsion or another balanced organic fertilizer will improve results. Apply after transplanting at intervals of 2–3 weeks. Protect crops from pests and extreme cold snaps with floating row covers. Low-growing cabbages can remain under cover until harvest. Prevent splitting by giving each head a quarter-turn twist a few weeks before harvesting. This will sever some roots and prevent the rapid growth that can occur after a sudden downpour. Slicing roots on one side of the plant by inserting a shovel into the soil will do the same thing.

Pests: In addition to cabbage worms and cutworms, cabbage maggots can be a problem, chewing roots and weakening plants. Row covers will keep adults away from young seedlings in the spring, and the insects are less active in the summer when the soil temperature has warmed. Parasitic wasps and roving ground beetles prey on larvae. Earwigs can hide inside heads and chew holes. Control with Spinosad.

Diseases: Black rot, leaf spot, and other fungal diseases can damage crops. Practice good rotation and avoid overwatering, especially if water splashes onto leaves.

Challenges: For small-scale farmers, the biggest challenge is battling pests and diseases for a long growing season and then getting low prices for the finished crop. Choosing good varieties will help.

Harvesting: Pick cabbages early in the morning, slicing the thick stems at ground level. If the leaves aren't showing insect damage, leave a layer of loose growth around the head.

Marketing Tips: Keeping attractive outer leaves will improve appearance, especially with savoy and red cabbages. Compact, conical cabbages like the heirloom Jersey Wakefield have a novelty appeal.

Carrots *Daucus carota* var. Sativus

Basics: Nothing says healthy like freshly picked carrots, their lacy green tops still lush and green. Carrots are relatively easy to grow, but they have a few basic needs. Without them, germination can be sporadic and roots disappointing—small, thin, and forked or twisted.

Details: Loose, sandy loam soil with lots of organic matter, ample nitrogen, and a pH of 5.8–6.8. Germination: 7–12 days at 50°F–75°F (10°C–24°C). Raised beds with high organic content will hold moisture and keep soil friable enough to allow unimpeded root growth. If your soil is heavy, double-dig to at least 15 in. (38 cm) and add compost and sand. Also remove rocks, which can twist roots. Seeds are viable for up to 3 years.

Ease of Growing: Moderate.

Varieties: There are myriad categories of carrots, but here are a few that are widely used by North American growers.

 Chantenay. Blocky and stocky, the cone-shaped roots are stronger, thicker, and better suited for compact soils—especially soils with rocks and pebbles. Red-Cored Chantenay (65 days) and Hercules hybrid (65 days) are two common varieties. Kuroda carrots are an improved Chantenay-style carrot with thick roots and vigorous growth. They hold moisture well and are superior for juicing: New Kuroda (68 days).

Carrots look fresher when they still have the tops attached. Keep them cool and out of the sun to prevent the greens from wilting.

Danvers. Similar to Chantenays but slightly less blocky (longer roots): Danvers Half-Long (70 days) and Danvers 126, an improved half-long hybrid (70 days).

Imperator. These are the foot-long beauties you'll typically find in the grocery store. Size and color are both excellent, but flavor is lacking from some varieties. Sugarsnax (68 days) is quite sweet. Nevis, Artist, and Interceptor are other Imperators.

Nantes. Roots are cigar-shaped with rounded or blunt tips. Flavor is exceptional, and the harvest is early. Good choices are Mokum (54 days), Nelson (56 days), Napoli (58 days), Necoras hybrid (68 days), and Yaya 60 days).

Mini-carrots. Sometimes called radish-carrots, they have small roots that feed snack-sized appetites. Mini-carrots are easy to grow and attractive. Varieties include: Adelaide Baby, Amsdor, Babette, Baby Ball, Parmex (an improved Baby Ball), and Thumbelina. They're good for containers and dense soils. Many ready-to-eat bagged tiny carrots that you buy at the supermarket are man-made minis—large, sweet carrots that are picked thin, then cut and ground into their conical shape by a machine.

Companions: Beans, peas, Brassicas, and allium crops. Avoid parsley, dill, coriander, and parsnips. Kale, chard, and lettuce can keep soil cool in the summer heat, making carrots more vigorous.

How to Grow: Sow in the spring, when the soil has warmed to at least 50°F (10°C). It's best to sow successive crops every 2–3 weeks from mid-spring to late summer, ensuring a steady harvest. Tiny seeds should be planted no more than a ¼-in. (6 mm) deep and spaced 1 in. (2.5 cm) apart in blocks or wide rows. That's nearly 150 seeds per square foot, but spotty germination will thin some plants; the others can be pulled as they grow. Keep seeds well watered but don't flood the soil. If the surface packs down and forms a crust, it can hurt germination. After 2 weeks, reseed bare spots and keep the entire bed watered with a gentle spray. As carrots grow, thin to about 2 in. (5 cm) apart. (Note: Some growers interplant carrots with radishes, but I don't like to do that. Fast-growing radishes will germinate quickly and mark the spots where carrots are growing, but they can quickly become bullies, crowding out the slower-growing carrots.)

Pests: The most serious pests are carrot sawfly larvae, which bore into roots and leave blackened tunnels. They also spread diseases such as rust. Use floating row covers over young carrots, especially when sown in the spring, when flies are most active. Harvesting carrots in the morning can attract sawflies, which are drawn to the scent of the foliage. (When you pull carrots, you inevitably disturb and possibly crush foliage.) Wireworms, the brownish orange segmented larvae of click beetles, will also tunnel into roots, and parsley worms, the black, green, and yellow larvae of black swallowtail butterflies, can defoliate the delicate leaves.

Diseases: Alternaria blight turns leaves yellow and brown, hurting overall vigor. Another fungal blight, Cercospora, leaves tiny black dots with a yellowish ring on new foliage. Prevent fungal disease by rotating for at least 3 years. Use cover crops—especially deep-rooted grasses—to loosen soil and provide beneficial fungi, which crowd out pathogens.

Challenges: Twisted, stunted, and misshapen roots are the biggest disappointments. Prevent them by double-digging the bed and ensuring loose, nonrocky soil for optimum root growth. Growing carrots in raised beds is ideal. Raised beds can also deter carrot rust flies, which cruise low to the ground looking for carrot foliage. A surer control is to use floating row covers over hoops, which provide air space and allow the tops to grow tall.

Harvesting: Snack on thinnings, or sell them as baby carrots, but don't market carrots that are spindly or pale in color (they won't have much flavor).

Marketing Tips: Bundle multiple colors of carrots together: yellow, white, purple, red, and orange. For the freshest-looking carrots, leave the tops attached. For storage in a cooler or root cellar, trim the tops, leaving about 1 in. (2.5 cm) of stubble.

Cauliflower *Brassica oleracea* var. Botrytis

Basics: Like its Brassica cousin broccoli, cauliflower is grown for its crispy flowering head, which is picked while buds are still compact. It can be fussy, and it takes a lot of space; one cauliflower plant produces just one harvestable head, but freshly picked cauliflower, especially the colorful varieties that are hard to find in grocery stores, commands a handsome price. Fast-maturing white-curd varieties can provide a second harvest in the fall, and striking new hybrids in purple and orange are popular at farmers' markets. Cauliflower is susceptible to many of the same Brassica pests and diseases and can be protected with similar techniques.

Details: Soil temperature: 70°F–80°F (21°C–27°C) for germination, cooler for growth; fertile, well-drained sandy loam with a pH of 6.5–7.5. Germination: 5–7 days. Seeds are viable for up to 3 years.

Ease of Growing: Difficult.

Varieties: Snow Crown hybrid (50 days) has creamy white heads. Snowball (60 days), Fremont (62 days), and Skywalker (80 days) are other good whites. Violetta (85 days) is a purple Italian heirloom; Graffiti (80 days) is a strikingly purple hybrid. Cheddar (58–70 days) is an early-maturing orange cauliflower that turns darker when cooked. Panther hybrid is a green-headed cauliflower that's well suited for fall crops.

Companions: Beets, bush beans, carrot-family crops, alliums, cucumbers, and potatoes. Avoid pole beans and strawberries.

Cheddar cauliflower, a variety with deep orange-colored heads, looks as if it has already been dipped in cheese sauce. Sunset is another yellow-orange variety.

How to Grow: Start seeds indoors in early spring, about 5–6 weeks before transplanting. Timing is important, since seedlings kept in pots for more than 5–6 weeks can suffer cramped roots, resulting in tiny heads. Cauliflower grows best in cool conditions, so get it outdoors and into the ground as soon as the soil can be worked. Young plants will tolerate frost, but adult-sized heads will be damaged by a freeze.

Many cauliflowers have leaves that don't fully envelop and shade the developing heads. These will need to be blanched. When the heads are small, turn the leaves upward to create a loose envelope around the head. Use a rubber band or twine to secure the leaves at the top. Even varieties that are self-blanching can benefit from having their leaves wrapped to ensure good protection from direct sun, especially for plants sown in spring and harvested in summer.

Pests: Protect seedlings against cutworms, and use floating row covers to keep flying pests away (cabbage butterflies and moths). Cauliflowers can reach more than 2 ft. (60 cm) tall, making it difficult to protect mature plants with floating row covers. If necessary, spray with Bt or Spinosad once the plants grow tall.

Diseases: In addition to clubroot, some strains of cauliflower are susceptible to bacterial rot. Rotate crops and dispose of residue instead of composting it. (Feed it to the chickens if you have them.)

Challenges: Growing cauliflower can be time-consuming and produce small heads if the weather is hot or if seedlings get pot-bound before transplanting. But there's no joy quite like harvesting an 8-in. (20.5-cm) bright purple or vibrant orange head and hearing oohs and aahs from customers.

Harvesting: Watch developing heads carefully, especially during warm weather. If the heads start to bulge and the tight curd begins separating, the head can lose its sweet nutty flavor.

Marketing Tips: Grow orange-hued Cheddar cauliflower and Graffiti purple cauliflower for an attractive harvest.

Collards *Brassica oleracea*

Basics: A traditional food in the South, collard greens are now finding a welcome home in northern kitchens as well. This cabbage-family crop is both heat and cold tolerant but is best when you pick it after a light frost. Northern growers should plant it in the spring or summer for a fall crop (picked into winter). Southern growers can sow 2 crops: one in late winter for a mid-June harvest and again in late summer or early fall for harvesting once the weather cools.

Details: Rich loamy soil with a pH of 6.0–7.0. Germination: 7–14 days at 70°F–85°F (21°C–29°C). Seeds are viable for up to 4 years.

Ease of Growing: Easy.

Varieties: Blue Max (68 days), Champion (65 days), Georgia Southern (70 days), and Top Bunch (70 days).

Companions: Tomatoes. Dill and fennel attract good insects that are cabbage pest parasites.

How to Grow: In the North, sow ½ in. (12.5 mm) deep, spacing plants 3 in. (8 cm) apart and then thinning to 6 in. (15 cm), 12 in. (30 cm), and even 18 in. (45 cm) if the plants become crowded. Keep the plants watered and side-dress with an organic fertilizer for extra nitrogen. Cover young collards with floating row material and watch for bugs when the plants get taller and you have to remove the covering.

Pests: Harlequin bugs are the biggest pests for southern growers, sucking the juices from leaves and causing unsightly damage. Handpick adults, and crush the nymphs and clusters

Collard greens taste best when harvested after an autumn frost. This leafy Brassica is finding a wider appeal among northern cooks.

of keg-shaped eggs. Destroy crop residue to curb recurrent problems. Grow mustard or horseradish as a trap crop, then kill the insects on the trap plants. The organic pesticide PyGanic is a last resort, but be aware that it can damage beneficial insects. In the North, cabbage butterflies and cabbage worms do the damage. Cover with floating row covers until the plants get too tall, then spray with Bt. Green cabbage aphids are a problem when weather cools; spray with a forceful hose to knock them away. Ladybugs and lacewings will help with aphids.

Diseases: Downy mildew and alternaria leaf spot are common fungal diseases. Clean up old crops before winter, avoid overhead watering, and rotate crops.

Challenges: Despite the insects, collards are easy. In a nutrient-rich, biologically active soil, healthy plants will weather insect and disease problems. Harlequin bugs have extended their range northward and are now a problem among organic growers in Pennsylvania.

Harvesting: Pick individual leaves or cut down the entire plant. Harvest in early morning and keep cool and moist.

Marketing Tips: Bundle leaves, using a strong rubber band or a wrap of natural jute. Fresh collards look beautiful in comparison to older leaves sometimes found on produce counters. Keep the leaves fresh at a farmers' market by covering with iced burlap.

Corn *Zea mays*

Basics: Corn is a high-demand crop, but one that requires ample space. Pollination is best when planted in blocks with at least 10 plants spaced 10 in. (25 cm) apart in rows 2 ft. (60 cm) apart. It's also a heavy feeder and uses lots of nitrogen; plant in a rotation after legumes.

Details: Full sun, fertile soil, with high organic content, ample nitrogen, and a pH of 6.0–7.0. Germination: best at 70°F–90°F (21°C–32°C). Plant organic corn when the soil has warmed to above 65°F (18°C). Seeds are viable for up to 2 years.

Ease of Growing: Easy to moderate.

Varieties: Supersweet and sugar-enhanced hybrids hold their sugars long after harvest instead of converting to starch. Geronimo (63 days) is an early bicolor supersweet hybrid. Spring Treat (66 days) is an easy-to-harvest hybrid. Luscious (75 days) is a yellow-and-white midseason corn, and Silver Queen (80 days) is an all-white corn that arrives later in the season. Golden Bantam (70–85 days) is an old-fashioned, yellow-kernel variety with classic corn taste. Country Gentleman (92 days) is a late shoepeg variety with deep white kernels that are not aligned in straight rows. Popcorn, grain corn, and ornamental corn have slightly different growing and spacing requirements; check seed literature for instructions.

Companions: Beans and soybeans add nitrogen and may repel earworms and armyworms. Native Americans used the classic combination of beans, corn, and squash, called the Three Sisters. Tall sunflowers at the border can repel fall armyworms and block the spread of weeds. Some weeds can harbor pests and hurt growth, especially perennial quackgrass, which limits nitrogen for corn.

How to Grow: Sow seeds directly when the soil is warm (65°F/18°C) and dry enough to be worked. In cool areas, soil can be warmed with black plastic. Because wireworms and fungal diseases can damage corn that is planted in cool, wet soil, commercial seed is often treated with fungicides and insecticides. Avoid chemically treated seeds if you want to be organic. (There is an organic-approved treatment called Agricoat Natural II that provides a microbe- and nutrient-rich coating to boost growth.) Sow corn seeds 1½–2 in. (4–5 cm) deep in rows 2 ft. (60 cm) apart. Plant seeds 5–6 in. (13–15 cm) apart and thin every other seedling. Seedlings can be dug with soil and transplanted to dead spaces in the row. For best pollination, grow corn in a block instead of a few long rows. Keep corn watered with at least an inch of water per week and side-dress with organic fertilizers when corn is knee-high and again when the stalks show tassels. Don't cultivate the soil near stalks or you can damage shallow roots.

Pests: Corn earworms and European corn borers are two destructive insects. Earworm caterpillars eat young growth, tassels, and ears. Stop them by growing earworm-resistant varieties. Country Gentleman (92 days), Silvergent (87 days), and Staygold (88 days) have tight husks that resist penetration. Setting a clothespin over each ear at the point where silks enter will keep larvae from reaching the kernels. Corn borers eat leaves, tassels, stalks, and ears, and then begin a second and more damaging generation. Parasitic wasps and green lacewings eat larvae. The biological pesticide Bt works on borers but not as well on earworms. To deter both pests, shred and hot-compost stalks, and till the stubble into the ground in the spring. Control nearby weeds where borer moths mate. Spraying with Bt will kill caterpillars. There are other pests, too. Contact area farmers or check a local university website for specific solutions.

Diseases: There are multiple diseases, many specific to geographical regions and soil conditions. Check with fellow farmers, your cooperative extension office, or state-run website. News changes each year. In 2011 and 2013 a bacterial disease called Goss's wilt cut harvests in Midwest corn-growing states. Scientists think genetically modified corn that was engineered for production traits also developed a weakness to this disease.

Challenges: Space, soil nutrition, and pests can make organic corn culture difficult for small farmers. Uneven pollination results in "nubbins," stunted ears with missing kernels. Cross-pollination of supersweet hybrids and other varieties can result in tough kernels. Grow varieties at least 25 ft. (7.5 m) apart or time your plantings to ensure that 2 varieties aren't flowering at the same time.

Harvesting: Pick corn when the ears feel full (pinch the tops lightly). The silks should be brown and dry at the tips. Pick 1 ear and bite into a kernel; it should be sweet and milky. Popcorn and ornamental varieties, as well as grain corn, need more time to mature since the husks should be dry and the kernels hard when harvested.

Marketing Tips: Sell corn immediately after picking. Older varieties can turn starchy in less than a day. Newer hybrids hold their sugar longer. Popcorn, a different variety with smaller, denser kernels, can be a fun option. The variety Strawberry is an heirloom with small ears and kernels that look like tiny pomegranate seeds; it's attractive, ornamental, and edible.

Corn needs lots of space, along with steady water and high-nitrogen soil. When conditions are good, the reward of sweet, fresh-picked ears is unmatched.

Cucumbers *Cucumis sativus*

Basics: The first crisp and crunchy cucumber of the season reminds me of why I take the trouble to grow my own. (Of course, that's the same reason I share my harvest with others.) The snap and flavor are beyond compare. Cucumbers can be finicky to get started. They need lots of water and nutrients and are prone to some insect and disease problems, but seeing the small fruits hanging on a trellis and swelling to harvest size in just 2–3 days brings a joy to any grower. The harvest is more than worth the effort, especially if you grow some of the specialty cukes that can't be found in mainstream markets.

Details: Nutrient-rich, well drained loamy soil is mandatory for good cucumber growth, with a pH of 6.0–7.0. Germination: Seeds won't germinate until the soil warms up to more than 70°F (21°C), with optimum germination in 3–5 days at 80°F–95°F (27°C–35°C). Growing temperature is slightly cooler at 70°F–80°F (21°C–27°C). Seeds are viable for up to 5 years.

Ease of Growing: Easy to moderate.

Varieties: For starters, there are pickling cukes, slicing cukes, serpentine cukes, heirloom cukes, and bitter melons, which are gaining in popularity for their peculiar appearance and flavor. No matter which varieties I plant, I prefer long, trailing vines because I train my cucumbers up trellises, and the vertical space simply provides me with more area for vines to catch the sunlight and manufacture food. For slicers, Diva hybrid (58 days) is exceptional, with thin skin, a crisp, sweet honeydew flavor, and no seeds. The vigorous vines also produce less of the cucurbit odor that lures insects to the garden, so cucumber beetles are less of a problem. It can also be kept under a floating row cover because all-female plants don't need pollinators. Other good slicers are General Lee (52 days), Tasty Jade (54 days), a slim-fruiting Japanese cucumber that needs no pollinator. Marketmore (52 days) is the standard open-pollinated slicer, and Straight Eight (65 days) is an old variety (1935), with vigorous vines, uniform 8-in. (20.5 cm) fruit, and small seed cavities. Some top picklers are: Southern Homemade Pickles (55 days) and Picklebush (52 days), producing crisp pickling cukes on short vines. Boston Pickling cucumber (57 days) is an heirloom with resistance to mosaic disease; prolific vines reach up to 8 ft. (244 cm) and produce 5–6-in. (13–15 cm) 15 cukes over a long period. Armenian cucumber (60–70 days) produces 12- to 18-in. (mild-flavored and pale-skinned fruit on trailing vines. Lemon cucumber (65 days) is an heirloom with round,

Easy Does It!

When growing slicing cucumbers on a trellis, the vines will be more productive if they are pruned. At each fruiting node, you will see a leaf, a flower (with a tiny cuke attached if it's a female flower), and a young branch. Prune off the branch, leaving the leaf and the flower.

Whether you grow spiny picklers or long, sleek slicers, cucumbers are a midsummer treat. Beat pests naturally by growing parthenocarpic varieties, which don't require pollination to produce fruit. You can keep them covered with insect–proof floating row covers from seed to harvest. Parthenocarpic choices include Diva for slicers and Little Leaf for picklers.

lemon-yellow fruit that is sweet tasting but has a large seed cavity. Boothby Blonde (63 days) is another popular heirloom with pale yellow fruit that has bumpy skin and black spines; the flavor is sweet, and the vine produces over a long period.

Companions: Cucumbers like radishes, peas, dill, and nasturtiums. Avoid potatoes, tomatoes, and other cucurbits.

How to Grow: Sow spring cucumbers indoors in large plug trays or cell packs about 4 weeks before the last frost date. Prepare outdoor beds by adding a spade of compost for every cucumber vine, mixing it in to the soil to a depth of about 1 ft. (30 cm). (Note: Cucumbers do not like to be transplanted but will survive if treated gingerly.) Harden transplants outdoors, then carefully remove the seedlings from containers and place into wide holes, gently backfilling the holes to avoid disturbing the roots. Transplant on a cloudy day or in the evening and finish by fertilizing with liquid fish emulsion. For direct-seeding wait until the soil has warmed to 70°F (21°C); covering the bed with black plastic will increase the soil temperature. Sow the seeds ½ in. (1.2 cm) deep and 3 in. (7.6 cm) apart along a trellis or in groups of three on a mound of soil. Thin the vines to at least 6 in. (15 cm) apart once the first true leaves appear. Keep the soil moist and coax the vines onto the trellis until the tendrils wrap around and hold them in place.

Pests: Cucumber beetles are serious pests that damage young seedlings and the roots of adult plants. They can also transmit wilt disease, a fatal disease that causes the vines to die in a matter of days. Other pests include cutworms, squash bugs, and squash beetles, which look like Mexican bean beetles but are slightly larger and more orange in color. Remove the adult insects before they lay eggs. Sticky yellow traps can catch cucumber beetles. PyGanic spray, an organic contact pesticide, can also be effective against cucumber beetles, but it has to contact them directly. It also kills beneficial insects, so watch drift and take care not to use it when bees are visiting cucumber blossoms. Cucumber beetles flock to some winter squashes, so plant a few Blue Hubbard squashes as trap plants. When the Blue Hubbard vines are overrun with cucumber beetles, spray them with PyGanic. Don't plant trap crops without taking care of the pests that infest them.

Diseases: Cucumber mosaic is a virus that leaves yellow spots on foliage and fruit, especially when the first true leaves are developing. Plucking off affected leaves and planting resistant varieties will usually allow plants to regain strength. Always wash your hands to avoid spreading mosaic and other diseases. Wilt disease is more serious. If the plants droop in the midday sun and don't come back after watering, break off a piece of stem near the base of the vine. If it exudes a milky sap, pull up the entire vine and throw it in the garbage. Do not compost it or you may reintroduce the disease next season. Prevent wilt disease by controlling cucumber beetles. Fungal diseases include powdery mildew, anthracnose, and alternaria. To minimize the diseases, practice good crop rotation, don't compost diseased plants, and control cucumber beetles, which spread the diseases.

To ensure a long harvest, sow a second crop of cucumbers by direct-seeding as soon as the first crop begins to set blossoms (about 1 month after transplanting). In some areas there may be time for 3 cucumber crops.

Challenges: Cold weather after transplanting can leave young vines stunted. Cucumber beetles are far-ranging fast fliers that can be difficult to control. (Read more about insect strategies in chapter 4.) Lack of water can result in undesirable cucumbers that swell at one end and curl up at the tips. Diseases can shorten the harvest, especially if you plant only one crop.

Harvesting: Small cucumbers can double their size in a day, so watch the vines carefully and pick fruit when it's slim and firm, not swollen and beginning to grow seeds. Use scissors to snip the stems, leaving a tiny stub on the fruit. It will help the cucumber hold water and stay crisp.

Marketing Tips: Plant dill to grow at the same time as cucumbers, and market the two together for a winning combination.

Eggplant *Solanum melongena*

Basics: Eggplants have become a healthy alternative to starchy side dishes, and as such they are gaining new appreciation from chefs at restaurants and in homes. In the field, they are an undeniable cash crop, producing strong harvests with limited space. Starting eggplants, however, takes some effort, and they are not immune from diseases or insect pests. As with so many other farm-fresh vegetables, local growers will benefit from trying something new. There are heirlooms and hybrids in multiple shapes and sizes in purple, orange-green, and rose—all of which contrast nicely with the traditional pear-shaped black-skinned varieties we've grown accustomed to.

Details: Rich and friable soil, with ample phosphorus and a pH of 5.5–6.5. Heat-loving eggplants perform best when daytime temperatures are above 80°F (27°C). Seeds are viable for up to 5 years.

Ease of Growing: Easy.

Varieties: Rosa Bianca (75 days) has oval-shaped fruit in pastel purple, streaked with white. Kermit (60 days) is a hybrid with small, round fruit that is frog green, streaked with white. Raveena (70 days) is a banana-shaped eggplant with jade-green fruit. Fairy Tale (63 days), one of my favorite new hybrids, is a diminutive pear-shaped eggplant with medium purple skin and longitudinal streaks of white; it's pretty, tasty, and prolific. For something completely different, try Turkish Orange (80 days), a tomato-shaped eggplant that's pumpkin orange with blackish green streaks.

Companions: Pole and bush beans and garlic. But don't let pole beans shade heat-loving eggplants.

How to Grow: Start seeds indoors about 6 weeks before the last frost. Most varieties germinate in about 10 days, and they grow slowly—about the same speed as peppers. Harden the seedlings, then set them outdoors once the daytime temperature is consistently above 70°F (21°C). Sturdy plants won't need trellising but might benefit from a dark-colored perforated mulch in northern climates.

Eggplants look more enchanting with streaks of white across their skins. Try heirlooms Udumalpet and Listada de Gandia or the hybrid miniature eggplant Fairy Tale.

Pests: Watch for Colorado potato beetles and their eggs. Pick off the clumsy striped bugs and smash their orange eggs before the larvae hatch. Once at work, they can quickly defoliate plants, which have a hard time leafing out again. Spray with Bt for potato beetles.

Diseases: Several fungal and wilt diseases affect eggplants, but most can be prevented with good crop rotation and planting in soil that drains easily.

Challenges: The growing season is long for some older varieties. Cool, damp weather can make the growing time even slower.

Harvesting: Pick black varieties when the skin develops a satiny shine; otherwise, squeeze the fruit. It should give slightly but not be mushy. Look out for sharp spines on the calyx.

Marketing Tips: Eggplants in multiple shapes and colors sell themselves. Thin-skinned small varieties are great for summer grilling. If your customers don't know this, let them know how easy it is to do: Slice and baste the open side with olive oil, then grill until tender.

Garlic *Allium sativum*

Basics: One of the easiest crops to care for, garlic is a must for market gardeners and CSA managers. It doesn't take up much space, and it has a high value at harvest time. In most of the country except in the Deep South, garlic is planted in the fall for harvest in early summer. Thinnings can be used as green garlic, giving growers a jump on the harvest and another option for early-season farmers' markets and CSA shares.

Details: Garlic likes loose, well-drained soil with lots of organic matter, with a pH of 6.0–7.0. The plants aren't fussy but will need mulch in cold winters, along with weeding in the spring and steady water during dry spells. Soil temperature: Hard-neck varieties require about a month of temperatures below 50°F (10°C) to trigger new growth. For that reason, it's best to plant in the fall and allow the garlic to overwinter in the ground. Growers with warm soil temperatures can chill hard-neck varieties in the refrigerator for about a month prior to sowing.

Hard-neck garlic varieties produce gracefully curled flower stalks called scapes. Cut them before they bloom and add to salads, stir-fries, and pasta primavera.

Ease of Growing: Easy.

Varieties: One of the biggest reasons for growing garlic is to enjoy the variety, which is far greater than the simple white garlic you find in most grocery stores. There are more than a dozen different types of garlics—Porcelains, Artichokes, Silverskins, Rocamboles, Asiatics, and others—each with varying characteristics. For most growers, however, the biggest distinction is that some garlics have soft necks and others have stiff necks. Hard-neck garlics produce woody stalks surrounded by a small number of large, easy-to-peel cloves. Hard-necks are generally more cold-hardy than soft-necks, and they sprout flower stems called scapes, which can be harvested in the late spring.

Hard-Neck Garlics. These varieties include several sub-varieties: Creole types are heat and drought resistant, often with brightly colored skins and a strong flavor. Porcelain garlics have extra-large cloves and are extremely cold-hardy. Purple-stripe varieties are named for the purplish streaks over off-white skins. Rocamboles are the most popular hard-neck garlics grown and include Spanish Roja, Argentine Red Stripe, and Italian Purple.

Soft-Neck Garlics. Soft-neck varieties prefer warmer climates and produce a bulb with large cloves on the outside and smaller cloves inside. They include the subgroups Artichoke and Silverskin. Artichoke varieties grow quickly and are easy to care for. Silverskins are stronger flavored, take

Healthy garlic is one of the easiest crops you can grow. Break each bulb into individual cloves and plant in the fall, as you would do for flowering bulbs. Your harvest will arrive in early summer.

longer to harvest, and are good for storage; they are the best varieties for weaving into a handsome braid for gifts.

Companions: Garlic repels slugs, aphids, and cabbage worms, making it a good companion for lettuces, Asian greens, and cole crops. In return, shallow-rooted, low-growing plants between rows of garlic will prevent weeds.

How to Grow: For the best quality and advice on planting, buy hard- or soft-neck garlic from an established supplier, not from the grocery store. In most of the United States, garlic is best planted in late fall. I like to wait until a few weeks after the first frost, when most of the beds have been cleaned up and cover crops sown. Planting about a month before the ground starts to freeze allows roots to grow but keeps top growth slow, until the warmer weather arrives. To sow, break garlic bulbs into individual cloves. Plant the biggest cloves about 5–6 in. (13–15 cm) apart, holding a clove between your thumb and forefinger, pointy-side up. Push your finger down to the second knuckle, which should put the bottom of the clove about 4 in. (10 cm) belowground. Smaller cloves can be used in the kitchen or sown for green garlic.

After sowing, water the bed and mulch with 2–4 in. (5–10 cm) of weed-free straw or shredded leaves. The mulch will stabilize soil temperatures and prevent late-winter thawing and refreezing, which can heave cloves out of the ground. In the early spring, shoots will sprout and grow quickly. (Don't worry if they sprout early and get a little freezer burn on the tips.) Keep the bed weeded and fertilize lightly, but keep nitrogen to a minimum. (It can promote fast foliage growth at the expense of the bulb.)

Pests: Garlic is relatively pest free, since the plant produces a natural compound called acillin, which repels insects. A few pests that also attack other alliums can affect garlic, so avoid planting in the same spot for at least 2 years.

Diseases: Garlic is relatively care free, but soilborne fungi can become a problem, especially after several seasons. Garlic rot turns foliage yellow and slowly takes over the plants. White rot, rusts, and mildews are also problems—all them caused by pathogenic fungi. To avoid diseases, never plant garlic where onions, garlic, or other alliums have been grown in the past 2 years. Also, watch and remove any plants that show signs of rust patches or sudden lack of vigor accompanied by yellowing leaves. Don't compost old allium foliage if you plan to use the compost in your garden.

Challenges: The biggest problem most growers face is small bulbs. Poor-quality seed cloves or sowing in springtime are often the reasons for stunted plant growth because the bulbs sprout top growth before the roots become fully established. Soil that is compacted and too wet can also harm growth. Make sure the soil is friable and well drained, with lots of compost.

Harvesting: Pick garlic when the stalks begin to turn yellow but before the stalks wilt. Bulbs left too long in the ground can start to separate and regrow. Stop watering about a week before picking time, and then use a spading fork to loosen soil as you pull straight up on the stalks. Brush off the dirt and allow garlic to cure for several days out of the direct sun. Cut the dried stalks, leaving about 2 in. (5 cm) of stem, and store in a cool, dry place. A root cellar or refrigerator is ideal.

Marketing Tips: Use smaller cloves to sow green garlic, which can be harvested in early June, before the main crop is picked. Plant the smaller cloves between larger ones, and thin for a harvest when the stalks are as thick as your pinky finger. Rinse off soil, trim roots, and bunch the stalks together like scallions.

Hard-neck garlic varieties produce scapes, which should be allowed to grow at least 1 ft. (30 cm) long (usually curling into 1 or 2 loops). Then cut them and bundle in a bunch for sale. Gourmet chefs prize garlic scapes for their deep green color and mild garlicky flavor.

Greens

One of the fastest-growing and most profitable crops for small-scale growers is salad greens—lettuce, beets, mustards, kale, spinach, and other leafy vegetables or herbs that are picked and marketed at different stages of maturity. The earlier they're harvested, the higher a price they command—from Asian greens and microgreens to baby greens and full-sized salad ingredients.

Asian Greens *Multiple Species*

Basics: Asian greens are almost all Brassica family crops, sharing many of the same needs and pests as broccoli, kale, cauliflower, cabbage, radishes, kohlrabi, and mustards. Spicy, crunchy, sweet, and mustardy, Asian greens are so healthy, so varied, so easy to grow, and so undiscovered that it pays to plant multiple crops. Experiment with a few new items each season to settle on your favorites. Asian greens make sense for small-scale organic growers because pest protection is easy, the yield is extremely high from a small plot, and the turnaround time from planting to harvesting is fast. The optimum growing time is spring and fall, providing valuable crops during the shoulder seasons when summer staples are unavailable. Finally, the joy of introducing others to new culinary experiences is unparalleled. Most cooks will recognize a few Asian greens—Chinese cabbage and Bok Choy—but will probably need to be introduced to others such as Hon Tsai Tai, Tatsoi, Senposai, Komatsuna, and so on.

Details: Most thrive in rich, well-drained soil, with a pH of 6.0–7.0 and lots of organic matter. Soil temperature: 65°F–75°F (18°C–24°C). Germination: 3–7 days. Seeds are viable for up to 4 years.

Ease of Growing: Easy to moderate.

Varieties: Unlike varieties of lettuces, tomatoes, or peppers, the Asian greens described here are completely different crops—individual species, not cultivars of the same plant. Some of the more progressive seed catalogs carry good selections of Asian greens and others are catching up. Look to Johnny's Selected Seeds, Fedco, Baker Creek Heirloom Seeds, and Southern Exposure Seed Exchange. The list below describes some of my favorites.

Red and green Bok Choy are fast and easy to grow. Protect them from insects with row covers and handle the harvest carefully so leaves don't get tattered. Stalks are filled with water and are surprisingly easy to break.

Bok Choy. Crunchy, celerylike stalks with soft, mustardy foliage. Good in stir fries and soups. Red Choy (44 days) has striking burgundy leaves with green ribs and stems; Black Summer (45 days) has pale green stems with dark green leaves; Joi Choy (50 days) has dark green leaves with pudgy white stems and white ribs; San Fan hybrid has green stems and green leaves and tolerates heat better than other varieties.

Chinese Cabbage. Heads can be loose leaved or tightly packed, in a torpedo, oval, or round shape. Napa varieties form tighter heads. All Chinese cabbages are best planted in late summer and grown through the fall for late harvests. Plants that are sown in early spring have a strong tendency to bolt to seed, especially if they experience cold temperatures. Chinese cabbages are great for stir-fries and soups and can be used for mild-flavored, less-crunchy coleslaws. Early-maturing varieties include Orient Express (43 days) and Minuet (48 days); slightly later are Rubicon (52 days) and Blues (55 days). Yellow King (77 days) has deep yellow interior leaves and prefers cooler weather. Green rocket (70 days) forms tall torpedo-shaped heads up to 4 lb. (2 kg.) Kingdom (80 days) is disease resistant, slow to bolt, and grows up to 10 lb (4.5 kg).

Fun Jen. Plants are actually a type of Chinese cabbage, but the leaves are frilly and soft, more like an upright escarole or loose-leaf lettuce. Fun Jen grows quickly and is ready to harvest in just 35–45 days. Sow successive crops all season long, but watch plants closely for quick bolting in hot weather. Also, earwigs can be a problem in warmer months, and slugs will chew holes in leaves during cool rainy spells. Fun Jen can be spaced just 6 in. (15 cm) apart, providing a fast and heavy harvest in a little less than a month.

Tatsoi has deep green spoon-shaped leaves with white ribs. It adds a mild peppery flavor to salads and stir-fry dishes.

Hon Tsai Tai. Green leaves, purple stems, yellow flowers; all of it is edible. No wonder this attractive green is getting attention from discriminating chefs. All parts are used in stir-fries and soups, but the leaves and stems are best when picked young. Sow directly in cool weather, and harvest by cutting outside leaves. Hon Tsai Tai will produce fairly well in spring and summer but is best sown as a fall crop or almost year-round in the Pacific Northwest.

Komatsuna. Sometimes called Japanese mustard spinach, this Brassica is actually closer to turnip greens, with an appearance like tender Bok Choy. Young leaves can add a spice to salads, with full-sized plants used in soups and stir-fries. There are numerous varieties, some reaching harvest size in just 30 days. Very easy to grow, especially in fall months. Like many mustards, the leaves can become bitter in hot weather.

Mustards. Young mustard leaves can be snipped for spicy salad green mixes, while mature leaves are best to sell in a bunch for stir-fries, soups, or steamed like spinach. Most will grow through the summer months in cool areas but perform best as season-ending crops. Japanese Red Giant is impressive in size and color, with purplish red leaves and striking off-white veins. Miike Giant has similar color, but its ribs and veins are green; size can be over 5 lb. (2 kg.)Mizuna mustard has a mild arugula-like flavor, with jagged-edged leaves that resemble the shape of frost on a windowpane; it is widely used in spicy mesclun plantings. Ruby Streak is similar in size and shape to Mizuna but is flecked with reddish purple streaks. Peacock Tail is a green mustard, with mild flavor and narrow upright growth.

Senposai. This versatile and flavorful green is gaining in popularity because of its long-lasting harvest. When planted in the spring, the leaves can be picked in as little as 30 days, and the harvest persists into summer—sometimes into autumn before the plant goes to flower. Senposai is a recently developed hybrid between Komatsuna (Japanese mustard) and cabbage. The broad leaves, shaped like ping-pong paddles, are tender enough to use in salads. Plants grow about 18 in. (45 cm) wide and about 2 ft. (60 cm) high but much higher when flowering. You can pick Senposai one leaf at a time, since the foliage regenerates quickly in rich soil.

Tatsoi. Picture Bok Choy that has been flattened out and pressed into a pancake-shaped rosette. Tatsoi has handsome white stems and ribs, with rich forest-green leaves, and a delicate mustardy-radish flavor. Use young leaves in salads and mature plants in soups and stir-fries. Use row covers to protect against flea beetles in the spring.

Companions: Garlic, onions, and strong-flavored aromatic herbs (rosemary and thyme) help repel slugs and some insect pests. Avoid beans, strawberries, tomatoes, and crops in the carrot/parsley family.

How to Grow: In most of the country, Asian greens perform best at two times of the year: as spring crops that are harvested in late May or early June and as fall crops that can be sown in late summer for picking in October or beyond. Seeds can also be started indoors and transplanted to a greenhouse, cold frame, or protected tunnel to grow into the winter months. Don't sow in late fall for a spring crop because most Asian greens will bolt quickly to seed after they've been exposed to temperatures near 40°F (4°C).

For May and June harvests, start seeds indoors about 6 weeks before the last frost date. Sow in six-packs, plug trays, or flats. Seeds germinate quickly, in some cases as early as 4 days. Give them at least 12 hours of light to prevent spindly growth. In about 3 weeks, when greens are transplant size, move them to the garden. Raised beds with hoops and a floating

Komatsuna or Japanese mustard spinach is a wonderful green that's easy and productive. It has a delicate mustardy flavor and is perfect for salads when harvested young. Mature plants can be steamed or stir-fried.

row cover will protect them from a frost (and prevent later bolting), and the covers will serve double-duty, blocking out flea beetles, which chew tiny holes in young Brassicas. Because Asian greens don't need pollination, you can leave the row covers on until harvesttime.

In the fall, start Asian greens in August and September if winters are late or if you have protection. Sow seeds directly in the bed or in seed flats or plug trays. Use shade cloth to protect young seedlings from late-summer sun and transplant in about 3 weeks.

Pests: Because so many pests feed on Asian greens, it's difficult to list them all, but here are a few: Flea beetles, usually a bigger problem in the spring than in the fall; slugs, which are bigger problems in long rainy spells; and earwigs, which emerge on summer nights to chew holes in tender growth and hide deep in the folds of the leaves during daylight hours. Brassica-eating caterpillars, specifically the larvae of white cabbage butterflies, cabbage looper moths, diamondback moths, and cross-striped caterpillar moths can cause serious damage. Fortunately, organic protection is easy. The same floating row cover that protects against frost becomes an impenetrable barrier for hungry insects. Stretching a medium-lightweight cover over steel or PVC hoops will protect seedlings and allow ample space for growth. If plants are exposed, spray with Spinosad as a last resort to control earwigs and flea beetles; use Bt or Spinosad for cabbage caterpillars.

Diseases: Asian greens aren't susceptible to a wide range of diseases, but the one disease you need to guard against is clubroot. It's a fungus that stunts the growth of Brassica roots, causing them to bunch up into a knot or club. Avoid it by rotating crops regularly and keeping the soil pH close to 7.0, a neutral state where fungal diseases falter.

Challenges: Planting too early in the season can cause bolting. Growing Asian greens without floating row covers can lead to a season-long battle with insect pests. Woodchucks, if you have them, love to munch on the milder Asian greens, along with other Brassicas. Voles love to chew the tops off of young seedlings.

Crisp Pak Choi and other Asian greens are susceptible to flea beetles, but protecting them is manageable if you use a floating row cover supported by hoops of plastic or metal.

Harvesting: Many Asian greens can be picked when immature for baby Bok Choy, tender Tatsoi, or salad greens. Mustards and other leafy greens can be harvested by plucking individual leaves, but Bok Choy and Chinese cabbage are best when the heads are cut off at soil level and the stalks are trimmed flush with the bottom. Pick in the morning and plunge into cold water to prevent wilting.

Marketing Tips: Red and green Bok Choys look appealing both in a stir-fry and in display. Take care not to damage the delicate leaves on mustards and other greens. A mixed bundle of red and green mustards looks attractive for market. Bundle the leaves and stems from several large-growing Asian greens, such as Senposai, Yokatta-Na, and Komatsuna for a stir-fry assortment. Package the greens with fresh garlic or scapes and include a simple stir-fry recipe.

Red Tatsoi and Bok Choy add color and spice to salads when picked as baby greens.

Baby Greens

Leafy salad greens that are picked early (but not as early as microgreens) are often marketed as baby greens. Typically, the leaves are very soft and tender and no bigger than a flattened soup spoon. Spinach, arugula, Tatsoi, Pak Choi, kale, and lettuce mixes are all good choices. Baby greens generally have enough root development to produce subsequent cuttings if you're careful about harvesting. Snip them with sharp scissors or a bread knife, slicing just above the crowns—the node of young growth where new leaves are waiting to sprout. To be sure of the right level, get down on your knees, grab a fistful of leaves, and look closely at where you're making the cut. If your soil is good and the young plants get adequate water, several cuttings can come from a single sowing.

After cutting, give the plants plenty of water. You can feed them with a liquid fertilizer but be careful with fish emulsion. Allow at least a week between application and harvest and give your greens a taste test to be sure there is no lingering aroma. Do not use compost tea, especially if it has been made with compost that contains manure. Also be careful about cuttings taken later in the summer, since the leaves can grow bitter. Taste them before packaging for sale.

Baby Cos lettuce can be cut several times before the lettuce becomes tough and bitter. Blend with red salad bowl lettuce and frisee (curly endive) for a summer salad mix.

Lettuces *Lactuca sativa*

Basics: Technically, this family is in the Compositae family, named for the blossoms, which are actually groupings of multiple flowers. The family includes sunflowers, artichokes, chickory, and escarole, but lettuce is certainly the most important food crop of the bunch. There are hundreds of varieties—some that head up like cabbages; others that form loose clumps of frilled leaves. There are tiny and sweet Bibb lettuces and Cos varieties with pointed heads like romaine. You'll want to grow multiple varieties of lettuce with different tolerances for heat and cold so you can keep your harvest coming all summer long.

Details: As cool weather crops, many lettuces bolt quickly to seed in warmer temperatures. Sow indoors to protect seedlings from either extreme heat or cold. Fertile, friable, and well-drained soil, with ample nitrogen for good leaf growth; pH is best between 6.0–7.0. Seeds are viable for up to 5 years.

Ease of Growing: Easy.

Varieties: There are hundreds to list, with new hybrids and rediscovered heirlooms becoming available every year. Some standouts: Red Fire (55 days), large ruffled heads with green on the bottom and maroon at the top. Red Oak Leaf (60 days), similar to Red Fire, but slightly smaller with a tighter head. Black-Seeded Simpson and Simpson Elite (45 and 53 days): large, frilly leaf lettuce with bright green color. Rouge d'Hiver (60 days), a red and gold romaine well-suited to cold weather. Butter Crunch (46 days), a popular Bibb lettuce. Deer Tongue (50 days), Red Deer Tongue (60 days), and Really Red Deer Tongue (48 days), each one darker than its predecessor, an Amish heirloom. Paris Island Cos (58 days), an heirloom romaine lettuce with large, compact green heads. One of my favorites, Les Oreilles du Diable or "Devil's Ears" lettuce (50 days), a red-, green-, and gold-tinged romaine lettuce with large heads that stand up to the heat.

Companions: Beans, carrots, cucumbers, and onions. I like sowing small heads in the space between rows of garlic. Avoid cabbage.

How to Grow: Sow in flats or directly as soon as the soil can be worked. Keep planting every 2 weeks for a continuous harvest. For mesclun, sprinkle seeds onto smoothly raked beds and cover lightly with less than ¼ in. (0.6 cm) of fine soil. Plant a new section every 2 weeks for a continued harvest. Lettuce mixes can be cut once and allowed to regrow; however, in warmer weather, second cuttings can get bitter.

Pests: Slugs and earwigs are problems, chewing holes in the leaves and leaving waste behind. Control both with physical traps. Bait traps with beer or yeast and water for slugs and olive oil for earwigs.

Diseases: Fungal diseases can ruin plants if the weather is damp and the soil doesn't drain well. Provide adequate air space between heads or rows.

Challenges: Heat is probably the biggest reason for crop failure. Head lettuces grow rapidly and the leaves become twisted, while leaf lettuces send up quick flowers in extreme weather. Grow heat-resistant varieties and protect with shade cloth or the shade of nearby plants during hot spells.

Harvesting: For leaf lettuce, cut loose leaves or the entire head. To protect the heads from wilting, pick them in the early morning and plunge the heads into a pail of cold water.

Marketing Tips: Colorful lettuce heads sell well when packaged together. In the summer, when lettuce has a tendency to bolt, pick smaller-sized heads. Make them into an attractive package by bundling them together—reds, greens, and multicolored varieties, bound by a single rubber band.

Easy Does It!

Fast-growing weeds are often a problem with salad greens. Because plants are harvested by cutting them down in a group, it can be hard to distinguish desirable greens and tender herbs from undesirable weeds. Here's how to let the fast-growing weeds beat themselves.

After smoothing the bed, water it and wait for a week to 10 days. Then go back with a sharp hoe and cut down all the weeds that have sprouted, scraping them off at the surface of the soil. Don't dig in and disrupt the soil or you'll expose new weed seeds. Do this in the early morning on a sunny day so the tiny weeds will dry out in the midday sun. If you have time, you can repeat the process to eliminate all but a few weed seeds. Or just sow your greens the next day, cover them with a ¼ in. (6 mm) of weed-free compost or topsoil. This technique is called stale seed bedding, and it's used by organic farmers large and small. (Note: Stale seed bedding works best in sandy-loam soils that are loose and friable. Heavy clay soils can become crusty after stale seed bedding.)

Some growers use a weed flamer to kill young weeds, avoiding any possibility of disturbing the soil. A weed flamer is essentially an extension and nozzle that attaches to a propane tank. It has a valve to control the flow of gas and is either towed in a hand cart or carried inside a backpack. You can find weed flamers at farm and garden-supply stores.

Microgreens

Basics: Tiny, flavor-packed microgreens are a little bit like sprouts grown in soil. They're fast growing, nutritious, and in high demand from both home cooks and restaurant chefs. For growers, the turnaround is almost instant; young seedlings of some greens are ready to pick as soon as 2 weeks after sowing—when they are about 2 in. (5 cm) tall and have developed their second set of leaves (not cotyledons).

Details: Microgreens are often grown in 1020 trays with a lightweight growing mix or soil-less medium that encourages uniform germination. This makes them excellent candidates for greenhouse production or growing indoors under lights in the winter months.

Ease of Growing: Easy.

Varieties: Popular microgreen choices include cabbage, radishes, cress Asian greens, kales, and even such annual herbs as basil and cilantro.

How to Grow: Unlike cut-and-come-again salad greens, however, microgreens won't have enough energy in their young roots to produce a second crop. You'll just have to sow them again. Keep sowing them on a schedule (about every week, depending on your needs).

Harvesting: To harvest microgreens, cut the plants (stems and all) at ground level; do not pull out the roots. Some growers use an electric knife to lop them off evenly.

Marketing Tips: Start slowly with microgreens and take notes on the timing and success of different crops. After you've become reliable and productive, you can begin to market them to customers.

Organic microgreens are easy to grow in the winter using trays of soil–less growing mix under a light table. Cut them off at ground level when plants begin developing leaves. Beets, chard, radishes, and Asian greens can all be used to produce microgreens.

Salad Greens

Basics: Greens that are left to grow a week or two longer than baby stage will provide more bulk and will generally be easier to work with. Often the second cutting of baby greens will be more mature and leafy. Lettuce and chicory mixes and spicy combinations of arugula and mustards are typically grown as cut-and-come-again crops, offering several harvests. Harvest salad greens in the same way as baby greens, taking care not to damage new growth.

Details: Many salad greens will bolt to seed in warm temperatures. In regions where the summers are hot, protect the crop from sun with a layer of shade cloth. Or sow greens in the early spring for harvest before the weather warms, then sow again in late summer for picking in autumn. Some hardy salad greens such as kale, cabbages, and winter radishes can be grown into winter under the protection of a low tunnel covered with plastic sheeting or a floating row cover.

Turnip or mustard greens are used in salads when small or for cooking when they grow larger. Sell them in bunches tied with jute or secured with a rubber band.

Ease of Growing: Easy.

Varieties: Kale, Swiss chard, turnip greens, beets, and mustards are tender enough for salad greens when young but are better suited for cooking when they grow to full size. If given the space and time to reach maturity, Mizuna, Tatsoi, endive, chicories, and radicchio will produce individual heads or grow into mounding clumps or rosettes. Mixes that contain kales, mustards, and other peppery-flavored Brassicas are often sold as braising greens. They can be picked at baby stage for salads and used in stir-fry dishes when harvested later.

How to Grow: When you're planting baby greens and salad greens, prepare the soil in advance by mixing compost and organic matter to a depth of 6 in. (15 cm). Add organic fertilizers as recommended by your last soil test. Rake the bed smooth and level so that the seeds can be spread evenly. Sow seeds according to directions for the type of greens you're growing. Brassicas (arugula, mustards, and most Asian greens) are covered with just ¼ in. (6 mm) of soil; beets, Swiss chard, spinach, and radishes should be sown about ½ in. (12.5 mm) deep. Space them about ½ in. (12.5 mm) apart for greens grown outdoors. For microgreens sow about 5 seeds per sq. in. For a continuous harvest, sow new greens every 2 weeks.

Pests: All salad greens should be protected from pests with floating row covers. Spring-grown Brassicas, such as arugula, mustards, radishes, and mustards, are susceptible to flea beetles, which chew tiny holes in the leaves, marring the harvest. Spinach, beet, and chard greens are a favorite target of leaf miners, which tunnel through leaves and leave squiggly, brown trails. Covering the young crops with a floating row cover is the most effective and least environmentally disruptive way to prevent damage.

Harvesting: After harvesting salad greens, spread them out on a large table or counter indoors or in a cool, shady spot. Quickly pick out any grass or weeds, then get the greens into a deep sink or basin filled with cold water. After rinsing, scoop out the greens, spin them in a salad spinner, then bag and keep them refrigerated. For the freshest salad ingredients, greens should be harvested early in the morning and sold the same day.

Easy Does It!

Planting greens from a premade seed mix will produce a ready-to-cut assortment. But because lettuces, herbs, kales, and chicories grow at different rates, cutting the mix can damage taller plants, preventing an even second harvest. Experienced growers sow each salad ingredient in its own location, then mix the greens after harvest. This allows plants of the same variety to be harvested at the same height and development without cutting the new growth. Once greens have been harvested, they can be custom mixed to create the right colors, flavors, and textures.

If you want to grow salad greens from premixed seed packets, one way to sow them evenly is to pour the contents of a seed packet into a spice jar with a shaker lid. Add a scoop of horticultural-grade vermiculite and/or fine brick sand to help the seeds flow through the shaker top.

Some Choices for Greens:

Arugula *(Eruca sativa)*. Tangy arugula (also called Rocket) is popular in a mix or sold on its own. Many chefs use baby arugula as the only green in some salads. It's very easy to grow, germinating in less than a week and reaching cutting size for baby arugula in three weeks; 40–50 days for full size. Arugula varieties: Astro is heat tolerant, wild arugula (a different species) has more deeply serrated foliage and a spicier taste. When grown in the fall, full-sized arugula can be harvested for a longer period without worries that it will bolt to seed. It can take several hard freezes without harm.

Beets *(Beta vulgaris)*. Beet greens are good candidates for microgreens but can be harvested at baby stage; thinnings from root-crop beets can be added to mixes. Beets are slower than other greens, but most are ready for microgreens in about 3 weeks. Baby-leaf beet greens can be picked at 5 weeks with full-sized greens available a week or two later. Try Bull's Blood for deep red leaves and Early Wonder Tall Top for deep green leaves with red veins. Swiss chards have similar seeds and growth habits when used for salad greens. Rainbow chard is often used in salad mixes for its colorful stems and leaf veins.

Kale *(Brassica napus* Pabularia Group). The best kale for microgreens, baby greens, and mixed braising greens is Red Russian kale, which has a delicately serrated flat leaf that's a pale bluish green with purplish veins. Tender leaves can be cut in about 3 weeks and full size in 7 weeks. If allowed to grow to full size, the leaves grow in a clump and don't develop a tall, single-stem growth habit like other kales.

Lettuce *(Lactuca Sativa)*. If you're planning to mix lettuce with faster-growing mustards, Asian greens, or arugula, sow the lettuce a week earlier so it will be ready to harvest at the same time. A few good lettuce choices for salad mixes are: Black Seeded Simpson, Simpson Elite, Salad Bowl, and Tango (all green varieties); Red Salad Bowl, Red Oak Leaf, and dark Lollo Rosa (reds); and speckled Trout's Back, Red Sails, and romaine varieties (red and green).

Mustard *(Brassica juncea* and *Brassica rapa)*. Not just for southern growers, mustards add a tangy flavor along with color and texture to salad mixes. Most reach baby size in 3 weeks and full size in 6 weeks. Mizuna, a Japanese mustard, has a mild flavor and delicate shape, like the swirls of frost on a windowpane. There's also a Dark Purple variety, which is green with purple tips. If you want darker purple, try Ruby Streaks, a variety of Mizuna with lacy leaves that are streaked in burgundy. Red Giant and Red Rain are flavorful and colorful Asian mustards. Komatsuna (red and green) has a delicate flavor and growth habit like a thin-stemmed Pak Choi.

Kale *Brassica oleracea*

Basics: "Eat more kale!" should be the mantra for dietitians, farmers, and vegetable lovers everywhere. Kale is nutritious, versatile, easy to grow, and can be harvested from late spring into early winter. One planting in April can provide a steady and almost effortless harvest all year long. I know that I felt like cheering when several of my CSA customers said their biggest reasons for signing up is that they wanted a steady supply of fresh-grown kale every week.

Of course, not all customers are so informed about the benefits of kale. Part of your job as the supplier is to educate customers about the benefits of this prized vegetable. Try kale recipes yourself and share some of them with your customers through a CSA e-newsletter or include a printed recipe with your shares.

Details: Full sun and rich well-drained soil, with a pH of 6.0–7.2. Germination: 5–7 days at soil temperatures of 45°F–90°F; prefers cooler soil temperatures for growing. Seeds are viable for 4 years.

Ease of Growing: Easy.

Varieties: Most people know only of curly-leaf blue-green kale or the ornamental kale sold at garden centers in the fall. To encourage more people to eat kale, grow varieties that are better suited to the table. Nero di Toscana kale (50–60 days), also called Black kale, Lacinato kale, Black Palm kale and Dinosaur kale, has deep blue-green leaves that are smooth-edged

Kale is healthy, delicious, and very easy to grow organically. So it's your duty as a farmer to spread the word about this underused vegetable.

Vegetables and Herbs: The Spice of Life and Farming

Easy Does It!

Are you tired of kale sautéed, kale steamed, or kale in soup? Try this easy recipe using raw kale, wilted with your favorite dressing. Start with fresh-cut organic Lacinato kale. Remove the stems and ribs, then roll leaves and cut into a chiffonade. Add chopped fruit, dried cranberries, red onions, and diced nuts or sunflower seeds. Drizzle with your favorite vinaigrette dressing; toss well to coat the leaves and refrigerate for an hour before serving.

but dimpled like savoy cabbage. It's one of the best kales for eating. Dwarf Blue (55 days) is a compact blue-green kale with frilly leaves. Winterbor is a good tall-growing, blue-green variety with curled leaves; Redbor (50–60 days) is a purple version of Winterbor that makes a striking edible garnish (but it can get bitter in hot weather). Rainbow Lacinato (61 days) is an open-pollinated cross between Redbor and Nero di Toscana kale that has dark violet stems and frilly, dimpled leaves in hues of purple and green; vigor and flavor are good, and the color turns a deeper red when cool weather arrives. Frilliness and color can vary in different plants. Portuguese kale is growing in popularity at markets; it's a fast grower, providing large, flat leaves that look more like collard greens or outer cabbage leaves than kale.

Companions: Kale likes onions, beans, mint, radishes, and tomatoes. Avoid pole beans.

How to Grow: Start spring crops indoors about 6–8 weeks before the last frost. Set frost-tolerant plants outside as soon as the soil can be worked. Protect from cutworms and feed with fish emulsion. Use row covers to guard against insects and extreme early-season cold. Keep weeded when the plants are small and mulch the ground or plant lettuce seedlings below tall plants to smother weeds and keep the soil cool.

Pests: Not as susceptible to pest damage as other Brassicas, but kale does attract cabbage-eating caterpillars. Floating row covers work well when the plants are short. For taller plants, spray with Spinosad and Bt to control caterpillars.

Diseases: Clubroot can be a problem with this and all Brassicas. Rotate crops and keep soil pH near or above 7.0.

Challenges: Few challenges other than insect pests.

Harvesting: Dwarf Blue kale is often clear cut and sold whole, but most growers pluck the individual leaves and bundle together with rubber bands.

Marketing Tips: Plant multiple varieties and mix several leaves of each in a thick bunch, bundling with rubber bands. Educate your customers about such kale uses as baked kale chips and easy cold kale salad, wilted with lemon vinaigrette dressing.

Organic melons need warm temperatures, rich, loamy soil, and steady water. But when harvest time is near, cut the watering for a week to ensure firm, sweet fruit.

Melons

Cantaloupes, Honeydews, and Specialty Melons

Sweet cantaloupes, honeydews, and watermelons are a late-summer treat, but growing them can be tricky in northern climates with short growing seasons. They require lots of space and steady water for successful crops.

Cantaloupe *Cucumis melo*

Basics: Black plastic with drip irrigation works well to warm the soil and ensure steady water. Growers with limited land can try trellising the vines, but the fruit will need support in the form of a hammock made from lightweight netting.

Details: Fertile, well-drained soil with lots of compost and other organic matter and a pH of 6.5–7.5. Seeds germinate in less than a week when the soil temperature is 75°F–85°F (24°C–29°C), and vines grow best when the soil is 75°F (24°C) or warmer. Seeds are viable for up to 4 years.

Ease of Growing: Moderate to difficult.

Varieties: Ambrosia (86 days), Burpee hybrid (82 days), Fastbreak (65 days), and Iroquois (75 days) are good muskmelons; Charantais (90 days) and Jenny Lind (80 days) are small-sized heirloom melons with superb flavor. Early Dew (80 days) and Passport (83 days) are excellent honeydew choices for short seasons. Galia (85 days) and Banana (90 days) are some attractive and flavorful specialty melons.

Companions: Avoid places where cucumbers, squash, and other cucurbits have been grown in past years. Oregano and marigolds can help deter pests.

How to Grow: Southern growers can direct-seed in hills, then thin to 3 vines per hill. Northern growers may need to start seeds indoors and transplant into rows covered by a black plastic or dark-blue mulching film. Melons and other cucurbits do not like their roots disturbed, so transplant carefully and do it when the weather is cool and preferably cloudy. Keep vines watered regularly throughout the season—at least until the fruit becomes baseball sized. As summer fades, place the fruit on dark-colored boards or plastic sheeting to attract heat and ripen them. The layer will also protect the fruit from insects.

Pests: Cucumber beetles are the most serious pests, attacking young vines and chewing cotyledons to nothing. Larvae eat roots and weaken vines. Cucumber beetles also spread diseases among most of the cucurbits. Use floating row covers to protect young plants, but remove the covers when the vines start flowering to ensure pollination.

Diseases: Wilt disease, which is spread by cucumber beetles, can quickly destroy a crop. There are also many fungal diseases, including powdery mildew. Don't overcrowd the vines, and don't splash water and soil onto the foliage when irrigating, especially in the evening.

Challenges: Inadequate space, cool soil temperatures, and a short growing season are the biggest challenges for northern growers.

Harvesting: Don't water your melons for several days before picking or the fruit can become mushy and lacking in flavor. Melons are ripe when the stems start to dry and you can pick them just by sliding the stem off the fruit with your thumb.

Marketing Tips: Grow small, personal-sized and flavorful melon varieties like Jenny Lind and Charantais.

Watermelons *Citrullus lanatus*

Basics: Sweet, juicy watermelon has long been a favorite for southern growers. But northern growers have good options as well, thanks to newer personal-sized melons that mature and ripen within 90 days. Watermelons perform best in hot growing conditions with plenty of space for vines to spread out.

Details: Likes deep, loamy soil with high organic matter, with a pH of 6.0. Germination: 7–10 days, with best temperature from 75°F–90°F (24°C–32°C).

Ease of Growing: Moderate.

Varieties: Full-sized melons are still a classic crop, but many small-scale growers get better results from personal-sized (6 lb./3 kg) and ice-box–sized melons (6–15 lb./3–7 kg), especially seedless varieties with high sugar content and unique qualities. (Some watermelons have yellow rinds with red fruit or traditional green rinds with orange or yellow flesh.) A few favorites for full-sized "picnic" melons (reaching about 25 lb./11 kg) include: Charleston Gray (85 days), which has oval fruit with pale green rind and red flesh; Sweet Favorite (80 days) with light oval fruit, green-striped rinds, and red flesh; and Crimson Sweet (85 days) with round to oval fruit and red flesh. Smaller, seedless hybrids that for northern growers include: Treasure Chest (80 days), extremely sweet with green-striped rinds and bright yellow flesh, 10–11 lb. (4.5–5 kg); Troubadour (87 days), dark green-striped rinds with red flesh, resists anthracnose; 15 lb. (7 kg) fruit; Harvest Moon (90 days), a triploid hybrid with red flesh and rind that looks like the heirloom Moon and Star watermelon (dark green with splotches in yellow). Another popular non-seedless watermelon is Sorbet Swirl (77–84 days), green-striped rinds with sweet yellow and orange flesh and 10-lb. (4.5 kg) fruit.

Companions: Watermelon can trail through corn plants, but not if the corn is planted so thickly that it blocks out the sun. Radishes and nasturtium also complement watermelon. Avoid potatoes.

Big oval watermelons are perfect for picnics and weekend barbecues, but smaller icebox watermelons may be more practical for everyday use. Ask local farmers for a variety that ripens well in your area.

Solarizing Your Soil

All you need is 6 weeks of sunshine and a sheet of clear plastic to kill soil pests and diseases naturally. Soil solarization involves stretching a layer of plastic film over your planting area during the hot summer, allowing the sun's energy to cook the soil and eliminate insects, surface weed seeds, fungal diseases, and other pathogens. The goal in solarization is to raise soil temperatures to 110°F–125°F (43°C–52°C) in the top 6–8 in. (15–20 cm) of the soil (the plow depth) without the use of chemicals that can persist in the soil. To use the technique, follow these steps:

- Till soil in the bed you are treating, then rake it smooth and level. Remove any sticks or sharp surface debris that might puncture your sheet of plastic.

- Water the area thoroughly. If possible, use drip irrigation tape or a soaker hose beneath the material to keep soil moist.

- Spread a sheet of clear polyethylene or PVC over the bed, with at least 6 in. (15 cm) overlapping the area. Use 2 mil clear plastic for good results. Thinner sheeting can puncture, and thicker material won't let in as much sunlight. (Note: 1 mil = 0.001 in. or 0.025 mm.) In general, black plastic does not trap as much solar energy as clear sheeting, but it may be effective in cool, coastal areas where the summer sun is weaker and weeds grow underneath clear material.

- Dig a trench around the bed roughly 6 in. (15 cm) deep and tuck the overlapping edges of your sheet into the trench. Use rocks or U-pins to hold the sheet inside the trench. Fill the trench on one side and pack the soil down.

- Stretch the material snugly across the bed and secure the edges in the trench with U-pins or stones. Fill in the trench, and pack the soil down, making sure the sheeting is tight so no air can penetrate.

- Leave the tarp in place for up to 6 weeks during the hottest part of the summer. If weather is cloudy and cool, you may need to leave it in place for up to 8 weeks.

- Remove the tarp and plant your crops directly, trying not to disturb the soil. Don't till or you'll bring weed seeds to the surface.

Solarization kills good soil organisms as well as the bad, but beneficial microbes will quickly repopulate the soil because the heat also breaks down organic material. The process also increases nitrogen content in the soil and weakens pathogenic fungi, allowing beneficial bacteria to attack and finish the job.

How to Grow: Prepare the bed before planting by amending the soil with compost or rotted manure. Plant seeds in flattened hills, about 4–6 ft. (1–2 m) apart, depending on the variety. Sow about 4 seeds per hill, 1 in. (2.5 cm) deep. Thin to 2 plants per hill when the plants are strong. Northern growers should start watermelon seeds indoors about 2 weeks before the last frost (and set them out about 2 weeks after the last frost).

Seedless watermelons also should be started indoors. Hybrid plants are produced by tetraploid crosses, which result in triploid hybrids that do not produce seeds. For growers, it means that seedless watermelons need to be planted near a different "seeding" variety for pollination. Seed packets should contain seeds for both the hybrids and the pollinators.

Sow watermelon seeds in biodegradable pots made from peat, manure, or coir—just 1 seedling per pot (because the seeds are expensive). Use a light, well-draining mix and keep soil above 75°F (24°C). Transplant to the garden when seedlings have 3 or 4 true leaves and soil has warmed to 70°F (21°C). Watermelons don't like to be transplanted, so handle them carefully. Hint: You're more likely to disturb roots if you allow the seedlings to get large in the pot.

Pests: Cucumber beetles and squash bugs can be a problem, but they usually attack other cucurbits (squash, cucumbers, honeydew, and muskmelons) first. Cutworms will destroy young seedlings, and a number of caterpillars can chew rinds late in the season. Watch for insects and hand pick. Prevent cutworms with a small ring of diatomaceous earth around plants. Aphids spread the virus that causes fusarium wilt disease. Encourage beneficial insects that prey on aphids. Root knot nematodes can attack roots, stunting growth. A healthy, diverse population of microbes will reduce the number of destructive nematodes. Solutions: Rotate crops; leave a planting area fallow for one season to starve nematodes; plant a cover crop of French marigolds or sorghum (Sudan grass) and till into soil; or solarize the soil with a sheet of plastic to kill nematodes in the top foot of soil. (Note: Solarization also kills beneficial microbes, and nematodes can return the next season.)

Diseases: Watermelon can be damaged by bacterial, fungal, and viral diseases. Keep plants and soil healthy; rotate crops. If a problem develops, destroy diseased plants and grow resistant varieties next time.

Harvesting: Thumping watermelons is not a reliable indicator of ripeness. Look for visual signs instead. The tendrils around fruit should be starting to shrivel; rinds will lose their shine and look dull. Finally, the underbellies—where the fruit rests against the ground—will turn from white to yellow.

Marketing Tips: Small, icebox-sized melons are popular for everyday sales. Cut open an orange-flesh or yellow-flesh watermelon for customers to see.

Okra *Abelmoschus esculentus*

Basics: Heat-loving plants are a staple for southern growers, as well as for southern cooks who use tender okra pods fried, sautéed, grilled, and mixed in spicy gumbo soups. North or South, okra is a striking plant that can reach 6 ft. (2 m) tall, with creamy yellow or white hibiscus-like flowers.

Details: Loamy soil with high organic content and a pH of 6.5–7.5. Germination: best at 75°F–90°F (24°C–32°C). Seeds are viable for up to 2 years.

Ease of Growing: Easy.

Okra is a beautiful organic plant to grow, with tasty pods that are sautéed, grilled, or used in soups. Pick them when pods are less than 4 in. (10 cm) long for best flavor and texture.

Varieties: Clemson Spineless (60 days) and Annie Oakley (52 days) are green-podded okras, without the caustic hairs that irritate skin during picking. Burgundy (70–80 days) and Carmine Splendor (51 days) produce colorful red pods with purplish foliage.

Companions: Peppers and eggplants. Grow in a rotation after peas or another legume. Low-growing crops will be shaded by taller okra during the summer months.

How to Grow: In the South, okra can be direct-seeded; northern growers should start seeds indoors about 4 weeks before the last frost. Soak the seeds overnight before planting in pots or in large cell packs and use a germination mat to keep the seed flats warm. Wait until the soil temperature is 70°F (21°C) before setting the plants outside.

Pests: Not usually a buggy plant. Stinkbugs, corn earworms, cucumber beetles, and flea beetles are possible, depending on the region. Root knot nematodes are damaging but can be controlled by rotating crops. For larger insect pests, handpick what you can and spray pyrethrum as a last resort.

Diseases: Several fungal diseases, leaf spot, fusarium wilt, and damping off can occur— especially when the soil is cool and wet. Rotate crops and remove decaying plant material.

Challenges: Cool, wet soil can aid in fungal disease.

Harvesting: Pick okra when small; pods that are more than 4 in. (10 cm) tall begin to get woody.

Marketing Tips: Red pods are prolific and colorful. Okra doesn't have a long shelf life, so market it soon after picking.

Onions

Onions are easy if you follow a few rules. Plant the right varieties for your region and give them water, good soil, and time to set bulbs.

Full-Sized Onions *Allium cepa*

Basics: The most common of the allium family of vegetables, onions are easy to grow from onion sets (tiny bulbs), from seeds started indoors, or from field-dug transplants ordered from a supplier. Onions are divided into two types—each suited for different geographical areas. Short-day onions are for southern growers and long-day onions are for northern growers. Short-day onions produce bulbs when the days are shorter than 12 hours. They're planted in the fall in southern climates and do most of their growing during the mild winter months (when the days are short). Northern growers plant long-day onions in the spring—as early in the season as possible. For much of the northern United States, that means starting seeds indoors around Valentine's Day and transplanting outdoors in April. Some onion varieties are day-neutral, setting bulbs regardless of the daylight hours. They work in both the North and South. Adding to the choices—and sometimes the confusion for new growers—there are onions that are grown specifically for storage and others that must be used immediately because they don't keep well. In general, the stronger-flavored onions are better keepers and the larger, sweeter onions such as Walla Walla and Ailsa Craig are not good keepers.

Details: Rich, well-drained soil with a pH of 6.2–6.8. Soil temperature: 65°F–75°F (18°C–24°C). Shallow rooting plants require full sun and steady water.

Ease of Growing: Moderate.

Organic onions look like vegetable jewels when the outer skin is peeled away to reveal the shiny edible bulb. If you use field-grown transplants and you want to call your onions organic, purchase them from a certified organic supplier.

Varieties: Some long-day sweet onions: Ailsa Craig, Walla Walla, Candy, and White Sweet Spanish. Long-day red onions: Redwing, Cabernet, and Ruby Ring. Short-day sweet onions: Vidalia, Sierra Blanca (formerly Super Star), and Yellow Bermuda. Short-day red onions: Red Creole, Red Burgundy, and Red Grano.

Companions: Plant onions with Brassicas, beets, carrots, and strawberries. Avoid asparagus and legumes.

How to Grow: For best results, especially with large sweet onions, order organic field-grown onion transplants from a supplier. Plant as soon as the soil can be worked. When using onion sets, choose firm onions and sow as early in the season as possible. Smaller set onions will be more likely to produce large bulbs, while the larger sets can be harvested for green onions. When sowing onion sets for scallions, plant them 3–4 in. (7.6–10 cm) deep; this will result in a longer section of stem that is edible. From seeds, start plants in February or early March. Some sweet long-day onions like Walla Walla and Candy are best sown in the late summer, then wintered over for harvesting up to 300 days after sowing.

Pests: Onion thrips and onion maggots are potentially serious problems. Use floating row covers early in the season. Watch for silver streaks on the leaves and catch thrips with yellow sticky traps.

Diseases: Leaf blight, downy mildew, white rot, and neck rot are all fungal diseases. Practice good crop rotation and dispose of crop residue to guard against fungal outbreaks.

Challenges: Onions can lose out to weeds, especially when they're young. Also, make sure you choose the right varieties for your part of the country and get them started early so they have time to produce large bulbs.

Harvesting: Use thinnings for green onions. Also, small-sized bulbs can be picked and sold with clean greens attached before the foliage starts to fade—a good way to stretch the harvest time. Harvest full-sized onions when the stems start to wither and fold over. Brush off residue, but don't wash them in water. Dry for several days out of direct sunlight, then store in a root cellar or refrigerator.

Marketing Tips: Large sweet onions, my favorites, are more desirable than the long-storage varieties, which usually have a more pungent flavor. But having storage onions handy can be a benefit during times when the harvest is skimpy. Maximize your onion season by pulling scallions and near-mature green onions. To harvest and market near-mature onions, loosen the soil and pull the bulb, then peel back the brown leaves and mottled onion skin, leaving only green tops and shiny red or white bulbs. Trim the tops to an attractive length.

Leeks *Allium porrum*

Basics: Leeks are milder than onions and are simple to grow, harvest, and bundle for customers. They thrive in cool growing conditions, with full sun and copious water, but because they produce such a large amount of top growth, they need more nitrogen than other alliums.

Details: Fertile, well-drained soil, with a pH of 6.0–7.0. Harvest in 70–150 days, depending on varieties; tolerates frosts and can be picked into winter. Seeds are viable for up to 1 year.

Ease of Growing: Easy.

Varieties: There are numerous varieties and hybrids, some for earlier harvesting. King Richard (75–80 days) is tall and very early but does not tolerate hard freezes. American Flag (130 days) is better for late-season harvests. Giant Musselburgh (150 days) produces large, mild-tasting white stalks and lasts well into winter.

Companions: Like other alliums, leeks repel some pests, making them good partners with cabbage, carrots, and lettuce. However, in a production setting, leeks are spaced closely, so there's not much room for shared planting.

How to Grow: Start leek seeds indoors in February, planting about 20 seeds in a 4-in. (10 cm) pot. When the outdoor soil is dry enough to work, plant leeks with a dibble, pressing it deep into the ground and inserting the leek (roots down) so there's only 1–2 in. (2.5–5 cm) aboveground. Don't backfill the dibble hole; instead, allow water to slowly collapse it as the leek grows. Some growers start leeks in a trench and cover the stems with soil as they grow, but this can lead to soil in the leaves that eventually settles into the edible stem, making it harder to clean. Leek foliage grows to either side of the stalk, not in every direction. If you plant your leek seedlings in a row so the leaves grow into the space between rows, you can plant them more closely together.

Leeks are beautiful to grow, harvest, and cook. To keep the interiors from accumulating dirt, plant them in a deep, narrow hole and let the rain fill in the soil around them. Only the section of stem left belowground will develop the blanched (white) color.

Pests: Onion thrips leave silvery streaks on foliage. Use blue sticky traps or spray with Spinosad. Onion maggots bore into stems and can ruin entire crops. Use floating row covers in the spring to stop the first generation. Yellow sticky traps will catch adult flies.

Diseases: Rust can cause orange patches on the leaves; pink root causes the leaves to wilt (if you pull up the plant, roots appear pinkish). White rot and downy mildew also attack leeks. To avoid fungal diseases, remove any damaged leaves, taking care not to infect nearby leeks. Rotate crops and avoid composting leek foliage.

Challenges: Weeds can be a problem, especially in early spring when the seedlings are small.

Harvesting: Leeks can be harvested when they're ¾ in. (19 mm) wide (finger width). Loosen soil with a fork before pulling, especially for deeply planted leeks.

Marketing Tips: Thin leeks and sell the narrow summer leeks in a small bunch. Trim leaves to create a spearhead at the top. When trimming roots, leave 1 in. (2.5 cm) of stubble. (Clean, white roots tell buyers the produce is fresh.)

Scallions *Allium fistulosum*

Basics: Bunching onions have stiff upright stems and don't develop swollen bulbs like their allium cousins.

Details: Loamy soil, with a pH near 6.5.
Germination: best at 65°F (18°C). Seeds are viable for up to 1 year.

Ease of Growing: Easy.

Varieties: Evergreen Hardy White (65 days) can be sown in the fall and harvested the following season in many areas. Deep Purple (60 days) has reddish purple tips. White Spear (65 days) is a heat-resistant variety good for growers in warmer areas.

Companions: OK with beets and Brassicas. Avoid legumes.

Scallions or bunching onions can be grown from seeds, from transplants, or from onion sets. If planting from sets, push the tiny onions in a little deeper than normal so that more of the stalk remains underground, where it will develop the blanched white color.

How to Grow: Sow the first crop indoors in late winter or early spring, then transplant outdoors as soon as the soil can be worked. For longer blanched stems, plant the transplants several inches deep or pile soil around the sides as they grow.

Pests: Not susceptible to many pests. (*See also* Onions.)

Diseases: Relatively disease free if the plants are rotated and grown in biologically active soil.

Challenges: Weeding and watering scallions is critical when they're young.

Harvesting: Begin picking scallions when the tops are 8–10 in. (20.5–25 cm) tall.

Marketing Tips: When the plants become tall, trim off floppy stems. Mix red-stem and white-stem varieties in a bunch.

Shallots *Allium ascalonicum*

Basics: Shallots are a type of multiplier onion, prized by chefs for their mild, sweet flavor. They can be started by seeds if you sow them indoors in late winter and get the transplants into the ground early enough for a 100-day season. The easiest way to grow shallots is to order bulbs for spring planting in the North and fall planting in the South. Bulbs planted in the fall can be susceptible to bolting; spring-sown shallots grown from bulbs are ready for harvest in about 75 days.

Details: Sandy loam with good organic matter, with a pH of 5.5–7.0. Soil temperature: 35°F–85°F (2°C–29°C). Best in full sun but can grow in partial shade. Ensure ample water and balanced nutrients for shallow roots.

Ease of Growing: Easy.

One shallot bulb will divide and multiply into new bulbs, with minimal care. Plant in spring and harvest in late summer, when the tops die back.

Varieties: Conservor, Picador, Pikant, Red Gourmet, and Red Sun are copper-skinned shallots. Golden Gourmet is a yellow-skinned variety.

Companions: Alliums like Brassicas, carrots, and beets but not peas, beans, or asparagus.

How to Grow: Sow bulbs in springtime, about a month before the final frost date. For best results, shallots need at least 4 weeks with soil temperatures below 50°F (10°C). Southern growers can sow in autumn for a late spring harvest. Add lots of organic matter to the soil, for nutrients as well as to keep the soil loose and friable. Push the bulbs into the ground so the tops are just below the surface. Space shallot bulbs 4–6 in. (10–15 cm) apart in rows that are about 1 ft. (30 cm) apart. Or plant them in a block with 4–6 in. (10–15 cm) between bulbs. Closer planting can result in smaller bulbs.

Pests: Onion thrips, tiny insects that scour and suck fluids from leaves, can be a problem. Keep the patch well weeded. Sticky traps will both control pests and help you monitor pests. If you see silver streaks on the foliage, spray with Spinosad.

Diseases: Shallots are relatively disease free but can succumb to white rot if the soil is compacted and doesn't drain well.

Challenges: Keep shallots well watered and free of weeds for the biggest, best-quality bulbs. Small bulbs can be frustrating to peel and use.

Harvesting: Pull shallots when the tops start to dry back, usually in midsummer. Spread the plants onto a screen or grid where they will be exposed to the air. Leave them to dry for about a week in a shaded location. Trim dead tops and store in a cool, dry place.

Marketing Tips: Green shallots can be picked and trimmed like scallions or green garlic but the tops are generally not as large. Dried shallots keep well once cured and can be added to fall and winter CSA shares.

Peas *Pisum sativum*

Basics: Peas are one of my favorite early summer crops—especially edible-pod varieties, which are picked and eaten like pole beans, albeit for only a 2–3 week window. Pea greens are also finding more fans and are an especially easy crop to grow. Many farmers plant 2 crops of edible-pod peas—one for picking in the late spring and a season-ending crop, timed for harvest just before the first killing frost.

Details: Fertile, well-drained soil, with a pH of 6.0–6.8. Soil temperature: germination good from 45°F–85°F (7°C–29°C) but best around 75°F (240°C). Seeds are viable for up to 3 years (although I find germination drops off after 2 years).

Ease of Growing: Easy.

Varieties: Snow peas, with flat, translucent pods that reveal the growing peas inside, include Oregon Giant and Oregon Sugar Pod II, both with large but tender pods on 30-in. (76 cm) vines (ready to pick in 60–70 days). Snow Sweet (60–70 days) has more uniform pods on compact 2-ft. (60 cm) vines; Avalanche (61 days) is prolific, resistant to powdery mildew, and grows on semi-leafless vines, making for an easy harvest. Snap-pea choices include Super Sugar Snap (60 days), which is sweet and prolific with 5-ft. (1.5 m) vines. More compact snap peas (2–3-ft./60–90-cm vines) are Sugar Sprint (60 days), Sugar Star (55 days), and Cascadia (66 days). The variety Dwarf Grey Sugar, is a good choice for pea greens, ready to pick less than 2 weeks from sowing, depending on the weather.

Crisp snow peas are more versatile than shell peas and offer greater value because both peas and the pod are edible.

Companions: Peas grow well with almost any other vegetable in the neighborhood, but avoid alliums, especially onions and garlic.

How to Grow: Sow seeds as soon as the soil can be worked in the spring. I like to soak them overnight in a small bowl, allowing the peas to swell. Don't soak them any longer or they will start to sprout, and you may risk breaking the brittle white roots. I also sprinkle inoculant on the moistened seeds before planting. It provides beneficial bacteria that colonize the root zone. Compact pea varieties can be sown in a shallow trench about 3–5 in. (7.6–13 cm) wide, spacing seeds about 1–2 in. (2.5–5 cm) apart. They will support themselves, or you could insert pea sticks into the wide row for support. To make pea sticks, just cut some thin limbs with multiple twiggy branches. Push the single stem into the ground and let the branches touch each other about 2–3 ft. (60–90cm) above the soil. To grow pea greens, sow as you would normally sow fruiting peas, but cut off the tops when the vines are about 6–8 in. (15–20.5 cm) tall. You can take a second cutting or just pull up the peas and sow another crop.

Pests: Peas are relatively insect-free, but the young greens are favored by deer, woodchucks, and even birds. Use fencing to keep critters away.

Diseases: Powdery mildew can leave pea vines looking ragged, but it usually doesn't strike until the weather turns hot and the spring-planted harvest is almost over. Pea root rot turns the entire vines brown; avoid it by rotating crops and planting in well-drained soil.

Challenges: A strong burst of hot weather can shorten the length of your harvest. It can be difficult to time your second crop. But count backwards 60 days from the first frost and sow your fall crop 2 weeks earlier.

Harvesting: For best flavor, pick peas when the pods are plump but not ready to burst. Once the peas inside get too big, the pods become tough, and the peas grow starchy. Taller varieties growing on a trellis are easier to pick than low growers.

Marketing Tips: Sell with green garlic, spring Pak Choi, and other Asian greens for stir-fry meals. Sell cut pea shoots in plastic bags or plastic clamshell cases. Let customers know how to prepare them—raw in salads or wilted with cooking greens or in stir-fries.

Peppers *Capsicum annuum*

Basics: Sweet or hot, peppers are a late-summer favorite for customers who like salads, salsa, relishes, or dishes that are stir-fried, roasted, or grilled. (That pretty much takes care of everyone, doesn't it?) To keep the customer satisfied, mix colors, shapes, and heat (measured in Scoville units).

Details: Loamy soil, with a pH of 5.5–7.0. Needs adequate moisture and ample phosphorus and calcium (calcium in the soil helps prevent blossom end rot).

Ease of Growing: Easy.

Varieties: Growers today have myriad pepper choices, both sweet and hot, in an assortment of colors, shapes, and fire power.

> **Sweet Peppers.** Ace (50 days green) is a prolific, early maturing pepper that tolerates cool climates and ripens to red (70 days). California Wonder and California Wonder Gold (60 days) are dependable bells that ripen to red or gold (75–80 days). Sweet Banana (58 days yellow, 70 days to red) is a frying pepper that's early and prolific. Giant Aconcagua

Sweet bell peppers are available in a rainbow of colors. If you're patient, let some green peppers stay on the plant for another two to three weeks until they ripen to their natural color—chocolate, yellow, purple, orange, or red.

How Hot Is Your Pepper?

The sizzle in hot peppers is caused by the presence of capsacin, which fires up the palate, puts sweat on your cheekbones, and sends you scrambling for the nearest water spigot. It's measured in units on the Scoville Scale, invented by pharmacist Wilbur Scoville in the early 1900s and not always uniform or precise because it relies on human taste. The scale is determined by how much sugar solution it takes to dilute capsacin to the point that it becomes undetectable to a panel of tasters. Standard sweet bell peppers, used raw in salads, have zero Scoville Units (SU). This is how some other popular peppers stack up

Pepper Name	Scoville Units
New Mexico and Anaheim	100–1,000
Jalapeño	2,500–5,000
Serrano	4,000–15,000
Cayenne	25,000–50,000
Habanero and Scotch Bonnet	100,000–350,000
Carolina Reaper	More than 1.5 million

Carolina Reaper set the Guinness Record in 2013.

If you ever bite off more pepper than you can handle, turn to the freezer for some ice cream. Sugar and dairy products will put out the fire better than water.

(70 days to green) is a frying pepper that exceeds 10 in. (25 cm) in length and ripens to red. Chinese Giant (75 days) is a blocky sweet pepper that ripens to red. Sweet Chocolate (58 days) ripens to a chocolate brown (78 days). Snapper (62–78 days to red) is resistant to many diseases.

Hot Peppers. Early Hot Jalapeño (60 days) is a prolific pepper with large, spicy fruit. Serrano del Sol (55 days). Hot Lantern (70 days to green; 90 days to red) is a habanero-style pepper that grows well in cooler climates. Anaheim peppers (numerous varieties are mildly hot and good for grilling or roasting): Numex Joe E. Parker (75 days) and Goliath Griller (73 days).

Companions: Peppers like onions, parsley, Swiss chard, and tomatoes. Avoid Brassicas and beans.

How to Grow: Peppers are the latest spring seedlings to go into my beds—about 1–2 weeks after tomato seedlings have been transplanted. Since they take longer to develop from seed, I typically sow them indoors at the same time I sow tomatoes. Like tomatoes, they need heat to germinate but will grow stockier if you remove the heat and bathe them in bright light after germination (14 hours per day ought to do it).

Harden transplants after tomatoes have been planted and give peppers a warm, sunny place in the field. Northern gardeners may want to use plastic mulch to warm the soil. Even though peppers are associated with heat and dryness, keep the plants well watered throughout the summer.

Pests: Cutworm collars or soil nematodes can stop cutworms, but there are plenty of other pests. Pepper maggot flies infest the stem end of peppers and can destroy the fruit. Use cherry pepper plants as a trap crop around the perimeter of your other peppers. European corn borers have also become a serious problem in many areas. Spray foliage with Bt to control earworms and corn borers. Neem can slow down tarnished plant bugs. Also watch plants for damage by potato beetles and tomato hornworms; handpicking can keep them in check.

Diseases: Fungal diseases and wilts can affect pepper plants, but good crop rotation will help control problems. If you've had problems with peppers in the past, choose disease-resistant varieties.

Hot peppers vary in their intensity, from relatively mild Anaheim and Jalapeño peppers to scorching Habaneros. The heat will also vary by climate, growing more intense in dry, hot regions.

Challenges: Failing to provide steady water all summer will reduce your harvest. A lack of calcium in the soil can cause blossom end rot.

Harvesting: Sweet peppers and larger-fruiting hot peppers typically start setting fruit when plants are quite small. It's a good idea to pick off any fruit that sets while the plants are still immature. Fruit pruning actually improves the harvest by allowing the plants to grow larger. But don't pick all fruit early. Sweet peppers eventually turn red, yellow, chocolate, orange, or purple when they ripen. Most seed descriptions include two harvest times: the first for green peppers and the second date for fully ripe colorful peppers.

Marketing Tips: Sell multiple-colored peppers together (usually late in the season.) Don't grow too many of the famously hot habanero peppers. Even though they don't reach their full heat potential in cooler climates, the peppers are still too hot for most palates. We find that a few plants are enough to provide us with peppers for jams and sauces. Most people don't eat fresh habaneros, so for consistent sales, stick with jalapeños, serrano chilies, and sweet or mildly hot peppers.

Potatoes *Solanum tuberosum*

Basics: The biggest reason to grow fresh potatoes for yourself or for market is to experiment with beautiful but hard-to-find varieties. Colorful heirlooms with rich flavor and histories are found in the potato world, too. So it makes sense to plant something other than ordinary russets. Potatoes can be grown in the summer and saved in a root cellar for use in a fall CSA, when other crops have ended their harvests for the season.

Details: Potatoes like loose, sandy soil with a pH range of 5.0–6.5. Growing them at the lower (more acidic) range may help prevent scab, a disease that leaves lesions on potato skins.

Ease of Growing: Moderate to difficult.

Varieties: Fingerling potatoes such as Russian Banana, Ruby Crescent, and Rose Finn are small, cigar-shaped spuds with thin skins and rich, waxy texture that keeps them firm in potato salads. Other eye-catching spuds include All Blue, a violet-skinned gourmet potato with an inside that looks like a purple tie-died T-shirt; Cranberry Red, a red-skinned potato with pinkish insides, and Caribe (pronounced car-EE-bay), a blue-skinned, white-fleshed potato that keeps well and is superb for baking. The best varieties will probably be the ones that other farmers in your area have had success with; ask around.

Companions: Potatoes like horseradish, cabbage, carrots, peas, and beans. Avoid asparagus, cucumbers, squash, and turnips.

How to Grow: Time your seed potatoes for delivery early in the season—as soon as the soil is dry and warm enough to be worked. Before planting potatoes, dig a trench about 1 ft. deep and fill the bottom 6 in. (15 cm) with a mixture of compost and soil. Add a couple of tablespoons of granular organic fertilizer for each plant and place the seed potatoes about 1 ft. (30 cm) apart in the trench, then cover with 2–3 in. (5–7.6 cm) of soil. As the plants grow, fill the trench in around them. After the

Why grow boring, brown potatoes when there are purple, pink, and yellow–skinned tubers—sometimes with different coloration inside? If you sell spuds, your customers will appreciate the variety of shapes and hues.

trench is full, hill-up nearby soil around the growing plants. Potatoes set tubers above the spot where the original seed potato was sown, so the extra soil will ensure more growing space for bigger spuds. And it will help conserve moisture.

Pests: Whiteflies, thrips, and Colorado potato beetles are common pests. Scout leaves for clumsy-looking brown- and white-striped potato beetles and their fluorescent orange eggs. Crush the eggs and handpick the adult beetles. Once pinkish, sluglike larvae appear, it's too late for handpicking. Spray the foliage with the biological insecticide Bt, formulated specifically for potato beetles. Control whiteflies and thrips with sticky traps—yellow for whiteflies and blue for thrips. Wireworms can also cause damage but usually only when spuds are left in the ground for too long.

Diseases: Prevent the most damaging diseases by planting only certified disease-free potatoes from a reputable source. Don't plant grocery store potatoes. Order certified seed potatoes from a specialty grower who is familiar with plant diseases as well as the best practices for organic potato production. Jim Gerritsen of Wood Prairie Farm in Aroostook County, Maine, says his seed potatoes are certified twice—once by the state, which finds them disease free, and again by his organic certifying agent to make sure that all organic requirements are met. Keeping the soil pH close to 5.0 can fight scab.

Challenges: Keep plants adequately watered throughout the growing season for best tuber production. Insects can stress plants, causing a smaller crop, even if the plants are not destroyed.

Harvesting: For summer markets, steal a few new potatoes while the vines are still growing, but save the bulk of the harvest until after the tops have died back. Dig up potatoes with a fork and not a shovel, which will likely damage tubers.

Marketing Tips: Sell fingerling and colorful heirloom varieties, which are in higher demand than white- or brown-skinned potatoes.

Radishes *Raphanus sativus*

Basics: They're so easy a child could grow them. But the trick to successful radish farming is to plant multiple varieties in succession and to keep insects, primarily flea beetles, off the tender tops. Because radishes are so quick to grow (25–35 days from seed), it's sometimes hard to remember where and when you last planted them. Wherever you sow radishes, plant early and often, since the seedlings can get leggy and the roots become intolerably spicy in the heat of summer. Fall plantings are preferred for larger icicle and Daikon radishes.

Details: Sandy or loose and loamy soil is best, with a pH of 6.0–7.0. Soil temperature: between 45°F–80°F (7°C–27°C). Seeds germinate in 4–10 days. Likes full sun or partial shade and ample water. Seeds are viable for up to 4 years.

Daikon radishes are sweet and big when planted in summer for harvesting in fall. Growers in hot climates plant successive crops of radishes in the spring, then wait until the midsummer heat has faded before planting a fall crop.

Ease of Growing: Very easy.

Varieties: Champion and Crimson Giant (25–28 days) are large, round, and red. Cherry Belle (22 days) is fast growing, round, and red. Purple Plum (25 days) has purplish red skin with white interior. Snow Belle (30 days) is white inside and out. Heirloom French Breakfast (23 days) is torpedo-shaped and red, with an attractive white tip on each root. Icicle (25 days) is white and carrot-shaped. China Rose (52 days) is a good keeper, as is Round Black Spanish (55 days), which has rough-textured charcoal-colored skin and white interior. Watermelon radish (50 days) has large, round roots that are pale green or white on the outside and a stunning hue of magenta inside; the color sometimes varies in intensity, looking like a tie-dyed T-shirt.

Companions: Radishes grow well with just about anything for neighbors. Some growers sow a few radish seeds with carrots to mark the spot and to loosen crusty topsoil. If flea beetles are a problem, consider planting arugula in a faraway corner of your property as a trap crop to lure the insects away. Spray insects on the trap crop instead of your main crop.

How to Grow: Sow radishes directly in blocks or wide rows, spacing seeds about 1 in. (2.5 cm) apart and covering with ½ in. (12.5 mm) of soil. (I plant seeds ¾ in./19 mm deep in warmer months, which I think gives me less bolting and better roots.) Keep young seeds well watered. For a continuous harvest, sow a new crop every 2 weeks until the weather gets too hot, and then sow more crops in the fall.

Pests: Wireworms can damage roots that are left in the ground too long, but the most common pests are flea beetles, which are especially problematic in the spring. Flea beetles chew tiny holes in the leaves, leaving them tattered and unattractive for selling with the tops attached. Control with row covers or Spinosad.

Diseases: Few diseases, since you'll most likely harvest the crop before you notice any problems.

Challenges: Hot weather, close spacing, and not enough water can be a problem during the summer months. Flea beetle damage can sometimes be severe enough to prevent you from selling radishes at market.

Harvesting: Pull radishes before they reach full size, picking them in the morning and dropping both roots and tops into a bucket of cold water. (This will keep the tops green and turgid.)

Marketing Tips: Bundle red, purple, and white radishes together for a rainbow mix.

Spinach *Spinacia oleracea*

Basics: The window for growing spinach is short for many growers, since spinach performs best in cool seasons—spring and fall. Sow seeds directly as soon as the soil temperature reaches 50°F (10°C).

Details: Loose, loamy soil, with near neutral pH (6.5–7.5). Germination: 50°F–75°F (10°C–24°C). Seeds germinate in 5–10 days.

Ease of Growing: Easy.

Varieties: Red Cardinal (32 days) has purplish red stems and veins with green leaves. Tyee (40 days) is a hybrid with savoyed (crinkled) leaves and good heat resistance. Bloomsdale Long Standing (50 days) is an old, open-pollinated variety with large leaves that hold up well to hot weather.

Companions: Brassicas, legumes, and onions. Not potatoes.

How to Grow: Sow seeds as soon as the soil warms to 50°F (10°C). I prepare a bed, rake away ½ in. (12.5 mm) of soil from a block about 8 in. (20.5 cm) wide and 3 ft. (90 cm) long, then scatter the seeds 1–2 in. (2.5–5.0 cm) apart, lightly pack them down, and cover with the ½ in. (1.2 cm) of soil. After they show their first leaves, I thin to about 3–4 in. (7.6–10.0 cm) between plants. When the weather is warmer and germination is more consistent, I use a homemade planting grid that leaves ½-in. (12.5 mm) impressions every 2 in. (5 cm) in a 1-ft. square. I sow 1 seed per indentation, and the spinach grows in perfect spacing. Then I pluck the seedlings for salads, leaving more space for the remaining plants.

Pests: Leaf miners, the larvae of a tiny fly, can be a problem. Adults lay eggs on the undersides of leaves as soon as the warm, sunny weather arrives. Eggs hatch, and the worms tunnel into the leaves, causing a squiggly trail of brown. A high number can ruin your crop. Stop adults with floating row covers. Don't plant spinach in the same place every year. Use blue sticky traps to catch adult flies.

I grow baby spinach under floating row covers to protect the young plants from an early spring freeze or from insect damage in the form of leaf miners.

Diseases: Not prone to many diseases.

Challenges: Cold, soggy soil can cause poor germination. A week of hot weather in the late spring can make spinach shift gears and suddenly bolt to seed. Choose bolt-resistant varieties.

Harvesting: Pick spinach by pulling selected leaves or by cutting the entire plant 1 in. (2.5 cm) above the ground. A full cut will stimulate a second crop of leaves if the weather isn't too hot.

Marketing Tips: Red Cardinal is an attractive variety. Savoy-type spinach has deeply crinkled leaves. Baby spinach is highly sought after for gourmet salads.

Squashes

Winter squash, summer squash, and pumpkins are colorful, fast-growing, and nutritious. They're also extremely productive. Successive summer squash crops produce from early summer to the frost, while pumpkins and winter squashes save their bounty for season's end.

Pumpkins *Cucurbita maxima*

Basics: I once grew giant pumpkins and entered them at the local fair (not even close to winning). I've also grown petite pumpkins and sizes in between. While pumpkins signal the start of cool weather, the vines like it hot. Don't start pumpkin seeds outdoors until well after your last frost date. Most varieties also need plenty of room to sprawl.

Details: Rich, well-drained sandy loam with high organic content. Soil temperature: 65°F–95°F (18°C–35°C), with a pH of 6.5–7.0. Germination: 7 days. Seeds are viable for up to 4 years.

Ease of Growing: Moderate.

Varieties: New England Pie, Sugar Sweet, and Winter Luxury (all 100–110 days) are small-sized pumpkins (2–5 lb./1–2 kg) with thick flesh and sweet flavor. Sugar Treat is a hybrid sweet pumpkin with compact vines. Autumn Gold and Jack O' Lantern (both 100–110 days) are medium-sized carvers reaching up to 15 lb. (7 kg). Big Autumn, Connecticut Field, and Howden (all 110 days) are large-sized carving pumpkins, reaching up to 25 lb. (11 kg). Those who want to shoot for a world record can try Howard Dill's Atlantic Giant (130 days). Monster pumpkin growers have selectively crossed different strains of Atlantic Giant to produce pumpkins weighing more than 1 ton (907 kg). A "smaller" giant pumpkin (over 200 lb./91 kg) with a nice shape and deep reddish orange skin can be grown from Burpee's Prizewinner seed (120 days).

Companions: Beans and mint. Trailing pumpkin vines can meander through corn and block out weeds. Avoid potatoes.

Easy Does It!

What are sticky traps? Sticky traps are cardboard or plastic rectangles in colors that attract bugs. They're coated with a gluey material that catches the flying insect pests as soon as they make contact. Growers who practice Integrated Pest Management (IPM) use the traps to count populations and decide the best time to spray pesticides. But the traps alone can control bugs in small plots.

Hang them from a stick or secure them to a stiff wire so they stand a few inches above the height of the crop you're trying to protect. Yellow sticky traps are the most widely used and attract cucumber beetles, whiteflies, flea beetles, and other pests. Blue sticky traps lure thrips and beet leaf miner flies. White and red traps are also available for specific pests.

How to Grow: Cultivate pumpkins much like winter squashes, as the plants are very closely related. Seeds can be started indoors under light, but vines grow rapidly and will need to be transplanted in about 3 weeks. Like squashes, pumpkins don't take kindly to having their roots disturbed, so grow them in peat pots or handle the transplanting task very carefully. Plants will need at least 1 in. (2.5 cm) of water per week, more when setting fruit. If starting outdoors, prepare a mound of soil and plant 3 or 4 seeds on the mound. After the vines develop, thin to 1 or 2 seeds per hill, depending on the variety. (The bigger the pumpkin, the more space it will need.) Protect young plants from cold and insects with floating row covers. Keep the vines well watered throughout the growing season, and mulch with straw to conserve moisture and prepare a clean growing spot for developing fruit.

Pests: Cucumber beetles, squash bugs, squash vine borers, and squash beetles. (*See also* Winter Squash for controls.)

Diseases: Powdery mildew and other fungal diseases. (*See also* Winter Squash for treatment and prevention.)

Challenges: Space and time are the biggest obstacles to growing pumpkins, especially for northern growers.

Harvesting: Time your harvest so that pumpkins aren't early or late for early October sales. Pick them by cutting the stem and leaving a nice, decorative handle.

Marketing Tips: Small baking pumpkins have a longer season of usefulness. Carving pumpkins are not in high demand after Halloween.

Summer Squash *Cucurbita pepo*

Basics: By mid-July there are more zucchinis than I can handle, but in June, when that first tiny squash is forming under its yet-to-unfurl female blossom, I get as excited as I do about the season's first tomato. There's something about a new zucchini that ushers in the real start of my summer harvest.

So how do you make summer squashes that exciting for customers, too? Sometimes it's not enough to know they're versatile, delicious, and low in calories. To make summer squash special, you need to grow several different varieties. And you need to pick all of them while still young—some with flowers still attached. Grow green-and-yellow pattypan squashes, golden zucchinis, and flavorful, prolific Italian heirlooms with color, shape, and textures that make supermarket squashes seem like soggy sponge cakes.

Details: Grow in rich, well-drained sandy loam, with ample organic matter, good nitrogen, and adequate phosphorus, potassium, and micronutrients. Soil pH should range from 6.0–6.8. Germination: 5–10 days at 70°F–90°F (21°C–32°C); plants grow best at about 70°F (21°C).

Easy Does It!

Plant in full sunlight and provide copious water. Seeds are viable for up to 4 years.

Ease of Growing: Easy.

Varieties: It's easy to grow a variety of squashes, from golden zucchinis and green-and-yellow pattypans to flavorful, prolific Italian heirlooms with color, shape, and textures that make supermarket squashes seem like soggy sponge cakes. Here are some of my favorites.

Pattypan Squash. Another variety of summer squash is the pattypan, which is shaped like flying saucers in shades of green and yellow. Scallopini hybrid (55 days) is golden yellow, with a dot of green at the blossom end; the flavor is superb, and the vines resist powdery mildew. Benning's Green Tint (53 days) has pale green fruit on vigorous, bushy vines.

Yellow Summer Squash. These squashes include straight-neck and crookneck varieties; crooknecks have a potbelly end with a larger seed cavity. Saffron (42 days) is an early, open-pollinated semi-straight-neck with smooth, buff-yellow skin, and superb flavor. Southern growers will like Gentry (43 days), a hybrid crookneck with smooth skin and a sweet, mild flavor.

Zucchini. Costata Romanesco (54 days), with its trailing vine that yields pale-green squashes with raised ribs, is one of my favorite zucchinis. This Italian heirloom is crisp and sweet with a nutty, pistachio-like aftertaste. But it's not as prolific as other varieties. 'Black Beauty' (48 days) is an All-America selection from 1957 that was introduced in the 1920s; its fruit is deep green and firm, with a large harvest from compact vines. Embassy is dark green and prolific on almost spineless vines. Gold Rush is a bright yellow zucchini with a firm and nutty flavor (55 days). Ronde de Nice (52 days) is a French heirloom with pale green round fruit that is perfect for single-sized stuffed zucchinis.

Companions: Summer squashes like corn, onions, and radishes but not potatoes or tomatoes.

How to Grow: Summer squashes are all heavy feeders, so amend the soil with a bucket full of compost or aged manure per plant, working it about 12 in. (30 cm) deep. For an extra-early harvest, sow 2 or 3 seeds in 8-oz. (240 ml) pots about 4 weeks before the first frost date. At 2 weeks, thin to the strongest seedling per pot. Warm the planting bed with black plastic and set out hardened seedlings about the same time you would plant tomatoes. Cover with floating row covers to protect from wind, insects, and the cold. I also sow summer squashes directly into the soil when it has warmed to at least 70°F (21°C), spacing plants 15–18 in. (38–46 cm) apart. Keep soil moist and watch for insects. If the soil has been well prepared, it won't need extra fertilizers for the life of the vine.

For a continuous harvest, sow a second crop about 1 month after the first planting. It should start to produce about the time the first sowing peters out. If you have the space, start summer squashes at the end of May, end of June, and end of July for three separate harvests that will keep you buried in squashes all season long.

Pests: Cucumber beetles and foul-smelling dark brown squash bugs chew foliage and suck plant juices, weakening the plants. Another pest, the squash vine borer, can be especially destructive with larvae that tunnel into squash stems and eat the vines from the insides out. Look for vines that suddenly droop and show a pile of sawdust around the stems. Home gardeners can slice stems to cut out borers, but that kind of attention isn't practical for large-scale growers with dozens of plants. It's better to protect young squash vines with row covers. Later plantings grown under covers will avoid squash vine borers. By the time the vines bloom and you remove the row covers, the adult pests (clear-winged red and black moths) will be out of season.

Diseases: Powdery mildew seems to overtake summer squashes no matter what precautions you take, which is why successive plantings are essential. Dispose of crop residue to prevent the spread of both diseases and insect pests.

Challenges: Squashes are generally easy to grow, but can become difficult if insects get the upper hand. Scout plants regularly, looking for

Pattypan squashes are a delicious departure from zucchini and yellow summer squashes. Pick them when they're young and small (about 4 in./10 cm), before the center seed cavity swells and the skin gets tough.

cinnamon-colored squash-bug eggs on the undersides of leaves. Squash beetles, which look like large Mexican bean beetles, will chew the leaves and breed yellow fuzzy larvae that skeletonize the foliage. Avoid overcrowding to keep the harvesting and cleanup easy.

Harvesting: Pick young summer squashes when they're small. (Bigger is not better!) To keep fruit fresh, cut it off the vines with a sharp knife, leaving at least 1 in. (2.5 cm) of stem still attached. It will help conserve water and keep the fruit crisp. Wear long sleeves if your arms are sensitive to the hairs on zucchini stems.

Marketing Tips: Mix green, yellow, and bicolor squashes together. Sell along with basil and other stir-fry ingredients. Late summer crops can be sold with tomatoes and eggplants. This is also a great time to supply your customers with recipes.

Winter Squash *Cucurbita pepo, C. moschata, C. maxima*

Basics: With a few exceptions, most winter squashes are derived from the same species as summer squashes, *Cucurbita pepo,* and tending to them is a lot like caring for summer squashes, except you'll need more patience. While summer squashes produce fruit in 40–50 days, winter squashes need at least double that time. The name doesn't mean they like winter weather; they don't. But the long growing season means they're harvested at the end of the season, and their thick skins allow them to be stored through much of the winter without spoiling.

Details: Sandy loam with lots of organic matter, substantial nitrogen, with a pH range of 5.5–6.8. Germination: best at 70°F–85°F (21°C–29°C), and plants grow best when soil remains above 70°F (21°C).

Ease of Growing: Easy.

Varieties: Winter squashes come in shapes round and elongated, scalloped and pear-shaped and have flesh that ranges from golden yellow to brilliant orange. Here are some of my favorites.

Acorn Squash. These medium-sized, yellow-fleshed squashes have tough rinds and are usually a dark shade of green with blotches of orange.

Blue Hubbard Squash. Large growing, Blue Hubbard squashes are another species, *C. maxima,* which includes pumpkins, Banana Squash, and Buttercup squashes. Hubbards and other *C. maxima* varieties do produce edible fruit, but they're also highly susceptible to insect damage. Their softer stems are an open invitation to squash vine borers, making them difficult to grow to maturity without pesticides. They also have a strong aroma that attracts cucumber beetles. In fact, many growers use *C. maxima* squash varieties as decoy

Delicata squash isn't just a delight to look at. It's one of the best-tasting winter squashes on the planet—with sweet flesh and a diminutive, one-serving size.

and trap crops to lure pests away from cucumbers, melons, and other squashes. When the Blue Hubbard vines are covered with pests, spray them with PyGanic, an organic contact pesticide, to keep the area population under control.

Butternut Squash. C. moschata is a customer favorite with its smooth beige skin, deep orange flesh, and a small seed cavity at one end. Waltham Butternut has long, trailing vines, while Burpee's Butterbush is more compact. Both acorn and butternut squashes have tough stems that resist squash vine borers.

Delicata Squash. Also called sweet potato squash and sometimes peanut squash, Delicata Squash is smallish (about the size of a small submarine sandwich) and has beige skin with dark longitudinal stripes. It's a good keeper and very flavorful.

Spaghetti Squash. Becoming more popular with health-conscious food lovers, Spaghetti Squash can serve as a substitute for pasta. The cooked fruit separates into spaghetti-like strings when scraped from the shell with a fork. Produced on a vigorous trailing vine, the football-sized fruit is pale yellow.

Companions: Winter squashes do well with pole beans, sweet corn, and radishes but not with potatoes, tomatoes, or eggplant.

How to Grow: Despite the long growing season, plants can't be placed outdoors until the ground has been thoroughly warmed. When daytime temperatures hover in the 70s and all frost danger has passed, it's safe to sow seeds in the warm earth. Northern growers may want to plant squashes through black plastic mulch, which warms the soil and helps retain moisture. I always start squash seeds indoors in individual 1-cup (240 ml) containers. When the first true leaves appear, I set the transplants into the soil, taking care not to disturb the root ball. Transplant on a cloudy day or in the evening, and give the plants a drink with liquid fish and seaweed emulsion to get them started.

Squashes that bear large fruit on trailing vines can be planted at the edges of your growing area and then trained to grow away from other plants. (Be aware that they will naturally grow southward, toward the sun.)

Pests: Young plants are often attacked by cucumber beetles, which can weaken and even kill the seedlings. Brown squash bugs also attack winter squashes, sucking plant juices and weakening the plant—especially during hot summer weather. Catch adult beetles in early summer, and drop them into a can of soapy water. Also crush egg clusters on the undersides of leaves.

Diseases: Fungal diseases can be a problem, especially during wet summers. Powdery mildew invariably arrives at the end of the season, regardless of the weather, but it usually doesn't kill the vines until the harvest is almost finished. Practice good hygiene and rotation, and make sure the soil has all the essential nutrients to help plants resist diseases.

Challenges: Space and time are the biggest obstacles to growing many winter squashes, especially for northern growers. Make sure you plant them in ample time to harvest before the first killing frost; otherwise, they're easy to grow.

Harvesting: Pick winter squashes when the fruit is ripe, and the vines begin to wilt. Always leave 1 in. (2.5 cm) of stem attached to the fruit; it will help the fruit keep longer in storage. Some winter squashes taste better if allowed to cure off the vine for a day or two in the field, but don't let them stay outdoors in a hard freeze.

Marketing Tips: Winter squashes are staples at autumn farmers' markets. For CSA members, they are a reason to keep people signed up for an optional late-season share. Choose varieties that produce medium- to small-sized fruit, and offer recipes.

Sweet Potatoes *Ipomoea batatas*

Basics: Sweet potatoes are a traditional southern crop, but the morning glory relatives perform remarkably well in northern climates, as long as the growing season is long enough (over 120 days). Even though sweet potatoes are sometimes called yams, the root vegetables grown and sold in North America are all sweet potatoes; yams are grown in tropical climates.

Sweet potatoes are easy to grow but take some effort to cure. They need warmth and humidity—not refrigeration.

A single plant will produce pounds of delicious sweet potatoes. Dig them up before a hard frost, which can damage the cold-sensitive roots and make them harder to cure.

Details: Plants like full sun, steady moisture and loose, sandy loam, which drains well and allows roots to swell. Ideal soil pH is around 6.5.

Ease of Growing: Moderate.

Varieties: Centennial, Georgia Jet, and Jewell (all 100 days) are easy-to-grow, orange-flesh varieties for northern growers. White Yam (90 days), which isn't really a yam, has light brown skin and white flesh. Violetta (120 days) has attractive purple skin and white flesh. All Purple (120 days) has purple skin and flesh and is a little more cold tolerant than other sweet potatoes.

Companions: Sweet potatoes will spread wide and thick once the heat of summer arrives, making it impractical to plant companions.

How to Grow: Wait until at least 2 weeks after your last frost, until the soil has warmed to over 65°F (18°C) before planting sweet potatoes. Order disease-free young plants called "slips" from a supplier with arrival timed for your planting area. If the soil is cool, a raised bed covered with black plastic mulch can warm faster than level, uncovered soil. It's easy to plant slips through slits cut in the plastic. The plastic mulch will also keep weeds from getting an edge on young vines. Plant slips about 3 in. (7.6 cm) deep and 10 in. (25 cm) apart. Keep the plants evenly watered throughout the growing season to ensure quality roots, which can crack or stay thin from inconsistent watering.

Pests: Large farmers may have to contend with insect pests, but for most gardeners and small-scale growers, the plants are relatively pest free. Nematodes can damage roots, but Centennial and Jewell are resistant varieties. Wireworms can burrow into roots, causing obvious damage; mice and voles can nibble roots, making them unmarketable. Deer will eat the foliage to the ground if they're hungry. Use fencing if deer are a persistent problem.

Diseases: In general, diseases can be avoided by ensuring healthy organic soil, giving plants adequate space, and keeping them evenly watered throughout the growing season (at least 1 in./2.5 cm per week).

Challenges: Several different fungal diseases and blights can affect foliage and roots of sweet potato plants.

Harvesting: Pick sweet potatoes before the first frost, since cold temperatures can damage roots and make them harder to cure. After harvesting, you'll need to cure the roots to help them convert starches to sugar. For the best flavor and quality, roots should be stored for 1 week at 85°F (29°C), with 90 percent humidity. This is difficult for a small-scale grower to do, but good results can be achieved if plants are stored inside a closed greenhouse. For smaller crops, try storing roots in a cooler kept inside a warm room (over 75°F/24°C) for 2 weeks. A seed germination mat can help increase the temperature.

Marketing Tips: Cured roots that are stored at 55°F (13°C) can last for up to 6 months. Do not refrigerate them. Growers may want to keep their harvest in a warm spot and sell at a late-fall market.

Swiss Chard *Beta vulgaris*

Basics: Swiss chard is such a versatile and consistent vegetable that it should be declared mandatory growing for anyone with a CSA or a regular table at a local farmers' market. One planting in early spring provides a constant harvest until well after the first frost, with little fuss. On the table, it's a summer alternative to spinach and a healthy green for stir-fries, salads, or steaming. It's adaptable to many recipes, including quiches. My favorite is Swiss chard golombkes (cabbage rolls), made with big multicolored chard leaves instead of cabbage.

Details: Loose, well-drained soil, with a pH above 6.0. Soil temperature: above 50°F (10°C). Seed germinates slowly, in 7–10 days or more, especially in cold soil. Seeds are viable for up to 4 years.

Ease of Growing: Easy

Varieties: Fordhook Giant is the standard green variety, but other good chards with green leaves and white ribs are Argentata (also called Bionda á Costa) and Lucullus. Rhubarb

Chard has green leaves with deep burgundy-red stems and veins. Golden Sunrise has bright yellow stems with green leaves. Orange Fantasia and Orange Chiffon have vibrant orange stems with green leaves. For multiple colors, try Bright Lights, also called Rainbow Chard, which is actually a mix of several chards with different colors.

Companions: Beans, Brassicas, and onions. Avoid cucumbers, muskmelons, and squashes.

How to Grow: Sow seeds directly, ½ in. (12.5 mm) deep in spring. Keep the bed weeded and well watered until the seedlings are strong. Because chard seeds are actually clusters of seeds, what seems like a single seed will often produce multiple plants. Thin to 1 plant every 4 in. (10 cm). As the plants grow larger, thin to 10–12 in. (25–30 cm) apart and use the thinnings for baby chard.

Pests: Chard is relatively pest-free, but beet leaf miners can destroy foliage, making the leaves unmarketable. Use floating row covers in the spring, spray with Spinosad, and hang blue sticky traps over young plants. (*See also* Beets for more details on leaf miners.) During wet summers, slugs and earwigs can also be a problem, hiding deep in the new growth by day, then emerging at night to chew holes in the leaves. Tender growth with tiny holes will look more like Swiss cheese than Swiss chard once the leaves increase in size. Spray foliage with Spinosad or set out slug traps baited with beer. Slug traps will also catch earwigs if you bait them with olive oil.

Diseases: Damping off can cause young seedlings to wilt and disappear, especially in cool, overly moist soil. Keep the seedlings moist but not soggy. Fungal diseases can damage foliage in warm, excessively wet weather, especially on plants damaged by leaf miners. Cercospora causes brown patches with a purplish ring around the edge. Prune and dispose of damaged leaves and ensure adequate spacing for airflow.

Challenges: Some chards can become harsh to the palate, especially in hot summer months. For people who are sensitive to low levels of oxalic acid, eating the vegetable raw can create a burning

Swiss chard is another organic staple—easy to grow, relatively pest free, and productive for the entire season. Mix green varieties with red, orange, pink, and yellow–stemmed chards.

sensation in the throat. Oxalic acid, the same poisonous substance found in rhubarb leaves, is also present in spinach and beet foliage but in smaller concentrations. To minimize harshness, keep chard mulched and well watered during the summer months and strip away the larger leaves. Advise customers to cook their chard when it is harvested in the summer months.

Harvesting: Young leaves can be cut with scissors or a knife. Chard can also be broken off at the stem by bending the leaves downward. If you harvest the lower leaves, the plant will keep growing and producing new leaves all season long. To prevent wilting, pick the leaves in the early morning or late evening and drop them directly into a bucket of cold water.

Marketing Tips: Bundle green leaves with multicolored varieties, either Bright Lights or your own favorite colors. When thinning Bright Lights chard, try to keep an even assortment of each color. To control the colors, buy individual varieties and sow them in rows, then mix after harvesting. Raw chard leaves make a delicious wrap for chicken or tuna salad.

Tomatillos *Physalis ixocarpa*

Basics: Extremely easy to grow and expensive to buy, these husk tomatoes are one of my favorite under-appreciated vegetables. They're great to skewer on the grill with eggplants, onions, and summer squashes, and they make the best salsa verde in either hemisphere.

Details: Full sun; loose, well-drained soil; tolerates almost any conditions. Soil pH is best in the range from 5.8–7.0. For fruiting, you'll need at least 2 plants.

Ease of Growing: Easy.

Varieties: Toma Verde (60 days) is the common green variety. Burpee's Tomatillo Gigante (90–100 days) produces larger green fruit (up to 4 in./10 cm across). Cisneros is also large fruiting, with purple and green shading. De Milpa (70 days) is an early-fruiting variety producing green fruit tinged with purple. Purple Tomatillo (65 days) is a deep, inky purple that grows darker when exposed to sunlight.

Companions: Similar to tomatoes and peppers.

Tomatillos grow like weeds, aren't bothered by many pests or diseases, and make terrific salsa verde (or in some cases purple salsa). Cisneros and Purple tomatillos turn darker when the fruit pops out of its paper husk and is exposed to the sun.

The season starts slowly for tomatillos, with small fruit on multiple branches. But once the harvest arrives, it keeps coming until the frost.

How to Grow: Start seeds indoors about 4 weeks before the last frost. Germination: Plants germinate quickly at indoor temperatures (70°F–85°F/21°C–29°C) and grow quickly without fussing. Harden for a week and transplant to a sunny spot when frost danger has passed. Tomatillo plants grow like giant umbrellas and will spread for several feet across if not contained. I grow mine in a wide, half-height tomato cage made from concrete mesh. Plants grow 2–3 ft. (60–90cm) tall, then branch out over the tops of the cages, helping to keep the fruit off the ground. Husks protect the fruit, but slugs and ants will damage any fruit that rests on soil.

Pests: Striped blister beetles sometimes chew holes in the leaves, but plants are so vigorous they don't seem to be bothered.

Diseases: Relatively disease free. Not harmed by tomato blights.

Challenges: The only challenge may be reining in the vigorous vines. Oh, and cleaning up the crop residue to keep the fallen fruit from reseeding next year.

Harvesting: Pick the fruit when the papery husk starts to crack open at the bottom. Leaving fruit exposed to the sun will darken the color in purple-hued varieties.

Marketing Tips: Sell Salsa Verde kits: Put in 10–15 tomatillos, 5 serrano chilies or jalapeño peppers, a big sweet onion, and a bunch of cilantro. (See page 323 for my favorite recipe.)

Tomatoes *Solanum lycopersicum*

Basics: America's favorite garden crop is also a hit for market growers. To put it bluntly, customers expect the best from your tomato patch, and you can't disappoint them. To get the top award for customer satisfaction, you'll need to grow a good assortment of tomatoes, including many high-demand heirlooms. Also try mixing up tomato colors and sizes. Green, yellow, black, marbled, and streaked varieties—cherry sized, salad sized, and beefsteak slicing size—add excitement to the harvest. Finally, you need to include an assortment of early-ripening and midseason tomatoes to complement those late-summer beefsteaks. Determinate tomatoes grow on vines that produce all their harvest at once, while indeterminate vines keep flowering and producing fruit all season long. (For more information see page 280)

Details: Full sun, fertile soil that drains well, with high phosphorous content and pH range of 5.5–7.5. Germination: 5–10 days. Soil temperature: 75°F–90°F (24°C–32°C). Extremely frost sensitive. Seeds are viable for up to 4 years.

Ease of Growing: Easy.

Varieties: Some good heirlooms: Black Krim (80 days); indeterminate; has deep mahogany-colored fruit, streaked with purple and red; flavor is about as good as a tomato gets.

Late summer brings a bounty of organic tomatoes—from sweet yellow-skinned varieties to deep mahogany heirlooms like Black Krim. This is the high point of your harvest.

Stupice—pronounced stu-PEECH-ka—(56 days) is a favorite early-season heirloom from Czechoslovakia with loads of golf-ball–sized salad tomatoes. Pruden's Purple (68 days) and Brandywine, (80 days) are two thin-skinned pinkish heirlooms with exceptional flavor. Juliet (62 days) is a superb hybrid cherry tomato with texture like a plum tomato; it's prolific and terrific for salads or salsas.

Companions: Basil may repel hornworms. Many other crops grow well with tomatoes, so it's easier to list what doesn't: cabbage, cauliflower, and kohlrabi. Also avoid potatoes, which share the same diseases.

How to Grow: Start tomatoes indoors about 6 weeks before your target date for outdoor transplanting. Use a lightweight seed-starting mix at temperatures between 75°F–90°F (24°C–32°C). A heated seedling mat works well. After plants germinate in 5–10 days, flood the seedlings with full-spectrum light for 14 hours per day and drop the temperature to 60°F–70°F (16°C–21°C).

I like to sow a dozen seeds of the same tomato cultivar in a single 4-in. (10 cm) pot, labeled by variety. I cover the seeds with ¼ in. (6 mm) of soil. Water thoroughly. After moving the pots to a light table for a week, the first true leaves become established (not the teardrop-shaped cotyledons). Once the plants have 2 sets of leaves, I water the seedlings thoroughly, then gently empty the container onto a tray and tease the seedlings apart. Holding the seedlings by the leaves—not the stems—I transplant them into individual 1-qt. (1 L) pots, burying the stems deeper to make shorter, stockier seedlings. I feed them weekly with liquid fish emulsion at half the recommended strength and return them to the light. I water the seedlings as soon as the soil starts to dry out, and I gently brush the tops of the plants with

my fingers. Lightly petting the plants bends the stems and keeps them from getting leggy.

About a week before transplanting time, I harden the seedlings in a cold frame, and I prepare the bed by mixing in composted manure, a granular organic fertilizer with ample phosphorus, and a fistful of rock dust with micronutrients.

To transplant, I dig deep holes and plant the seedlings at an angle, allowing the root ball to remain within the

Don't let the color fool you. Aunt Ruby's German Green tomatoes are one of the sweetest tomatoes I've ever grown. I never let a season go by without planting this family favorite.

topsoil, but keeping much of the stem underground. In a few days, the vine straightens out, and the plant is better for the deep, well-established roots. After transplanting, I give them one more feeding with fish emulsion and sprinkle a ring of diatomaceous earth around each seedling to repel cutworms.

Pests: Whiteflies can be a problem, especially on vines with lush, succulent growth—often caused by high-nitrogen fertilizers. Use balanced or higher-phosphate fertilizers for stronger, healthier growth. Trap whiteflies with yellow sticky traps placed between the vines.

The most damaging tomato pest might be the hornworm, a green-skinned caterpillar that grows as big as your thumb. If you see chewed leaves on the vines and pellets on the ground, look carefully on the undersides of leaves. Pluck worms off and drop them into a can of soapy water. If you find a hornworm that looks as if it has rice grains attached to its back, leave it alone. The rice grains are actually the cocoons of a parasitic Braconid wasp that preys on hornworms; the larvae eat the giant insects.

Mix a variety of red, yellow, green, and purple cherry tomatoes together. Sungold hybrid is a great-tasting yellow cherry tomato; Green Doctors is a prolific and sweet green cherry; and Black Cherry is a prodigious source of sweet purplish brown cherry–sized fruit.

Diseases: As the tomato vines grow larger, consider staking or caging the vines to keep them from sprawling on the ground. In either case, I pinch off the lower branches to increase air circulation and to prevent the leaves from touching the soil where fungal spores are located. Even with all of this TLC, the plants manage to get fungal diseases once the fruit starts swelling in late summer. To prevent the spread of diseases, I pinch off any yellowing leaves and toss them in the trash, taking care not to let them contact clean tomato foliage.

Fungal diseases can be checked to some degree by watering the roots, not the foliage, and with regular crop rotation. While most fungal diseases take a slow and steady toll on tomato vines, the presence of late blight, *Phytophthora infestans,* is a serious problem. Plants that turn gray and wilt in as little as 1 or 2 days must be pulled and destroyed (not composted). Unaffected plants can be sprayed with an approved organic fungicide. Biological fungicides contain living organisms that colonize tomato leaves and roots and help suppress the start of fungal diseases. Actinovate is one of several new choices available for controlling fungal diseases organically.

Challenges: Tomatoes that set fruit in cold weather can become cat faced. The fruit on the vines that don't receive adequate irrigation often split after a heavy rainfall. Tomato diseases can overtake plants in late summer and fall. Pruning off the lower leaves helps but don't prune foliage that shades fruit.

Harvesting: Don't pick tomatoes too early. Allowing them to ripen on the vine lets them reach their full flavor potential.

Marketing Tips: Mix colorful cherry tomatoes in a 1-pint (0.5 L) box or clamshell case. Sun Gold is a flavorful orange cherry tomato. Green Doctors is a green-when-ripe cherry tomato. Black Cherry, a blush mahogany cherry tomato, looks fantastic in a single container. Also mix colorful, flavorful heirlooms with proven hybrids. Sell large-sized beefsteak tomatoes with mozzarella cheese and basil for a ready-made appetizer.

Turnips *Brassica rapa*

Basics: These Brassicas are grown for their spicy, crisp roots, providing a late harvest for fall CSA shares or farmers' markets. Fast-growing salad turnips can be sown in the spring for an early summer harvest or in August for autumn picking. Larger turnips and rutabagas have longer growing seasons and are sown in the spring for an early fall harvest or late spring and early summer for deep into fall. Turnip tops are edible, too, including varieties with small woody roots that are specifically grown for their tasty foliage. All taste better when pulled fresh from the cool earth in the spring or late fall before the ground freezes.

Details: Full sun, loose, fertile soil, with a pH of 5.5–7.0. Germination: 3–10 days at 45°F–70°F 7°C–21°C). Seeds are viable for up to 4 years.

Ease of Growing: Easy to moderate.

Varieties: Fast-growing Hakurei salad turnips (38 days) have white globular roots that are better in quality when harvested at less than 2 in. (5 cm)in diameter. Another salad turnip, Scarlet Queen (43 days), has red stems and smooth, round roots in an unearthly shade of magenta. Purple-top White Globe turnips (50 days) reach 3–4 in. (8–10 cm) across and can be eaten crisp and raw, even though most are grown for cooking. Long-season Gilfeather Turnip (85 days) is actually an heirloom rutabaga with roots that reach softball

Hakurei salad turnips are sweet and fast growing. Plant them in early spring for harvesting in about a month. That's almost as fast as radishes.

size; the flavor is sweet and mild, especially when picked after a frost.

Companions: Peas add nitrogen and improve flavor, and alliums repel pests. Avoid potatoes and other root crops.

How to Grow: Sow salad turnip seeds directly in the spring as soon as the soil is warm. Space 1 in. (2.5 cm) apart and cover with ¼ in. (6 mm) of soil. Thin to 2 in. (5 cm) apart as the plants grow. Keep young turnips weeded and well watered. Fertilization isn't likely needed if the soil is loose and friable with balanced nutrients. Later-season turnips and rutabagas can be sown in late spring and early summer for an autumn harvest; just count backwards from your desired harvesttime and check that number with the seed packets.

Give purple-top turnips space for their roots to swell. Plant them in summer for a harvest in the fall.

Pests: Flea beetles and root maggots can be controlled with floating row covers. Wireworms can also damage roots. Destroy them by hand if you find any while planting or tilling. Woodchucks can also be pests, ruining crops at any stage; keep them away from the growing area with fencing.

Diseases: Leaf spot, alternaria, white spot, and other fungal diseases can cause problems, especially in overly damp conditions. Practice good rotation, use cover crops and organic material to encourage beneficial microorganisms, and avoid planting in soils where clubroot has been present.

Challenges: Relatively problem free, but insect damage and overcrowding can be problems. Thin plants properly and use thinnings for greens.

Harvesting: If the soil is friable, pull from close to the ground. If the soil is packed, loosen the soil with a fork first; root tops are usually exposed. Cut stems and keep cool and moist.

Marketing Tips: If the tops are attractive, leave them on the turnips for market. Educate your customers about the benefits of eating turnip greens. The exception is big winter turnips, which should be trimmed at stem and root ends. Rinse but don't scrub the roots before sale. Bundle red and white salad turnips and their handsome greens together for spring farmers' markets.

Perennial Vegetables

Some crops return year after year once established, eliminating the need for starting seeds and hardening and transplanting. . . well you get the picture.

Horseradish *Armoracia rusticana*

If you've ever had the experience of making your own horseradish, you can appreciate why a small jar of the ground root is so expensive. Even with a food processor on an outdoor table, the aroma will bring tears to your eyes. Less teary, however, is the ease of growing this popular root-condiment.

Order horseradish roots from a supplier or pick some up from a fellow grower. Plant them in the spring. Once growing, the plants develop large leathery green leaves that can be added to summer pickles to impart both spice and crunchiness. After midsummer pickle season has passed, pluck off all but a few of the leaves; doing this will encourage heavy root growth. Then dig the roots any time after the frost. You can even let the roots overwinter and harvest them in the spring. Dig up as much of the root as you can without breaking it, then wash it and sell it whole or make your own horseradish sauce for a value-added crop.

Whatever pieces of root you leave behind will sprout into a new plant.

Horseradish roots are a onetime investment that pays you back every year, but if you're not careful, horseradish can become invasive. Plant roots inside 5–gal. (19 L) pails sunk into the ground. Cut the bottoms out of the pails first.

The vigor of horseradish mandates this specific warning: Don't let it run wild in your beds or you will have a difficult time eradicating it. For greater control, plant the roots inside a deep plastic bucket with the bottom removed. (I grow mine in the spaces within the cinder blocks that support my raised beds. When it's time to harvest them, I pull away the cinder blocks and get clear access to the spicy roots.)

Rhubarb *Rheum rhabarbarum*

One of the tastiest spring treats I remember from my childhood is a fresh stalk of rhubarb, moistened and dipped into a cup of sugar. The thought of that intensely sour flavor still gives me cramps in the hinges of my jaw.

Rhubarb is one of the easiest crops to grow, and it comes back year after year, producing thick stalks for sauces, pies, and desserts. The leaves of rhubarb, however, contain high concentrations of toxic oxalic acid and should not be eaten. For a small-scale farm, rhubarb provides an early item for sale, coming to harvest in the late spring—about the same time as asparagus.

Rhubarb makes an easy and profitable addition to a small-scale growing operation. It's an edible perennial, meaning you plant it once and it returns year after year until you decide to divide it and pass the pleasure on to another gardener or just move a division to another spot, increasing your production. Because rhubarb is a perennial, it can remain in the same spot for a number of years. Finally, rhubarb has no serious pests, making it ideal for organic production.

Rhubarb is an ideal organic crop, with few pests. Divide plants every few years to increase vigor and to add more stock.

Rhubarb likes full sun and a cool climate, growing best in climates where the ground freezes solid for at least part of the winter. It tolerates a slightly acidic soil (a pH of 5.5 to 6.8) and thrives with minimal attention once planted.

Like lots of vegetable plants, rhubarb can be purchased in pots at a garden center. But it's quite expensive. Mail-order suppliers will often have better prices and varieties, but there's an even easier way still to obtain rhubarb. Most of my plants were started from root divisions passed along by a gardening friend. The plant is easy to divide and produces more consistent harvests if it gets divided every 4 to 5 years, so most growers who have been at it for a few years will be willing to part with some of their divisions—maybe for the promise of a dozen eggs, a jar of honey, or a nice, fat Brandywine tomato.

Sunchokes
Helianthus tuberosus

These sunflower relatives are often called Jerusalem artichokes, although they're unrelated to artichokes. When eaten raw, they taste crisp and nutty, like water chestnuts. When cooked, they take on an earthy, mushroom-like flavor.

Plant knobby white or smooth-and-slender red tubers in the spring, and

Sunchokes, also called Jerusalem artichokes, produce knobby white or smooth maroon roots that can be cooked or eaten raw for a nutty addition to salads. The taste of raw sunchokes is like an earthy water chestnut.

harvest after the first frost. Two things to beware of: First, the plants are extremely invasive and will grow out of control if allowed to roam through loose soil. Rein them in with a deep barrier or plant them in a grassy area and mow down the plants that spread outward. Second (and how can I say this delicately?), eating them can cause extreme gas. You might want to supplement them with the anti-flatulent herb epazote or recommend to your customers that they stock up on Beano.

Herbs

Herbs serve triple duty in an organic setting. Sharp-scented oregano, mint, and basil can repel insect pests. Cilantro, parsley, and dill have flowers that lure beneficial insects. In a CSA or at a market garden, bunches of basil or other herbs are natural companions for fresh vegetables.

In general, herbs are low-maintenance plants that are more flavorful when grown in lean soils. Don't overfertilize with nitrogen, which promotes lush growth but can leave plants weak in flavor and prone to diseases. I grow herbs in my main garden, as well as in big whiskey-barrel–sized pots around my patio, using a biologically active easy-draining mix of compost, perlite, loam, and builders sand. The plants only get fertilized once—with a foliar application of fish and seaweed emulsion early in the season. I've never had problems with insects or diseases.

Each week my wife, Valerie, and I pick a mix of herbs for our CSA customers, bundling them with a rubber band. Below are a few of the ones I like best—and the ones my customers like most, too.

Basil *Ocimum basilicum*

Fresh basil is a necessity when beefsteak tomatoes ripen. Finely cut leaves with fresh mozzarella cheese and extra-virgin olive oil over slices of tomato make for an unmatched late-summer appetizer. Come to think of it, basil is welcome throughout the growing year—with sautéed zucchini and garlic, chiffonade-cut basil in salads, and so on.

Basics: Sow the first basil crop indoors in containers to get a jump on the season. Tiny flea-sized seeds germinate in less than a week and reach transplant size in about 5 weeks. Harden and plant outdoors when the soil has warmed to 60°F (16°C) or more and all danger of frost has passed. That's also when you can start sowing basil directly in the garden. Plant seeds 1 to 2 in. (2.5–5.0 cm) apart and thin to 10 in. (25 cm) apart as plants grow.

Your basil isn't organic if you buy seedlings from a non–organic supplier. Start it from certified organic seeds sown every 6 weeks for a fresh crop with tender leaves. Genovese Basil and Sweet Basil are two common green-leaf varieties, but Purple Ruffled Basil and Lettuce Leaf Basil are other good choices.

Easy Does It!

Varieties: There are more than 150 varieties of basil available to growers, with variations in appearance and flavor. My No. 1 choice is still the old-world Italian sweet basil Genovese (60–90 days). Plants are strong and reliable, and it has true-basil flavor. Purple Ruffles is another top choice, with inky-purple foliage that's occasionally streaked in green. I grow it in pots where it fills out the space beneath a single Lacinato kale. Purple Ruffles grows more slowly than Genovese and resists flowering throughout the summer. Mammoth basil and Lettuce Leaf basil are large-leaf varieties. One or two leaves makes a great substitute for lettuce on a summer cheeseburger made with grass-fed beef.

Pests: Basil has a reputation for repelling some garden pests (aphids, hornworms, and asparagus beetles), but it attracts others. Japanese beetles and Oriental chafers chew leaves and ruin the appearance of basil. Whiteflies and slugs can also be a problem. Vigilant hand-picking will curtail beetle damage. Whiteflies can be controlled with yellow sticky traps.

Harvesting: To keep basil fresh and leaves tender, harvest it harshly, cutting stems down to a lower leaf node, which will encourage new branching. If you wait to pinch off just the flower buds, it's harder to keep up with the plant's will to seed. Invariably, the basil will start to change, and leaves will grow tough and bitter. I still plant successive crops of Genovese basil as a guarantee that I'll have fresh, sweet-tasting leaves at the end of summer when tomatoes come into season. A new variety called Bam is supposed to produce all season long without going to flower. I haven't tried it yet as it has not yet been released by Burpee.

Cilantro *Coriandrum sativum*

Cilantro is another herb that's useful all season long. In late summer, it is a mandatory ingredient in fresh salsas, but it has other uses, too. Try it in salads, in sandwich roll-ups, with seafood, and in Thai and Indian recipes.

Basics: Sow seeds directly, as cilantro has a tap root and doesn't like to be transplanted. It loves full sun, but will do fine in partial shade, especially in midsummer. Open-pollinated varieties will self-seed. In Zone 6a, I get my first cilantro of the season by shaking a tall, dead coriander plant over a small section of 1 bed in late autumn, letting seeds fall onto the soil. I cover them with a light (½-in./12.5 mm) top-dressing of compost and let them overwinter. Plants germinate in early spring and are ready to pick by May. In a few mild years, some have germinated before winter and survived the cold.

Cilantro and coriander are the same plant; coriander seeds are produced after cilantro bolts and flowers. Once cilantro starts to produce a central stem, it will quickly grow to 3 ft. (91 cm) tall and blossom, then set seed. Allow a few plants to flower, as they attract beneficial insects. To keep a fresh supply of cilantro leaves, make successive plantings every 2 weeks in summer and every 3 weeks during cooler weather.

Varieties: Some varieties that are slower to bolt include Calypso (50–55 days), Slow-Bolt (45 days), and Confetti (28–35 days), which has delicate carrotlike foliage and a milder flavor.

Pests: Watch out for white grubs and fungus gnats.

Harvesting: Harvest cilantro by pulling up the whole plant (roots and all). Sell the plants with roots attached, as they are an ingredient in many Thai recipes (rahk pak chee). Wash roots thoroughly, then cut within an inch of the stem.

Cilantro is easy to grow from seeds. Successive sowings every few weeks will keep you supplied with this versatile herb throughout the growing season. Plan a larger crop for late summer when tomatoes and tomatillos are available to make salsa fresca.

Easy Does It!

One way to get an early cilantro crop is to start seeds indoors. Cilantro has a large taproot and doesn't like to be transplanted. But you can work around that issue. Start seeds under a light table in early spring, sowing 3 or 4 seeds in a small biodegradable pot made from compressed peat, manure, or shredded coconut husk (coir). You can also use homemade newspaper pots—anything that can be transplanted, pot and all, into the garden. Thin seedlings to 1 plant per pot, and transplant to the garden as soon as soil is workable. Small plants will transplant easier, and cilantro can survive a frost.

Fern Leaf and Dukat Leafy dill are two varieties that produce more foliage than seed heads. Time your harvest to coincide with the arrival of pickling cukes.

Dill *Anethum graveolens*

There's a reason they call it dill "weed." Annual dill tolerates myriad growing conditions and comes back every year if the seeds are allowed to develop and drop. I planted Long Island Mammoth about a dozen years ago, and it has come back every season since. I weed what I don't want, and I harvest others as they get in the way. But I always let a few plants flower and go to seed because they attract ladybugs, hoverflies, parasitic wasps, and other beneficial insects.

Dill is also handy in the kitchen. It's obviously a delight with cucumbers but can be used in soups, salads, and dips or in sauces with chicken and seafood. In companion plantings, dill does well with cabbage-family crops, lettuce, onions, squash, and cucumbers.

Basics: Sow dill in early spring, covering with ¼ in. (6 mm) of soil. Dill is drought tolerant once established, but it needs steady water while young. As dill grows, thin to about 6 in. (15.2 cm) apart and use the cuttings in your kitchen. Dill can also be grown indoors under lights and transplanted in the spring when the soil is dry enough to be worked.

Varieties: For a steady harvest of desirable dill (foliage, not seeds), sow successive crops every 2–3 weeks, using a variety that produces optimum leafy growth—Fern Leaf and Dukat Leafy are two good choices.

Pests: Parsley worms are the only pests I've noticed on my dill, but hornworms like the delicate foliage, too. Both caterpillars can cause damage when feeding. Handpick if they become a problem.

Harvesting: Harvest when leaves are full and plants are about 12 in. (30 cm) tall—before the plant flowers. Once harvested, dill quickly wilts. Keep it in water until you use it or sell it.

Oregano *Origanum vulgare*

There are multiple varieties of oregano, some for ornamental use and others with a strong flavor that complements tomato sauce and is popular on pizza, in spaghetti sauces, and sprinkled atop garlic bread. Much of the oregano that is commonly grown, however, lacks the zip of true Greek oregano. One way to test the flavor of oregano is to bite into it. If it's mild and doesn't make you take notice, it may be the common variety that's also called Wild Marjoram.

Basics: Oregano grows freely from minuscule seeds or it can be grown from divisions. The growing medium is sandy loam. I don't fertilize the plants.

Varieties: I grow *O. vulgare hirtum,* which I bought as a potted plant from a local herb and vegetable farm. It has a strong flavor and constitution, thriving with minimum care.

Pests: Oregano is relatively pest-free, and it can even keep insects away from other plants. But if aphids or spider mites become a problem, blast them with a strong spray from the hose or use an organic soap spray.

Harvesting: I chop them to within 3 in. (8 cm) of the ground. They respond with a flush of new growth. I do let some flower, so that bees and beneficial insects can swarm to the tiny, white blossoms. One way to sell oregano is to divide a larger plant into multiple potted plants for sale.

Oregano tolerates neglect, thriving in lean, dry soil. When plants flower, cut them back for a second harvest of leaves.

Parsley *Petroselinum crispum*

Italian flat-leaf and curly parsley are both "must-grow" items. Despite its reputation for being hard to germinate, I've always had good luck with seeds sown under fluorescent lights. The parsley plant contains a chemical that inhibits weed germination, which can also slow its own growth. At any rate, germination is slow (about 3 weeks or longer). Some growers soak their parsley seeds overnight in warm water for better germination.

Curly parsley can be slow to germinate, but it's worth growing from seeds. If you buy seedlings from a non-organic supplier, the plants are not truly organic until their second year, which is when the leafy growth fades after sending up a tall flowering stalk.

Easy Does It!

One common parsley pest is the brightly colored black, yellow, and green parsley worm. It's easy to spot on your plants, and identification is confirmed if you touch the worm and it responds by displaying a pair of bright orange horns. If these insects are so numerous that they're destroying your crop, handpick them. If there are only a few, consider letting them be. They are the larvae of majestic black swallowtail butterflies.

When my daughters were young, we took some tiny caterpillars from our dill and kept them in a terrarium, feeding the insects with fresh dill from the backyard every day. The girls watched the caterpillars grow, molt, pupate, and then emerge as butterflies. When the butterflies' wings dried out, we released them to the wild.

Basics: Parsley is a biennial and will survive the winter in much of the United States and parts of southern Canada. The following spring it will send up a tall stalk with a wide, flat cluster of blossoms. If you have the space, let it bloom; the flowers attract beneficial insects.

Start plants indoors 6–8 weeks before your first frost. Sow about 20 seeds in a 4-in. (10 cm) pot and cover with ¼ in. (6 mm) of seed-starting mix. Keep soil moist but not soggy. When seedlings become established (showing true leaves), thin them to about 4 plants per pot. Feed with half-strength fish emulsion and keep them under light. When plants grow large enough to handle, sow them outdoors (they can take a little frost). Grow them in pots on the patio, in your herb garden, or interplant with other crops wherever there is space. I plant all four seedlings together, and they thicken into what looks like one bushy plant. Parsley is a good companion for onions, peas, peppers, and tomatoes.

Varieties: Some good varieties: Krausa and Titan are acclaimed curly-leaf parsleys; Giant of Italy is a full-flavored, vigorous flat-leaf variety.

Pests: Watch for green parsley worms; handpick to remove.

Harvesting: Pick parsley when plants are bushy and tall—at least 6–8 in. (15–20 cm). Keep plants vigorous by cutting less than one-third of the plant each time you pick.

Rosemary *Rosmarinus officinalis*

Basics: I would grow rosemary just for the pleasure of rubbing my hands across the foliage and savoring the aroma. The fact that it's a useful kitchen herb is a bonus. I envy growers in warmer climates who can keep rosemary all year long. In my Massachusetts plot, even the hardy Madeline Hill rosemary variety won't survive most New England winters. I've grown several plants and a few have survived one winter and died the next.

Details: Rosemary likes light soil that drains well and is slightly acidic, with a pH of 6.0–6.8. If you keep it indoors over the winter, keep it in a cool room and allow the soil to dry out between waterings.

Ease of Growing: Easy.

Varieties: The aforementioned Madeline Hill is a good upright variety for northern growers. Arp is another somewhat hardy variety with pale blue flowers.

How to Grow: It's easiest to start rosemary plants from cuttings or a potted plant bought from a local nursery. I buy a few potted plants and put them at the center of a large whiskey-barrel–sized pot, then surround them with lower-growing herbs. I also grow rosemary in the garden, but I like to keep my herbs close to the door where I can run out and take a few snips as I'm cooking.

Pests: Rosemary doesn't have many pest or disease problems outdoors, but when brought inside, it can be bothered by spider mites and fungal diseases.

Harvesting: Use scissors or a knife to cut stems from the plant, cutting the stems near the base of the plant or ones closer to the top.

If you can't find organic rosemary, take cuttings from any rosemary plant that you raised organically the previous year. Snip fresh sprigs about 4 in. (10 cm) long and strip leaves off the bottom 3 in. (8 cm). Insert the sprigs into a pot of light growing medium and keep them watered. In 3 weeks, gently tug on the sprigs. If they rooted, they shouldn't move. Carefully transplant them into separate pots. (Note: Don't use synthetic rooting hormone, which is prohibited for organic growers. Instead, soak a fresh willow branch in water and use that to irrigate the young herbs.)

Spearmint *Mentha spicata*

Spearmint is superb for tea, jellies, in salads, or as garnishes. I like to mix 2 cups (450 ml) of spearmint leaves with 1–2 cups (240–475 ml) of parsley leaves and only a small amount of bulgur wheat for a lightweight, fresh tabbouleh.

Basics: Perennial mint spreads aggressively, so you have to be careful about where you plant it. But it's such a useful herb that no garden or farm should be without it. Buying one potted

Spearmint is great for teas, jellies, salads, and, of course, tabbouleh. Grow it in an area where it is not likely to run into your other crops. Allow some plants to bloom and provide nectar for bees and other beneficial insects.

plant from a nursery is an easy way to get started because the single plant will quickly spread via shallow underground stolons. Another reason to buy plants: Mint that is grown from seed can produce unpredictable strains.

To keep mint from running wild, grow it inside a deep pot or bucket with the bottom cut out. Deep underground roots do not spread, but stems that touch the ground can take root.

Varieties: There are dozens of varieties and flavors, from lemon to peppermint-chocolate, but the most useful is common spearmint. Kentucky Colonel and Best Mint are two flavorful choices.

Pests: Keep an eye out for spider mites, cabbage loopers, and flea beetles.

Diseases: Mint is a pretty tough plant, but it can develop rust, leaf spot, and other diseases. Spots on leaves, unhealthy-looking foliage, and stunted growth all point to problems. Rather than spraying fungicides, look to cultural remedies. Provide adequate spacing for airflow, remove and dispose of spent plants (don't compost anything that looks diseased), and make sure your soil has all the essential nutrients. A soil test will tell you what you need to know.

Harvesting: Harvest mint when the leaves are soft and flavorful—before it goes to flower. Cutting back hard stimulates new growth for another harvest. I always allow some of my mint to flower because it provides nectar for my honeybees and attracts beneficial insects. I also grow some peppermint, which is sharply aromatic and makes great tea but is not as useful in the kitchen.

Thyme *Thymus vulgaris*

Another mint relative, thyme is a woody-stemmed, low-growing herb with tiny aromatic leaves. I grow thyme around the rim of several large potted herb gardens, allowing it to hang over the edges.

Thyme has multiple uses in too many recipes to count. Fresh thyme cuttings can be added to salads and soups, mixed with butter for an herbal spread, or steeped in vinegar for salad or tomatoes. Fresh or dried thyme complements lamb, pork, beef, seafood, eggs, chicken (and stuffing) and myriad vegetable dishes.

Basics: Thyme is easy to grow from potted plants and likes full sun and light, well-drained soil. Thyme can be grown from seeds or propagated from cuttings and root divisions.

Varieties: In addition to common thyme, Lemon thyme (*T.* x *citriodorus*) is a good culinary variety.

Pests: Ants like to build their nests in thyme beds and can disrupt the roots.

Harvesting: Thyme is a magnet for bees when in bloom, and while flavor is at its peak just before blossom time, I take cuttings for the kitchen throughout the season.

   To dry thyme, snip 4-in. (10 cm) branches at bloom time and pull the stems backward between your fingers to remove the tiny leaves. Spread them on a paper towel on a baking tray and leave them in a dry place. Store in zip-sealed bags or an airtight container.

Thyme is a delightfully aromatic perennial herb that is useful in too many dishes to count. It's also easy to grow. If you buy thyme from a non-organic supplier, you can't call your cuttings organic until it has been grown organically for at least a year.

Starting herbs in pots from seeds or cuttings ensures that you control the fertilizers. Use only organically approved materials if you want to call your fresh-cut herbs "organic."

Fruits of the Earth

6

Adding berries and fruit trees to an organic hobby farm can be a three-way success story. First, the additions will bring more to the overall farm diversity, attracting birds, bees, and beneficial insects; second, fruits and berries can provide additional income to supplement the earnings from vegetable sales; and third, berry bushes and fruit trees make sensible use of land that is sometimes unsuitable for traditional row crops.

The Fruits of Your Labor

Ever wonder why so many apple orchards are planted on rolling hillsides? One reason is that the terrain may be too steep and rocky for vegetable production. Beyond that, the gentle slopes of a hillside are well suited to growing fruit. Rolling hills allow cold air to flow downward, preventing subfreezing air masses from settling in and damaging blossoms with a late-spring freeze.

Despite the higher altitude of the hilly terrain, the temperature may actually be warmer along the hillside than on the flat land in the valley below. This is especially true on cool, calm nights when cold air flows like a slow-moving river into low-lying areas.

Organic hobby farmers can benefit greatly by adding fruits and berries to their operations. Even if there's room for only half-a-dozen dwarf fruit trees, the effort of planting, pruning, and watering them can pay off. If the trees are well tended, the harvest can be enough to keep your family stocked with fresh, healthy fruit in summer and jams or jellies in winter. They can also bring in extra income—either at farmers' markets or as an optional fruit share sold under a Community-Supported Agriculture (CSA) program.

Berries can work the same way. And if you have a small CSA with responsible members, you might consider starting a pick-your-own option for the blueberries, strawberries, blackberries, and raspberries.

Blueberries are a fitting crop for small–scale organic growers. As with other fruiting trees, shrubs, and perennials, organic growers need to purchase certified organic stock or grow the stock under an organic production regimen for at least one year before the harvest can be called organic. Because blueberry bushes take 3–5 years before producing a strong harvest, there's plenty of time for organic culture.

Berries

Sweet, juicy berries are simply delicious and some of the easiest plants to grow. From highbush blueberries to ever-bearing strawberries, fresh-picked berries make welcome additions to your market baskets.

Blueberries

In my naturally acidic New England soil, there isn't a better organic berry crop than highbush blueberries. With minimal care and effort, blueberries provide a bountiful harvest that is healthy and delicious, command a premium price, and are rarely plagued by insects and diseases if soil and space requirements are met. However, in an imperfect world, there are no perfect crops, and blueberries do have some drawbacks:

- A newly planted blueberry patch will take 4–5 years before it provides an appreciable harvest.

- When blueberries finally begin producing a steady crop, you'll need to protect the fruit from marauding birds. (Where do they all come from?)

- If your soil isn't sharply acidic, you'll have to adjust the pH by adding sulfur, peat moss, and rotted wood chips. In areas with poor drainage, naturally alkaline soils, or heavy clay soils, doing this may not be practical.

Those considerations aside, blueberries are a popular food choice for their high level of antioxidants and their relative small calorie load in proportion to their explosively sweet and occasionally tart flavor. All things considered, the drawbacks aren't that serious.

It's true that waiting 5 years for a productive harvest is a stumbling block for someone growing nothing but blueberries; however, if blueberries are part of an overall diverse farming operation, the one-time expense of buying bare-root bushes from a commercial supplier can be offset by earnings from the rest of the farm. And once the bushes start producing, they may continue for another 25 years. The bird damage—especially from noisy catbirds, which have an obnoxious habit of pecking triangular holes into every berry just as it begins to ripen—can be controlled. Cats, scarecrows, reflective Mylar strings, and balloons don't seem to help much in my blueberry patch, but an inexpensive, improvised frame made of 2 x 4s (38 mm x 89 mm) or PVC pipes, draped with 1-in. (2.5-cm) plastic netting will keep birds outside, where they can only eye the protected harvest.

The most important consideration for growing great organic blueberries is giving them the proper space and soil conditions. Soil should be loose and sandy with lots of organic material and low calcium. The soil must be substantially more acidic than the growing conditions in your vegetable patch. For instance, the ideal pH reading for most fruits and vegetables is nearly neutral, with a slight tendency toward acidic (roughly 5.8–6.8 on the pH scale, where 7.0 is neutral). Blueberry bushes thrive in soil that is significantly more acidic—

from 4.5–5.0 on the pH scale. All the water and fertilizers in the world won't help the plants grow bigger unless the pH is adjusted first. If conditions aren't right for blueberries, it might be wiser to choose another crop than to attempt dramatic changes in pH and drainage.

Northern highbush blueberries, which are the most flavorful, thin-skinned and widely grown blueberries in North America, thrive from the Carolinas and Arkansas north to the Canadian border. In the past, southern growers had fewer options, relying on rabbit-eye blueberries that are a different species with thicker skins and larger seeds. However, more recently, hybrids have been developed between southern blueberries and northern highbush berries, allowing greater variety for southern and southwestern growers. Before ordering plants, check with the cooperative extension office or an experienced blueberry grower in your geographical area about choosing the right varieties.

Getting Started

Select a spot that receives at least 6–8 hours of direct sunlight per day. More sunlight is even better. Blueberry bushes will grow tall in partial shade, but the bushes won't fruit as productively when sunlight is diminished. Planting bushes on a slope will not be a problem, but consider that you will someday want to get around your bushes to pick. Adult highbush plants can grow to more than 6 ft. (2 m) tall and 4 ft. (1.2 m) wide. Give them ample space, and allow room between rows. Space also provides air drainage, which lowers the risk of damage from cold and helps prevent fungal diseases. The best place to buy plants is from a trusted garden center or mail-order supplier. Catalog companies that specialize in blueberries will ship dormant, bare-root bushes that are 1–2 years old, timing the delivery to coincide with the optimum planting season in your area. For most growers, that's early spring—before other deciduous trees and shrubs have finished leafing out. In early spring, the soil is still moist and cool, and the sun is warm but not scorching, affording ample time for plants to get accustomed to their new surroundings without becoming shocked.

Another option is buying potted blueberry bushes that are sold at garden centers. These are usually more expensive than bare-root bushes because potted shrubs require large plastic pots with at least a gallon of growing medium. That makes them costly to ship and difficult to display and keep watered. But the advantage of potted bushes is they can be planted into midsummer if you keep the young transplants well watered.

Prepping the Site

Before planting blueberries, have your soil professionally tested or test it yourself with a pH kit, available at most garden centers. (See "Testing Your Soil" on page 87.) Chances are the soil is not acidic enough. To increase the acidity (or lower the pH number), add peat moss, wood chips, and sulfur—or any combination of the three. The most effective course of action is to add sulfur to the planting area, mixing it with partly composted wood chips and working the mix into the top 6 in. (15 cm) of the soil. Then use that same mix inside the holes before planting each bush.

My organic blueberries thrive on copious amounts of rotted wood chips. The best chips are not the clean, golden chips used for landscaping or playgrounds, but the dark-colored

Best Picks for Blueberries

To spread out the harvest, choose early, midseason, and late-fruiting blueberries. The earliest berries start setting fruit around the end of June in my Zone 6 berry patch, and the latest will continue setting fruit into mid-September. For best pollination, plant at least 3 varieties.

Varieties to Stretch the Season
Below are some popular varieties, arranged from early- to late-season.

Variety	Season	Comments
Earliblue	Very early season	Avoid growing in low-lying areas where late frosts can damage blossoms. Production is moderate with medium-to-large fruit that is sweet and light blue in color.
Duke	Early season	Good all-around variety that is very productive with large, firm, and mild-flavored berries that keep well. Bushes can lose vigor after multiple years if not well cared for.
Reka	Early season	Extremely vigorous variety that tolerates soggy soils. Good yields of flavorful deep-blue berries and good winter hardiness. A fairly new choice for American growers that was bred in New Zealand.
Northland	Early to midseason	Vigorous spreading shrubs with exceptional tolerance to cold climates. Consistently good producer of mid-sized berries that are deep blue and unusually sweet. Easy to grow.
Bluecrop	Midseason	Consistently strong producer under varying conditions. Fruit is light blue but sometimes ripens inconsistently with a few tart, purple-colored berries in the mix. It tolerates frost well and is easy to grow.
Chandler	Midseason	If you want a big berry with an extended picking season, Chandler is your bush. Yield is good, with flavorful light blue berries that are simply enormous. (You might see Chandler bushes advertised in garden catalogs with a U.S. quarter pictured nearby to show their extreme size. Not advised for areas with extremely cold winters.
Elliott	Late season	Longtime standard for late harvests, it has good production of mid-sized berries that are light blue in color and sweet/tart in flavor. Berries produce for about 4 weeks. Some maintenance is required in the form of pruning away old, unproductive growth to retain vigor.
Jersey	Late season	Widely grown berry yielding medium-sized deep blue berries with high sugar content. Firm fruit is good for freezing. Easy to grow and very productive.
Nelson	Late season	Strong production of large medium blue berries with firm texture and good flavor. Easy to grow.
Aurora	Very late season	Strong harvest of large medium blue fruit that is firm but occasionally sweet/tart, especially if picked too early. Bushes are vigorous and cold hardy. An improvement over Elliott in most areas.

Fruit and Geography

It's easy to set unrealistic goals when looking through the pages of a slick, well-produced fruit tree catalog. But let's face it, you can't grow coconuts in New Hampshire.

Seriously, while even beginners know enough not to attempt growing tropical plants in northern climates without the protection of a greenhouse, there are subtle reasons why some fruit trees might not produce in your geographical region. Sweet cherries planted in the Northeast can have trouble with insects and diseases, making them difficult to grow organically. Rhubarb doesn't produce well unless the ground freezes in winter, making them a challenge for southern growers.

So before buying grape vines, apricot trees, berry bushes, or any other fruiting tree or vine, check with a local cooperative extension service. Experienced staffers can lead you to the best options for your climate and soil conditions. Or look to local pick-your-own farms. What fruits and which varieties do they grow?

Joining a regional organic farming group can put you in contact with dozens of growers who will be willing to share their advice. Also look for seminars and workshops offered at universities or local farms, where seasoned pros will reveal sage advice on the best techniques and varieties for novices.

natural chips that contain bits of leaves and small branches. If you know of a tree surgeon or landscaper who cleans up after storm damage, make friends... and promise him some blueberries if he sells you his chips. They're even better if stored for a few months, allowing some of the chips to partially compost or decompose. Fresh chips take a long time to break down and can rob nutrients from the soil in the process. But rotted chips already contain living fungi, which secrete acid at the tips of microscopic hyphae. The acid helps break down the tough cellulose in wood chips and in the process helps to slowly lower the soil's pH.

Planting

When the soil tests close to 5.0, it's prime for planting. You can continue to bring the soil pH down while the bushes are growing. For a good head start, mix rotted wood chips, compost, peat moss, and a very small amount of low-nitrogen organic fertilizer into each planting hole, which should be dug slightly deeper than the root ball and about three times its width. Don't add fresh manure or chemical fertilizers, or you'll burn the young blueberry roots. If soil is excessively high in clay or silt, dig your holes a foot deeper and fill the bottoms with a mix of soil, sand, compost, and rotted wood chips before planting. Spread the roots out evenly in the hole and backfill with the mix of chips, soil, compost, and peat. For optimum growth, plant blueberries at their original depth, then mulch above the root zone with 2–3 in. (5–8 cm) of rotted wood chips.

Care doesn't stop when the bushes are planted. I like to surround the young plants with a ring of wire fencing to protect them from dogs, deer, and my occasional clumsiness. Also take care to prune off all flowers during the first and second seasons so the bushes will concentrate on growing taller. Don't try to speed that growth by feeding plants with high-nitrogen fertilizers. Fertilizer that is applied during the first 2 months after planting can stimulate the growth of nonproductive suckers. Also avoid fertilizing in late summer and fall since that can spur late-season growth that won't harden off before the winter freeze.

Blueberries can get by without heavy fertilizing, but they do need water two or three times per week. If you're planting a lot of blueberry bushes, you might consider drip irrigation. An irrigation drip tape run over the roots but buried under wood chips will provide a steady supply of water, ensuring that roots will not dry out. Blueberries like soil that drains well, but the roots are shallow and require constant moisture.

Pruning isn't complicated. For optimum growth, each bush should contain a dozen or fewer canes. Remove crowded branches from the interior of the plant and prune off the small lower branches that don't produce fruit. In general, newer growth is more productive, so when doing maintenance pruning, cut away older branches that are more than 2 in. (5 cm) in diameter.

Pests

Insects vary with geography, but the most serious pests for growers in the northern United States and Canada are blueberry maggots. Tiny ¼-in. (6-mm) flies with black-banded wings lay eggs, and larvae eat the pulp inside the fruit—1 worm per berry. Damaged berries may drop from the bushes, but otherwise, the pests are hard to detect. Pick up and remove any fallen fruit. Don't leave unpicked berries on the bush. To monitor for flies, hang yellow sticky traps baited with ammonium acetate (common fruit fly bait). Controls include *Opius melleus*, a parasitic wasp, or spraying with neem (azadirachtin) or Spinosad.

Cranberry fruitworm is another pest, which is easier to spot. Larvae feed on clusters of berries, eventually spinning webs around them. Pick and destroy damaged berry clusters to prevent pests from spreading. Organic controls include garlic spray, Spinosad, or neem.

Diseases

Numerous fungal diseases can damage blueberry bushes. Prune branches to encourage upright growth, thin interior branches to increase airflow, control weeds (which can harbor diseases and pests), and remove and destroy diseased plant material. Make sure the soil conditions are hospitable to blueberries before you plant bushes. Also avoid high-nitrogen fertilizers, which can cause lush top growth that is susceptible to insect and disease damage.

Nothing beats the flavor of freshly picked organic strawberries.

Strawberries

There's no substitute for the flavor of fresh-picked organic strawberries. In my raised beds, the berries often disappear while I'm weeding, watering, or trimming runners—before the birds even know the fruit is ripe.

Depending on the varieties of berries you choose, the harvest can come all at once in June or stretch throughout the summer and fall. June-bearing berries provide a single harvest in (you guessed it) June, while day-neutral and ever-bearing strawberries spread their harvest throughout the growing season. Both have different requirements for care and culture.

Shoot the Moon in June

If you like more berries than you know what to do with—all arriving at the same time—then choose June-bearing strawberries. True to their name, June-bearers produce 1 big crop of strawberries from late May until the end of June. Most growers agree that June-bearers are also among the largest and sweetest of the berries you can grow.

The best way to buy strawberry plants is from a mail-order supplier who will ship disease-free, dormant, bare-root plants—typically sold in bundles of 25. Ordering bare-root strawberry plants is much more economical than buying individual potted berry plants from a garden center. The plants should arrive in early spring, which is optimum planting time. If you can't plant right away, keep your bare-root plants cool and moist until you can plant them.

Prepping the Site

Prepare the soil by adding lots of compost or aged manure. Also work in about ¼ cup (59 ml) of balanced organic fertilizer per plant, and adjust the pH with lime or sulfur to a slightly acidic 5.8–6.5. The ideal strawberry bed should be well drained, with full sun. Raised beds work well because the soil drains faster and can be worked earlier in the spring, allowing more growing time for young strawberries.

Don't plant strawberries in areas where you've grown roses, blackberries, raspberries, tomatoes, potatoes, eggplants, peppers, melons, mint, or okra (whew!) for at least 5 years, because they share the same susceptibility to Verticillium wilt and other diseases that can affect strawberries. Make sure your bed is large enough to accommodate your crop. Plants should be spaced 12–18 in. (30–45 cm) apart in rows that are 3 ft. (1 m) apart.

Planting

When it's time to plant, get yourself organized. Do the job on a cool, cloudy day or in the early morning or late evening, when there's less danger that the sun will dry out the roots of exposed plants. I like to unbundle the berries and place them in a shallow pail of cool water, keeping the roots moist, but with the leaves above the water. I pull out the individual plants as I place them into the ground, first pruning off dead roots and foliage with a sharp pair of utility scissors.

My favorite planting tool for strawberries is a small mason's trowel, also called a pointing trowel. It's used for finishing work with brick and mortar, but the flat diagonal blade makes an ideal strawberry dibble. Jab it into the soil and wiggle it back and forth to create a deep, flat hole. Then spread the pruned roots into a fan shape and slide the plant into its hole so the roots remain spread. Backfill, firm the soil around the plant's crown, and move on. It's very important to check the depth when planting strawberries, ensuring that the soil level is at the midpoint of the crown—that firm, stemlike portion of the plant located above the roots but before the leaves branch out. Plants that are too shallow will dry out, while plants that are too deep can rot in the cold, moist soil.

A few months after planting, your strawberry plants will look much larger and greener and might even send out flowers to produce fruit. Sadly, you'll have to prune off all blossoms from June-bearing strawberries for the first season. This means that you won't get any luscious fruit until the plants are bigger and stronger in the following year.

After the first harvest, allow a couple weeks' rest, then mow down the entire strawberry bed with clippers, a trimmer or mower, but take care not to damage the plants' crowns. Then apply a generous dose of balanced organic fertilizer, along with a foliar feed of fish and seaweed emulsion. (For more information, see "The Importance of Organic Matter" on page 72.) The plants will respond with strong growth of the new crown, which will determine the size of the next year's harvest.

Depending on the variety, strawberry plants will send out numerous runners, with new plants at the end of each. The plants can take root and quickly overtake the strawberry patch, resulting in an overcrowded bed susceptible to diseases and one that quickly loses vigor, resulting in a loss of production. Even if you give strawberry plants all the TLC and space they need, they will still lose production after 2 harvests. At that point, you can either order more berries or plant some of the runners.

In addition to producing fruit, strawberry plants will send out runners. In a matted-row system, some runners are allowed to grow, and older plants are removed to keep the bed productive.

Traditional culture requires trimming away the runners that create new plants. However under the "matted row" system, which is favored by many gardeners and pick-your-own farmers, you can allow some runners to set. The new plants will produce a harvest in the following year. In the third year, you can remove the original plants to keep the bed growing.

Many growers are also using plastic culture for strawberries—growing berries with drip irrigation tape underneath sheets of black plastic. Plastic keeps soil moist and warm, blocks out weeds, and keeps berries clean. Drip tape irrigation is low-cost plastic tubing with regularly spaced slits in the tubing to slowly release water.

Strawberry Fruits Forever

For a smaller harvest that lasts throughout the year, choose day-neutral or ever-bearing varieties. Ever-bearers produce a small harvest in June, then a few random berries throughout the summer with a large crop in the fall. Day-neutral strawberries fruit continuously throughout the summer. The berries are smaller, but the plants will give you a harvest in the first year. Bolero, Calypso, and Fort Laramie are ever-bearers; Selva, Tribute, and Tristar are day-neutrals.

Care is a little different for ever-bearers and day-neutrals. For starters, you don't have to pick off all the flowers. Just remove the first set of blossoms that appear in June, and allow subsequent flowers to pollinate and set fruit. The bonus is that you can pick strawberries the first summer and fall after planting; the drawback is that most ever-bearing berries are not as big or productive as June-bearing plants.

To maintain the ever-bearing patch, remove excess runners and any leaves that are spotted or discolored. Mulch all strawberry plants—both June-bearers and ever-bearing varieties—with a layer of straw in the winter to prevent repeated freezing and thawing, which can heave plants out of the ground.

Pests and Diseases

Strawberries are loved by all, including numerous insect pests. Watching plants closely and eliminating a habitat that encourages pests will help keep them under control. It's a philosophy of organic growing and Integrated Pest Management.

Insect pests include strawberry clipper, a tiny weevil with larvae that sever fruiting buds from plants; tarnished plant bugs, which suck sap and weaken

Sun-sweetened strawberries, still warm from the garden, provide a tasty addition to farmers' market offerings.

plants; root weevils, which produce larvae that burrow into crowns; and mites and slugs. They can't always be avoided, but cultural practices will help. Grow healthy plants in good soil and provide adequate water. Don't overcrowd berries, and remove damaged plants. For many small-scale producers, slugs are the major enemy, eating the fruit and leaving neatly scooped areas on your berries. Catch them in a slug trap baited with beer.

Numerous fungal diseases can also damage strawberry foliage, including leaf spot, leaf scorch, and leaf blight, all of which cause brown patches on the foliage. Botrytis blight, which turns berries into fuzzy, gray blobs of mold, is also a fungal disease. Remove diseased plants and maintain beds after plants fruit. Move the strawberry bed to a new location after several seasons and avoid overcrowding. Check with university extension specialists in your growing area for solutions to local problems.

Blackberries, Raspberries, and Black Raspberries

Blackberries, raspberries, and black raspberries are grouped together under the umbrella name of brambles—unkempt and ornery bushes with sharp thorns and fast-spreading canes that can quickly turn into an impenetrable thicket if left to the will of nature.

But you'd be a fool to leave them unkempt and unattended because the reward for a few hours of simple pruning will be a bountiful harvest of sweet, succulent berries that are unmatched in appearance, texture, and flavor. Yes, brambles can be difficult to harvest, and, yes, the soft berries can turn to mush if they aren't handled with the utmost care. But the difference between fresh-picked mouthwatering berries and fresh store-bought berries—shipped in refrigerated trucks from some faraway place—is astounding. You'll love the locally grown alternatives, your children will love them, your customers will love them, and the world will be a happier and more peaceful place—albeit with a few purple stains on fingers and mouths.

As with most fruiting plants, brambles grow best in full sun. The ideal soil is deep, well-draining sandy loam with lots of organic matter and a pH reading from 5.6–6.5. Avoid planting in sites that are prone to flooding, unless you can make raised beds that keep the roots above the sogginess. Also avoid planting brambles where strawberries, roses, or any of the potato-family (Solanaceae) crops were grown in the last 5 years. Also avoid growing

Best Picks for Blackberries

One group of bramble varieties I'd recommend are thornless blackberries since it's much easier to pick berries and prune the plants without worrying about cuts and scratches across your forearms.

Chester is a disease-resistant thornless blackberry with good production and sweet berries that ripen in early August; Natchez is another thornless variety with large, elongated berries and an early harvest that lasts up to 5 weeks; Triple Crown is semi-erect and thornless, with large, flavorful fruit that ripens early in the season.

What Is the Difference between Blackberries and Raspberries?

Blackberries are generally larger and darker in color, but the real difference is the way the berries come off the stems when you pick them. Raspberries leave the core behind, creating a cone–shaped hollow space inside the berries. When you pick blackberries, the core comes off with the fruit, creating a little more pulp with the sweetness.

in areas near wild brambles, which can transmit diseases. (Some recommend removing all wild brambles within 500 ft. [153 m] of your cultivated raspberries and blackberries to avoid harmful diseases.)

There are numerous bramble varieties. Pore over some berry catalogs to determine whether you want golden raspberries, black raspberries, summer-bearing or fall-bearing raspberries, or blackberries that grow on stout canes and don't require trellising. Some varieties are less hardy than others, and sometimes the fussiest plants are invariably the ones with the best-tasting fruit.

Planting potted brambles from a nursery or garden center is an easy way to go, but selection is often limited and the prices high. (Remember that you're paying for the pot and the soil and the purported care that the young plants receive while waiting on the shelves.) Prices are lower and quality is generally higher if you order bare-root plants from a trusted nursery. But the planting window may be short for bare-root plants, so prepare the growing area before their arrival: Dig the bed, adjust the pH, and add compost but no fertilizers.

When the young brambles are delivered, remove them from the box and soak the roots in water for 2 hours before planting. Also trim off the old canes, which may contain diseases.

Planting

To plant brambles, place them in holes that are at least three times the width of the root mass and deep enough to cover the original soil line with another inch of soil. Other than planting them an

Sweet thornless blackberries make planting, pruning, and picking easier on the arms.

inch deeper, the biggest consideration is spacing. The exact space needed for each plant will vary with the cultivar, but 3 ft. (90 cm) between plants is a safe distance for blackberries; 2 ft. (60 cm) works for raspberries. That spacing should allow ample room for getting around mature plants to prune, pick, and inspect for pests and diseases.

Once plants are established, they can take some fertilizers. Feed them every spring (before blossoms arrive) with a balanced organic fertilizer. Don't bother scratching it into the soil because brambles have shallow roots that can be damaged by digging or mixing.

Keep the plants watered well, especially when the berries are setting and growing. To avoid diseases, target the root area and not the foliage. Also avoid sprinklers and watering in the evening, which can invite fungal diseases.

Black raspberries are sweet and prolific, but they contain more seeds than red raspberries.

Pruning

Pruning brambles is a lot easier than most people make it out to be. The biggest trick is knowing which variety is planted, because midsummer varieties are pruned differently from bramble varieties that produce berries in late summer and fall from first-year canes.

For most varieties that provide a harvest in July and early August, the berries grow on side branches that sprout from the previous year's canes. When pruning these varieties, simply cut away all the canes with lateral branches that have already produced fruit and are now old and dying. Leave the younger primocanes that will produce lateral fruiting branches in the following year.

Many newer varieties produce fruit on first-year wood. To prune these primocane fruiting varieties, just cut all the canes down to the ground every fall. (It really doesn't get much simpler than that.) You can also lightly prune canes that are excessively long on both varieties to keep them in control and to stimulate more fruiting branches.

Easy Does It!

You can grow bramble varieties with "erect" growth habits without a trellis; the canes are generally sturdier and the plants a little more stout and easy to control. Semi–erect plants can be grown along a fence or trellised to keep the canes neat. To prevent weeds, mulch young plants for the first year; don't mulch plants in subsequent years.

Fruit Trees

There's almost nothing more frightening than the memories of my first attempts to grow apples and pears. It was before I switched to organic practices, and I followed the recommendations of knowledgeable mentors who advised using chemical sprays every 10 days, beginning in midspring and stretching until late August when I managed to harvest a few dozen fruit. I think back to those hot summer days, waiting for the wind to subside so I could wander outside dressed like a hazardous waste worker in a nuclear power plant. Sweating and barely able to breathe, but still smelling (and probably inhaling) toxic mist, I sprayed some concoction every other week until the moment of truth—when I bit into the first fruit of my yearlong labors.

Yuck. Despite the pesticides, the apple was still riddled with tiny brown streaks, telltale trails of the ubiquitous apple maggot fly larvae that had managed to feast on my fruit between sprayings, rendering the entire effort futile.

I say there was ALMOST nothing more frightening. The only thing that's scarier than breathing in all those fumes is thinking about all the pesticides that had accumulated on the fruit that I would have eaten or passed along to family members.

There is only one sure way to feel safe about the fruit you grow—produce it organically. It will take more understanding and patience because problems must be recognized and dealt with at

Dwarf and semi-dwarf fruit trees are good options for home orchards and small-scale organic growers. The trees are compact enough to allow easy pruning, inspection, fruit cleanup, and spraying of organically approved materials. For best results, choose disease-resistant varieties that are suited to your geographical area.

an early stage. And while organic solutions aren't always as immediate as chemical solutions, they work out in the long run. Besides, a random dimple, bug, or brown spot is infinitely preferable to the potential hazard of working with and eventually ingesting harmful chemicals. Make the job easier by choosing varieties that resist diseases. Then use physical barriers, natural materials, and Mother Nature's own ingredients to fight pests and diseases.

Apples

Many organic growers will tell you that apples are the most difficult fruit to grow without pesticides. I only know that they're extremely difficult to grow WITH pesticides.

The problem is especially true in rural areas where there are many wild and uncared for apple trees. The pests thrive in the unchecked orchards, and it takes no time at all for them to find your future crop.

Planting

You can make the task of growing apples considerably easier if you stick with dwarf and semi-dwarf trees in varieties like Enterprise, Goldrush, Jonafree, and Liberty, which have been bred to resist common apple diseases. Dwarf trees and semi-dwarf trees also make it easier to prune branches, spray for insects (with organic controls), and harvest the fruit without a ladder. Finally, dwarf trees also take up less space and produce fruit in just a few years after planting. The drawback is that the roots of dwarf trees are not as vigorous as the root systems on standard-sized trees. That means they need more consistent watering and may require staking in high-wind areas.

Semi-dwarf trees have advantages and drawbacks of both standard and dwarf trees. They also grow taller, meaning some parts of the trees may be out of reach. But their rootstocks are more vigorous, anchoring them to the ground and spreading farther for water.

Controlling Pests

The best way to control apple pests is a tried-and-true organic method—physical barriers. You can't put a floating row cover over an entire apple tree, but you can spray the fruit with Surround, which is made from fine kaolin clay—a general-purpose insect barrier that makes fruit look powdery and unappetizing.

Another method is to set traps and use physical barriers. Red spheres can be coated with a gooey-gluey material, like Tangle-Trap. When fruit flies or apple maggot flies land on the nice red apple to lay their eggs, they get stuck in the goop and die. Most traps are used as a method for counting and monitoring insect populations, but a larger concentration of traps spread among a small number of trees can effectively minimize damage from insect pests.

You can also bag the fruit in a process similar to protecting Asian pears. Cut a hole or a slit into a small paper bag (the kind used for old-fashioned penny candy), pull the bag over the tiny apples and their supporting branch, and staple it closed. Remove the bags 2 weeks before picking to let the apples ripen.

Asian Pears

Asian pears, those sweet and juicy fruits that sell for such a premium price at the produce market are among the best fruit trees for organic production. The key to growing Asian pears is to make sure the fruit gets enough water throughout the growing season.

Plant Asian pears in full sun in rich, well-drained soil with a pH of 6.0–6.5. If planting a group of trees, space them at least 10–15 ft. (3–4.5 m) apart. To ensure pollination, plant 2 different varieties.

Each Asian pear variety has its own subtle flavor—some with a melonlike quality and others imparting spicy or even buttery tastes. Beyond the flavors, growers should choose pears that resist diseases.

Asian pears are sweet and delicious. Standard–sized trees grafted onto *Pyrus betulaefolia* rootstock resist fire blight and have deeper roots to draw in the copious water needed for large, juicy fruit.

Planting

Once you choose your varieties, plant bare-root trees in early spring, digging a wide and shallow bowl-shaped hole, loosening soil inside so roots can expand. Make sure the graft union—that lump where the fruiting part of the tree is grafted to its rootstock—remains above soil level. After backfilling, watering, and lightly tamping down the soil, mulch above the roots with compost or wood chips, but keep the material from piling up around the tree's trunk where it can invite insect pests and spread diseases.

Pruning

Trees can be pruned to either an open-center or central-leader growth habit, with the heaviest pruning done in early spring. (See "Pruning and Training Fruit Trees" on page 270.)

Asian pears usually begin setting fruit within 3 years from planting. Once they start, you'll have to hand-thin the prolific trees to keep crops under control. Some varieties set up to a dozen pears per spur; leaving them all on the tree will result in smaller fruit and biennial production. Pinch off all but the biggest unblemished fruit on each spur, leaving 6–8 in. (15–20 cm) between pears.

Asian pears are sold on several different rootstocks, but *Pyrus betulaefolia* is the most vigorous and results in larger fruit and longer-lasting trees. The only downside is that it produces full-sized trees, so homeowners will have to prune to keep the trees manageable.

Best Picks for Asian Pears

Shinseiki is a popular yellow-skinned pear, prized for its juicy white flesh and sweet flavor. It has moderate resistance to fire blight. Another popular pear is Hosui, with gold skin and sweet, white flesh. Ya-Li is delicious, but it sets blossoms very early in the season. That means it's a variety best suited for warmer climates where there's little danger of a late spring frost that could hamper fruit set.

Controlling Pests and Diseases

Asian pears are relatively disease free when compared to apples or stone fruit. But one problem they have is fire blight, which is usually a bigger headache for large-scale growers than for small-scale orchardists. Olympic is a variety that resists fire blight. Others that are somewhat susceptible include Chojuro, Niitaka, 20th Century (Nijisseiki), and Shinseiki.

Pests include aphids, plum curculios, codling moths, and pear psylla. A careful eye and well-timed spraying with an organic control should keep problems under control. Many growers use Surround, which is made from fine kaolin clay. The clay coats the fruit with an unappetizing dust that repels pests. It's good for plum curculios and other insects but must be applied shortly after blossoms drop and several other times throughout the growing season.

Another technique that controls all types of insect pests is bagging the fruit. There are several ways to bag fruit to create an insect-proof environment inside. One method is to purchase small paper sacks from a supplier and poke a 1-in. (2.5-cm) wide hole in the bottom of each bag. Slip the hole over an immature pear and staple the bag closed with the pear inside. The pear will grow, and the bag will provide enough space for it to reach maturity without harm from hungry insects.

You can also use sections cut from an old pair of panty hose or small slings fashioned out of floating row cover material. It may seem like a lot of effort to bag an entire tree full of Asian pears. . . and it is. But once the work is done, you don't have to go out there every 10 days to do it again.

Peaches and Nectarines

Luscious, mouthwatering, tree-ripened peaches—still warm from the sun when you bite through their fuzzy skins—are superior in every way to fruit that is trucked in from distant farms. It's a wonder that more small-scale farmers don't grow peaches and their clean-shaven sisters—nectarines—because they can add variety to both the landscape and a farm's income.

One of the biggest keys to growing peaches is the overall climate. If winter temperatures regularly drop below –10°F (–23°C), flower buds can be damaged and fruit production diminished. Even if temperatures don't get that cold in winter, peach blossoms can still be damaged by a late-spring frost or by the chilly air that hangs in low-lying spots. So don't plant peaches in a hollow. But that still leaves much of the country—from Georgia, the Peach State, northward to New England—as suitable land for peach production.

Pruning and Training Fruit Trees

Pruning and training are both necessary tasks for fruit trees, but they're not interchangeable terms. Pruning will remove excessive, non-productive growth and stimulate new growth, while correcting structure problems. It's done throughout a tree's life and at several times during the season.

Training is the specific type of pruning and spacing that orchardists perform on young trees to achieve a desirable growth plan—specifically one that creates a sturdy structure for heavy, fruit-bearing branches and opens up the tree to sunlight.

There are two basic methods of training fruit trees for production—one that produces conical or pyramidal growth via a central leader and another that encourages outward growth, like a vase or martini glass, by opening up the center of the tree.

Central-Leader Training

This method is generally used for many apple, cherry, pear, and plum trees, which naturally produce a strong main trunk. Growers select healthy branches that spread outward from the trunk in an evenly spaced pattern at varying heights and prune away other branches that crowd the main branches or scaffolding. The first main branch should be about 3 ft. (1 m) aboveground—minimum spacing to allow you to get underneath it once the tree is large. Prune away other nearby branches, then move up the trunk another 10–12 in. (25–30 cm) to select the next main branch growing in a different direction from the first. Prune away nearby branches and move up another 10–12 in. (25–30 cm) for the third main branch. Eventually, you should have about five main branches at different heights, all spreading out in different directions from the main trunk. If you viewed them from directly above, they would look like spokes on a wheel.

Once the shape has been created, prune out any branches that grow inward and crowd the main limbs. When you're pruning branches for length, cut just above an outward growing bud; this will force branches to grow outward and allow space for air and sunlight to penetrate the tree.

Open-Center Pruning

This method is best for peaches, nectarines, and some plums, which don't have strong main trunk growth. It involves choosing four or five main branches, somewhere between knee-high and waist high, then pruning off the entire top of the tree. The result is a tree that grows strong lateral branches and retains a shape like a martini glass.

Open-center pruning is also used on some apple and pear trees to keep them at a manageable height. But to keep the shape, you'll have to prune off fast-growing water shoots that grow straight upward and create weak, overcrowded conditions.

There are improvisations to both techniques, and there is also espalier—a specialty variation of central leader training. For espalier training, you save only the branches that grow in a single plane to the left or right of the main trunk, then train those branches at 90-degree angles from the main trunk and follow along a wall or trellis.

However you decide to train your fruit trees, look to cut out weak limbs and branches that grow in a tight V-shaped angle with the trunk. Tight "crotch angles" accumulate bark inside the V and grow into weak branches that can easily break off under a heavy load of fruit. If branches are still young and supple, they can be encouraged to grow more outward than upward by clipping a clothespin to the section of trunk just above the branch. Clip the clothespin so that it bends the small branch downward. You can also make or buy spacers—a short section of light wood with a notch cut into each end. Place the notch against the tree so it forces the branch outward. Or use gravity to help train the growth. Fill a sock with sand, tie the open end into a knot to keep the sand from running out, and drape it over a branch. The weight should be enough to pull the branch downward. (Just make sure there are no holes to let the sand drain out.) The ideal branch angle is about 60 degrees—about the angle that your relaxed thumb makes with your index finger when you place your hand flat onto a table.

Easy Does It!

Chill Hours

Many fruit tree suppliers include the term "chill hours" in their descriptions for peach trees and other fruits. It means the number of hours the tree needs in dormancy to set fruit. Chill hour calculations can vary, with some models counting all hours below 45°F (7°C); others count only the hours between 32°F (0°C) (freezing) and 45°F (7°C). Check with your cooperative extension service to find out the number of chill hours in your area. (It's not a matter of simply multiplying the number of cold days by 24 hours, because on many days the temperature will be above 45°F [7°C] during daylight hours and below 45°F [7°C] at night.) In general, high chill hours are best for northern growers and lower chill hours better suited for southern climates.

Prepping the Site

Peach trees like soil that is slightly acidic, from 6.0–6.8 on the pH scale. And for best fruiting and overall health, they require ample nitrogen, phosphorus, potassium, and micronutrients. For best performance, get your soil tested and add the appropriate fertilizers to the entire growing area, not the planting holes.

Planting

Most suppliers ship bare-root trees for planting in the spring, as soon as the soil can be worked. Dig your planting hole about double the width of your root mass and deep enough so the tree can be planted at the same depth it was grown. Trim off broken or damaged roots. There's usually a visible bump where the tree was grafted, and that "graft union" should be facing north or northeast to prevent sunburning. The union should also be about 3–4 in. (8–10 cm) above the soil level. (After you mulch the tree and the soil settles a little, the union will be only an inch or two above the soil, which is perfect.)

If you're planting multiple trees, allow adequate spacing, depending on the variety and size—usually 10–15 ft. (3–4.5 m). After planting, prune the trees to encourage the best and most sensible

Peaches aren't just for Georgia. With a little TLC, they can be grown in northern areas, too.

Prune peach trees to create an open center, which improves airflow and helps prevent fungal diseases. It also makes picking the fruit easier.

growth. Peaches don't have a strong central leader and are generally trained to an open-center or vase-shape, which invites light and air into the tree, reducing the likelihood of disease. (See "Pruning and Training Fruit Trees" on page 270 for more information.)

Give your peach trees steady watering for the first year and mulch above the young tree's root zone to discourage weeds. Don't allow the mulch to pile up in a mound against the bark, or you can invite boring insects and fungal disease. It's a good idea to clear the mulch completely away from the tree's trunk before winter arrives, as it helps the tree harden and makes the area less hospitable to rodents. If the ground in your area is covered in snow during winter, you should also protect the bark from mice and voles by installing a trunk wrap.

Pruning

Peaches grow quickly and reach fruiting size in just a few years. They also produce multiple branches. You'll need to prune off much of the wood every year. As a rule, cut back all new growth that points inward (toward the open center of the tree). Don't be afraid to chop away; many growers remove up to 25–40 percent of the wood each year.

The best time of year for pruning is late winter or early spring, before the flower buds have opened. The flowers will help you to distinguish fruiting branches, which you'll want

There are multiple varieties of freestone and clingstone peaches that perform best in different climates. In general, freestone peaches—which break away easily from the center stones—are preferable for fresh eating, although clingstone peaches tend to be juicier. Some freestone choices include: Red Haven, an early variety with red and yellow fruit and nearly fuzz-free skin; Red Globe, which has red-and-yellow fruit on prolific trees; Suncrest, with mostly red fruit and cold-hardy trees; and Reliance, with its firm, yellow flesh and good cold hardiness.

Popular clingstone varieties include: Rich Lady, with deep red fruit, limited fuzz, and excellent flavor and Arctic Supreme, a white-fleshed peach with great flavor and good cold tolerance.

to keep until they set fruit. Once peaches have set fruit, you'll have to handpick some of the immature peaches when they're about the size of marbles. Thinning fruit to about 1 peach every 6 in. (15 cm) helps the remaining fruit grow larger. It's also a chance to get rid of peaches that have telltale C-shaped scars caused by plum curculios (see below).

Pests and Diseases

Insect pests like to eat peaches. (Can you blame them?) Plum curculios attack very young fruit leaving a C-shaped scar. The tiny, weevil-like insects can also be slowed down by planting away from other trees that provide winter shelter—in other words, not at the edge of woodland. Oriental fruit moth can damage fruit, and peach tree borers can damage the entire tree.

To stop most insects, remove fallen fruit and discard (not compost) the residue before insect larvae inside the fruit reach adult stage. Use Mylar balloons or a plastic owl to scare away birds and squirrels, but to keep them effective, move the decoys around every few days.

In the southern New England plot where I grow peaches and nectarines (and in most of the Eastern United States), black spot is extremely common. It looks like freckles on the fruit and leaves and mars the appearance of fruit and can weaken trees, which invites further problems. Black spot is caused by bacteria that favor warm and moist conditions. Keep trees pruned to improve airflow and choose resistant varieties like Cresthaven, Earliglo, Harbelle, Harbinger, and Redkist. In general, early varieties are less susceptible to bacterial spot. Brown rot is a fungal disease that may be even more devastating. It attacks fruit in the final weeks before harvest, quickly turning the entire fruit mushy and grayish brown, with white fuzz. Organic fungicides applied at fruit set and again just before harvest may help prevent the disease, along with regular pruning to improve air circulation. Good hygiene is the best prevention for brown rot. Remove all mummified fruit clinging to branches to prevent spores from infecting next year's crop. Also pick up and dispose of dead fruit beneath the trees.

Natural Remedies

Fight disease with preventive measures first—use resistant varieties, prune well, and keep soil fertile and well irrigated. If diseases still strike, use natural methods. Follow label directions carefully and check with your certifier.

Material	Use	Comments
Bicarbonate Mixes	Formulas of baking soda (sodium bicarbonate) with water and horticultural oil or a few drops of liquid soap can be effective on fungal diseases.	Note that accumulated sodium can also harm plants. Potassium bicarbonate, sold as Bi-Carb, is an alternative.
Bordeaux Mix	Lime (calcium hydroxide) and copper sulfate mix is a longtime natural control that works on anthracnose, bacterial leaf spot, fire blight, canker, and downy mildew.	A recipe can be found at the University of California, Davis, IPM website: ipm.ucdavis.edu/PMG/PESTNOTES/pn7481.html
Copper	Some treatments with copper are permitted in organic farming; they work on a wide range of fungal diseases but also kill bacteria.	Repeated copper use can leave harmful residue in the soil; it can also harm plants.
Horticultural Oil	Some applications of mineral oil are allowed in organic operations; on fruit trees, dormant oil is used to smother scale and insect eggs. Horticultural oil can also prevent fungal diseases on foliage.	Read instructions carefully because horticultural oil can cause phototoxicity in hot and humid conditions.
Neem Oil	Extract of neem seeds is used as both a fungicide and insecticide; it stops powdery mildew, along with the aphids and white flies that spread it.	———
Sulfur (liquid or wettable powder)	Use for powdery mildew and other fungal diseases; it doesn't kill an established fungus but prevents spores from colonizing.	Avoid using when temperature is above 80°F (27°C). Don't use with insecticides because it can injure plants in direct sun (phototoxicity). Also avoid using within 30 days of horticultural oil. Don't use at all on apricots, raspberries, or squash–family plants

"When chickens get to live like chickens, they'll taste like chickens, too."

—Michael Pollan, *The Omnivore's Dilemma: A Natural History of Four Meals*

Maybe it's the way they sneak up and peck at your bootlaces. Or perhaps it's how they charge en masse when you're toting a bundle of organic greens from the garden. Maybe it's the nervous cooing when you invade their henhouse at dusk—as if all of them are saying, "Uh-oh. . .uh-oh. . . . Wha-a-a-a-t?" Or perhaps it's the smell of freshly gathered eggs squeaking and sputtering in a cast-iron skillet with healthy bright orange yolks standing tall and firm. Whatever the reason, there's nothing that makes an organic hobby farm more complete than a flock of squawking biddies.

Keeping Chickens

Chickens are among the easiest of farm animals to care for and a perfect first step for someone who is moving from vegetable gardening to raising livestock. The ease of care is one reason for the current growth of urban and suburban poultry farmers—many of them nine-to-fivers at the office who sell backyard eggs to coworkers, neighbors, and friends. If you watch your expenses, supplement food with grazing and organic garden scraps, and sell your eggs for the premium price they deserve, you can turn a small flock into a profitable side business.

Are Chickens Right for You?

The truth is, almost anyone can raise chickens, but that doesn't mean the birds are care free. Chickens have a list of basic needs, so there is a commitment to make before getting started.
   Begin by checking your community's zoning laws to see if chickens are allowed. Then ask yourself if you can devote ten to fifteen minutes per day—every day—for chicken chores. Most days it will take less than ten minutes, but these chores can't be skipped even if you're tired, the weather is cold, or you want to go away for the weekend. In other words, if you can't picture yourself stepping outside in your slippers and a bathrobe at 6:00 A.M. to unlatch the chicken coop door, you might not be ready to start a flock.

Basic Chores

Below are some things you'll need to get used to:

- Releasing the chickens from their coop every morning, feeding them and changing their drinking water.

- Gathering eggs twice a day and keeping the nesting boxes clean. That includes shooing out the broody hens that want to sit on a clutch of eggs all day.

- At dusk, herding them back into a shelter. They usually go by themselves, but there are sometimes a few stragglers. Once they're all inside, you'll need to close them up tight for the night and make sure the coop is secure from predators.

Chickens are easy to care for, but they have important basic needs. You'll want to give them shelter from predators, along with organic feed, daily water, and grit or crushed oyster shell.

- Chasing the escape artists who fly the coop and always manage to stay two or three steps ahead of whoever is chasing them.

Beyond that, the chores include cleaning out the henhouse when it piles up with waste and feathers, keeping the birds cozy and warm, supplied with food and fresh water (not ice) in the winter, and providing supplementary light if you want them to continue laying eggs in the sun-starved months of November through March. And you may have to deal with occasional illnesses, parasites, predators and pests— but many of them can be tamed with good hygiene and preventive measures.

Fresh organic eggs are incomparable, especially when chickens add greens to their diet from your pasture (or front lawn).

Your Investment

You also have to be willing to accept the fact that your investment may take a little time to pay off. I often joke that my first egg—pale brown and roughly the size of a misshapen golf ball—was worth every penny of the $500 I spent for it.

Once you've weathered the initial expenses—the henhouse and the fenced-in run, along with the water containers, feeders, brooding equipment, optional mobile pen (or poultry tractor), and portable fencing—the chickens will continue to lay eggs. As the expenses quiet down to just a steady supply of feed and grit, the eggs will keep coming each day, providing food for you and your family, along with yet another product for your farm stand or neighborhood CSA (Community-Supported Agriculture) operation.

The Benefits

There are other benefits to keeping chickens, too:

- They are the ultimate organic farm recyclers, consuming bolted lettuce; crop residue; bugs; fallen apples, peaches, and pears; assorted kitchen scraps; fresh-pulled weeds; and all of your unwanted and sun-wilted greens. When all of that organically grown material has been "chickenized" and mixed with high-carbon bedding material, it becomes nitrogen-rich manure that will supercharge your compost piles.

- Chickens eradicate garden pests, feeding on almost any kind of bugs or slugs, including bothersome ticks. Guinea fowl are especially efficient at eliminating tiny black-legged deer ticks that transmit Lyme disease.

- Finally, chickens are a joy to have around the yard. They're funny to watch and, depending on the breed (and how frequently you handled them as young chicks), they can behave like pets.

The Mark of Egg-cellence!

All the other advantages aside, the obvious reason why you or anyone else would keep chickens is for their high-quality contributions to the breakfast table. Freshly gathered organic eggs from local grass-fed chickens are far superior to the mass-produced eggs found at grocery stores: They're better looking and better tasting—an assessment that can be made after just one bite of a supercharged, super-yellow omelet. Organic eggs also have less saturated fats, more vitamins A and E, four times the vitamin D, and up to seven times more beta-carotene than store-bought eggs. So naturally, the price for fresh organic eggs from healthy free-ranging chickens should fetch a proportionally higher price than supermarket eggs. How much higher? I used to sell a dozen extra-large eggs from my organically fed Barred Rocks for $4, but steady customers told me that mine were better than the eggs that sold for $5 a dozen at the organic supermarket. So, at my customers' suggestion, I raised the price to $5. And I still have a waiting list.

The quality of your eggs will depend on the health of your chickens and the way they are raised. But before you gather any eggs, you'll need some chickens. (Hey, didn't I just answer that eternal question?)

Getting Started

Before building a coop or ordering a flock of chicks, make sure you have a plan and know what you want. Ask yourself the following five questions:

1. How many chickens will you raise?

2. Where will you put the henhouse?

3. Will you have a fenced-in run?

4. Will the chickens be allowed to free-range around the yard?

5. Will you practice rotational grazing with a movable pen?

Housing

You'll need a level, easily accessible place to build a coop with convenient water, windows for light and vents for fresh air. The coop should be built to accommodate the number of hens you plan to acquire at peak production. If you're starting with a dozen or two birds the first year, but know that you'll be expanding, you'll need a coop that has enough room for the entire flock. The same goes for nesting boxes, roosting areas, and food and water dispensers.

Coop Construction

Raising chickens has become so popular that many garden stores and farm-supply centers offer chicken coops with a combined run/enclosure for sale. Most are intended for a few birds in an urban or suburban setting, and they're quite expensive. Making your own coop is a better (and more economical) option.

You'll find dozens—maybe hundreds—of websites offering chicken coops and chicken coop plans for sale. Buying plans is a good option, but just looking at the myriad styles may give you ideas on designing your own. Coops can also be fashioned from a large doghouse or a small shed.

Some designs are stationary, while others are movable "chicken tractors" that allow you to move the chickens to new locations so they can graze and scratch a new piece of turf every day (see page 293).

Below are some things to consider:

- Chickens need 2 sq. ft. of space per bird to be comfortable inside a coop.

- Nesting boxes, which are often built as an extension to your coop, should be located above the floor level so chickens don't hang around inside them. Make at least one box for every five hens and keep the boxes 1 ft. (30 cm) wide, 1 ft. (30cm) high, and 1 ft. (30 cm) deep.

- To keep chickens from roosting in the nesting boxes, give them a proper roosting area. They'll use it to rest during daylight hours as well as to sleep at night. Use two 2 x 4 studs at a 45-degree angle to one of the walls, creating a triangle with the floor. Then space sturdy saplings or straight branches (at least 2 in./5 cm thick) along the lumber about 12-15 in. (30-38 cm) apart to create a mini-ladder. Make the bottom rung 18 in. (45 cm) high so that submissive birds can jump up to steer clear of dominant hens.

- Chickens typically eat a large meal before retiring, and they create most of their waste while roosting overnight. Large-scale producers place a wire grid beneath the roosts, allowing manure to drop through and into a pit. You may want to create easy access to the roosting area to allow quicker cleaning.

- An earthen bottom is acceptable for the coop, but a heavy wooden bottom will keep the birds warmer during the winter. Predators won't be able to tunnel into a coop with a wooden bottom.

- Wood absorbs manure, so I found cheap linoleum tiles that I stuck onto the wood. My chickens now have a faux parquet floor that's easy to clean. I used a few extra tiles to line the bottom of their nesting boxes.

- Nesting boxes that extend outside will save space and allow you to build a hinged roof for easy access. That way you can collect eggs without going into the coop and disturbing the birds (or soiling your shoes). Make sure that any entrance to the coop has a hasp or tight hooks-and-eyes to secure the entrance from predators. Raccoons have the manual dexterity to flip open a loose-fitting latch.

- Venting is important. It's nice to keep birds snug in the winter, but it's an absolute necessity that they stay cool in summer months and ventilated all year long. I looked through classified ads and found four used windows with insulated glass and full screens for $25 each. The windows are shut tight during the winter but open to keep a breeze flowing through the coop in the summer.

- The coop should be watertight, with a roof that doesn't leak. Rolled roofing is often a good alternative to shingles.

- Build a full-sized door for yourself, but keep the chickens' door small—1 sq. ft. is plenty large. I have a chicken door on either end of the coop, one leading into a fixed run and the other to the open yard, where I can set up my portable mesh fence.

- The best site for a run is an area that drains well. For ideal conditions, dig out the soil beneath the run and replace it with layers of sand and gravel.

In addition to a coop for nighttime and inclement weather, chickens will need access to the outdoors. Fencing or an enclosed run will protect the flock from predators.

Space

It's unhealthy to keep chickens crowded into quarters that are too small for the entire flock. When birds are overcrowded, they become stressed and can exhibit aggressive behavior, going beyond establishing a pecking order and entering into the world of bullying. Overcrowding can prompt chickens to pull out the feathers and cannibalize the weakest members of the flock.

Coop Design

Bigger is better when it comes to coops, but small flocks in an excessively large coop during severe cold may not generate enough body heat to keep each other warm. If your coop is excessively large for a small number of birds—and especially if it has a 7-ft. (2 m) ceiling to accommodate the humans who have to clean the coop, it may need insulation and a heat source during extreme cold spells. But it's always better to have too much space than too little.

Minimum Requirements

The minimum spacing for chickens inside a coop is 2 sq. ft. per bird, with an additional 2 to 5 sq. ft. of outdoor space (or fenced-in pen). Allowing your chickens to roam around the yard will help keep down the bugs and make your birds happy. But if there are cats and dogs or raccoons and hawks around, you might lose some of your flock to predators.

The ideal situation is to provide a coop with ample space and an attached fenced area (also called a run) that will keep the birds contained and safe from predators. The run should have some shade from the sun and be sheltered from extreme winter winds. It should also be equipped with its own feed and water stations. (See "Coop Construction" on page 281.)

Fencing

To provide additional protected grazing space for your chickens, consider using a portable fence. Vinyl-coated electric fencing, often called electric poultry netting, is an option for small flocks. It's typically 42 (107 cm) or 48 in. (122 cm) tall and sold in 164-ft. (50 m) rolls, with lightweight poles attached. An entire 164-ft. (50 m) section weighs less than 25 lb. (11 kg) and can easily be set up or taken down and moved in about fifteen minutes. One person can do the job, but it's better with two—one holding the bundle while the other takes the poles and pushes them into the ground.

Moving the fenced-in area from one place in the yard to another will keep chickens from eating every piece of green growth, a process that can take just a week or two. Left in the same spot, chickens will denude an area of all vegetation and turn it into a dusty bowl that is pocked with holes where they burrow into the ground and give themselves dust baths. Moving them to a new location every week allows the grass to grow back and doesn't give them time to make too much of a mess.

Poultry netting can be rigged to a marine battery and solar collector to provide a small electrical charge, which is just enough to deter predators. Without electricity, the fences will confine chickens during daylight hours and allow you to practice rotational grazing techniques. It's not guaranteed to stop predators; hawks can still swoop down from above.

Choosing the Right Breed

You'll also, need to think about the reasons for keeping chickens: for eggs, for meat, or both. There are more than 200 breeds, divided between standard categories (large-sized birds) and bantams (miniature breeds). Some heritage breeds carry an impressive history, along with a proven track record. Others are new hybrids developed to lay bigger eggs or produce chicks that are easy to differentiate between males and females. There are heavy breeds that resist cold weather with big bodies, small combs, and thick feathers; others that lay colorful blue-green or deep brown eggs, and many that serve a dual purpose—as egg layers and meat birds. Some breeds are docile, making them easy to handle but prone to be picked on, while others are noisy or flighty.

Sizing Up and Selling Your Eggs

Egg size refers to the overall weight of a dozen eggs, not the size of individual eggs, which can vary in a given dozen. Weigh your eggs before labeling them.

Egg Size	Weight per Dozen		Minimum Weight (per egg)	
Jumbo	30 oz.	850 g	2.42 oz.	67 g
Extra Large	27 oz.	765 g	2.17 oz.	62 g
Large	24 oz.	680 g	1.97 oz.	56 g
Medium	21 oz.	595 g	1.92 oz.	54 g
Small	18 oz.	510 g	1.42 oz.	40 g
Peewee	15 oz.	425 g	—	

- Eggs don't need to be washed unless they are soiled. Clean under running water that is at least 10°F (6°C) warmer than the eggs. Do not soak eggs.
- Eggs should be stored at 33°F–45°F (0.5°C–7°C).
- Egg cartons should be unused and clean, with labels that include the sell date (30 days after packing); the producer's name, address and phone; and the FDA's Safe Handling Instructions: "To prevent illness from bacteria: keep eggs refrigerated, cook eggs until yolks are firm, and cook foods containing eggs thoroughly."
- Notify your local board of health before selling your eggs.

SOURCES: Pennsylvania State University Extension Service; Massachusetts Department of Agriculture.

Best Egg-Producing Breeds

Breed	Characteristics	Egg Color/Size
Black Sex Links (also called Black Stars)	This high-quality layer is a cross between Rhode Island Red and Plymouth Rock, bred to help hatcheries distinguish between male and female chicks, right out of the shell. The markings virtually guarantee that all of your birds will be hens, not roosters. Black Sex Links are a little flighty, but they're excellent layers and good cold-weather birds. Sex links are not true breeds because they don't produce like offspring.	Brown Extra-large
Golden Comets	These hybrids are superior layers, with coloration that makes sex identification easy as soon as they hatch. Comets begin laying at an earlier age than other breeds.	Brown Extra-large
Leghorns	They're noisy, fast moving, and more than a little bit skittish. They're about the best layers you can find. Leghorns are good flyers and are excellent at foraging. White Leghorns are the most common, but colors range from black to red to brown.	White Extra-large
Orpingtons	These big and gentle birds were bred for laying as well as for meat. They have exceptionally thick feathers, making them look even bigger than their large frames and providing good protection from the cold. Their disposition makes them a superb choice for families with children. Buff Orpingtons, which are big and pale golden yellow, are especially calm. Other plumage colors are black, blue (more like a slate gray), and white.	Brown Large
Plymouth Rocks (also called Barred Rocks)	This was the chicken of choice across America during the post-World War II boom, prized by farmers for both its prolific egg-laying ability and its meaty body. The traits are just as popular among backyard chicken keepers today. The original gray-and-white barred coloring is attractive. And the feathers are thick, making it a good choice for northern climates. Sometimes its thick plumage has a tendency to collect waste matter on the feathers around its vent; you'll need to clean the area off from time to time for clean eggs. Plymouth Rocks are also broody, having a tendency to sit on eggs. Disposition of Plymouth Rocks is calm.	Pale brown Extra-large
Rhode Island Reds	These versatile birds are extremely good layers in almost any conditions. They're active, noisy, and flighty and can be hard to catch when they escape from pens. (Trust me, I've been there.) They have good winter hardiness.	Brown Large to extra-large

Choose good egg–laying breeds for best production. From *left to right from top*: Black Sex Link, Golden Comet (another sex link cross), White Leghorn, Orpington, Plymouth Barred Rock, and Rhode Island Reds.

If you're just starting out, and you want to maximize your investment with the best supply of marketable eggs, stick with chickens that are known for prodigious laying. The best layers will produce extra-large eggs for 250 to 300 days a year. That works out to about five or six eggs a week per chicken or about 70 percent daily production. With ten such hens, you can expect about seven eggs per day during peak laying time. Twenty chickens should give you fourteen eggs daily, and thirty-five will produce about two dozen.

Egg size will start out small, when pullets begin laying (at eighteen to twenty weeks old) and increase for several months until the chickens mature. After that, expect the chickens to remain productive for two seasons before steadily declining. Egg laying can also drop off temporarily when birds enter a molting stage or during the winter months when sunlight is lower.

Laying Breeds

Different breeds and varieties within breeds vary in temperament and egg production. Use the table opposite as a guide when choosing the breed right for you.

Buying Your Birds

Once you've chosen your site and your breed of choice, it's time to determine the best option for starting a flock. There are several ways to purchase chickens.

Organic Requirements

Chickens raised for eggs that are marketed as organic must meet specific requirements and be treated with organic practices from the second day of their life (as soon as they arrive from the hatchery as day-old chicks). Some things to consider if you want your eggs to carry the organic label:

- Chickens must be fed with only certified organic feed. This also means that pastures in which they are grazing must not have been fertilized with chemical fertilizers or treated with unapproved insecticides, herbicides, or fungicides for at least three years. If you're feeding the chickens leftover fresh vegetable crops, they must be organic, too.

- Feed cannot contain animal drugs, including hormones, to promote growth. Supplements or additives, other than what is needed for adequate nutrition and health maintenance, are also prohibited. Feed must be free of antibiotics and mammal or poultry slaughter by-products.

- Chickens must have year-round access to the outdoors, with shade, shelter, exercise areas, clean water, and sunlight. Coops must have adequate ventilation and organic bedding material.

- Full-grown laying chickens are required to have at least 2 sq. ft. of indoor coop space per bird and 2 to 5 sq. ft. of outdoor space if enclosed in a run or pen.

- Animals must be kept in a healthy, low-stress environment that emphasizes the prevention of diseases and uses approved treatments (including vaccinations and limited synthetic remedies that are approved by the National Organic Program).

- Organic producers are not allowed to withhold treatment of their animals with non-organic means solely to retain their organic status.

This is only a summary of the major requirements. For more specific information on restrictions, talk to your certifier, or consult the National Organic Program. (See "Resources" on page 356.)

Pullets

The fastest way is to begin with young adults, called pullets. These adolescent hens, which are sold at sixteen to twenty weeks old, are usually close to adult size and just beginning to produce eggs. Buying pullets avoids the months of caring for young chickens inside a brooder, as well as keeping and feeding them during the five months it takes to reach laying age. Another advantage of pullets is their adolescence. Because most are already at the point where they're beginning to lay eggs, it effectively eliminates the possibility that you'll be stuck with an unwanted rooster.

But organic pullets are expensive. And when you introduce pullets to an existing flock, they can upset the pecking order. If you're not extremely careful, they can also introduce diseases. If you decide to add pullets to your poultry family, quarantine them in a separate area for at least thirty days to make sure they don't bring in some new parasite, bacteria, or virus that your birds aren't used to. And when it's time to introduce them to their sisters, do it in stages. Set up an adjacent pen next to the fenced-in run where your established chickens roam. Allow the two groups to see and interact with each other from behind fencing, so that nobody gets bullied. Then, after a week or two, put them into the coop together at night. The groups will assimilate better if they wake up and greet the new day as a unified flock.

Day-Old Chicks

The most popular way to acquire chickens is to order newly hatched chicks from a poultry supplier. You should prepare for the chicks before they arrive, making a brooding area and stocking up on food, feeders, water dispensers, bedding material, and other supplies. The day-old chicks will arrive in the mail or through your favorite feed-supply store as downy, little peepers—not much bigger than those marshmallow Easter chicks. You'll have to raise the youngsters inside a warm, secure climate-controlled brooding area and feed them a special diet until they feather out and can be moved outdoors. (For a listing of mail-order nurseries, see "Resources" on page 356.)

Chicks by Mail

How is it possible to buy live chickens by mail order? It works because young chicks absorb the last bit of the egg's yolk just before they hatch, providing at least two days' worth of food and water to sustain them while they nestle beneath their mother's feathers. That two-day food supply allows a nursery just enough time to ship the birds. Buying mail-order chicks is probably the most common and affordable way to get started with chickens, unless you have a nearby feed store that receives deliveries from a hatchery.

You'll have a choice of how you order chicks: straight run or sexed. Straight run means the chickens will likely be about half male and female. The price of straight-run chickens is cheaper, but you'll have to deal with the likelihood that half your layers might be crowers instead of layers. One option is to raise the cockerels (young males) for meat birds, sending them to a slaughterhouse when they are ten to twelve weeks old. The alternative is living with a bunch of noisy roosters crowing at dawn and angering your neighbors.

Buying sexed chickens isn't a guarantee that you won't have an occasional rooster. With most breeds of chickens, it's hard to distinguish the sex at just one day (when they are shipped), so suppliers usually guarantee about 90 percent female chicks in sexed lots. Incidentally, you don't need any roosters to make your flock lay eggs, but in a large flock, one or two roosters can keep the hens in order and protect them from small predators.

When the Chicks Arrive

Let your postal carrier know you are expecting a shipment of chicks, and ask that someone contact you to pick up the delivery as soon as it arrives. When you come to pick them up, open the box and check out the babies before you bring them home. In most cases, the chicks should be fine, but there may be casualties caused by cold weather, crowding, or other conditions. I had several perfect shipments in a row before receiving a batch of chicks with a number of weakened and dead birds. I reported the problem the next day and the nursery gave me credit.

Replacing Aging Flocks

If you're intending to stay in the egg-producing business for more than a few years, you'll ultimately need to replace your aging fowl with new stock. Most professionals recommend replacing the entire flock at once—all in and all out. This technique will help prevent the introduction of parasites and poultry diseases, and you won't have to determine which of your chickens are the ones who have stopped producing eggs.

Some backyard chicken keepers buy a new batch of chicks each year, and add them to the existing flock. If you go that route, you'll have to spend time introducing young chickens to the older flock members, ensuring that the youngsters don't get picked on. Understand that introducing new members to an existing flock often upsets the pecking order among hens and can lead to stress and fighting among all your birds.

Your brooding area must be warm, duplicating the heat that chicks find under their mother hen's feathers. A heating lamp with a red bulb should do the trick.

Making a Brooding Area

You don't have to spend big money on a state-of-the-art chicken brooder. Considering that you'll only need the brooder for the first two months of the birds' lives, you can easily rationalize using a big cardboard box lined with pine shavings. (Don't use cedar chips, which have an aroma that can irritate the chicks' respiratory systems.) When you've finished with it, you can compost it and put away the supplies until next year. I use collapsible moving boxes, starting with one box for the very young chicks and then in about two weeks doubling it in size by cutting one side of each box and taping the sides together with duct tape. A section of plastic deer netting placed over the top and secured with clothespins will keep the chicks from jumping to freedom once they get a little older.

Feeding and Care

You'll also need some basic supplies: chick food, chick feeder, and a small water dispenser, along with a heat lamp to keep the chicks warm. All of these supplies can be purchased at a feed store or ordered from a mail-order supplier.

Brooding

The brooding room should be a warm place, where doors can be closed to keep cats, dogs, or young children away. I set up my brooder in a closed room inside the basement, where the oil burner and hot water heater are located. In early spring, when the chickens typically arrive, the heat is in use, and the room is about 75°F (24°C).

It's a good idea to keep the brooding area away from your everyday living space because the chickens create and kick up dust that can be allergic. As they grow larger, even the cleanest birds can smell bad, especially during the last two to three weeks before you put them outside. When making a brooder, consider the following needs.

Heat. The brooder must duplicate the heat that young chicks receive under their mother's feathers, which is about 95°F (35°C) for the first week. After that, the temperature can be reduced by 5°F (3°C) per week as the chicks become more active and grow more feathers. A good heat source is a lamp with a wire guard and red lightbulb. A red bulb is preferable to a white light because it won't keep the babies awake and add stress, which could cause them to peck at each other. Hang the light with a chain over one corner of the brooding box so chicks will have a warm place to sleep, along with a cooler area to run around in. If you notice that the chicks are constantly huddled underneath the heat lamp, it's probably a little too far away so you'll want to lower it a few inches. If the chicks are spread out in the corners away from the lamp, it's too close. Raise it a few inches to cool things down. Happy birds will sleep in small groups but will spend their waking time eating, drinking, and hopping around inside their surroundings. They will also peck incessantly—at specks of dust, dots on the walls, at anything they see.

Raising your birds from day-old chicks lets you provide organic feed and care—mandatory steps if you want to call your eggs organic. Once hens are old enough to scratch on fresh pasture, the land must be under organic culture, too.

Bedding. Spread about 1 in. (2.5 cm.) of pine shavings on the bottom of the brooding pen to absorb the waste. The high-carbon material will keep the waste from stinking for a few days or more than a week, depending on the size and number of birds.

Water. To minimize stress and prevent diseases, check the water supply at least once a day (twice daily for larger numbers of chicks). Set a plastic bell-shaped water dispenser made especially for young chicks on the floor of the brooder. I like to set mine on a shallow plate to raise it slightly above the pine shavings and to keep the chicks from splashing water onto their feathers. You'll need to place round, smooth stones or marbles inside the basin to prevent very young chicks from falling asleep with their heads in the water and drowning. Yes, it happens. To introduce your chicks to the concept of drinking, dip their beaks in the water so they'll know where to find it. Clumsy chicks often spill water, so be prepared to place fresh shavings around the dispenser when you change the water.

Purchased pullets—laying-age hens—must have been raised organically from day 2 of their lives if you want to call their eggs organic.

Food. Feed your babies only chick food or "starter" feed. It's different in formulation from adult chicken feed and layer pellets, with more protein to support the rapid growth that chicks undergo before they reach adult size. Starter feed (along with grower feed for young pullets that have not yet begun to lay) has much less calcium than the feed for layer chickens. Layers need the extra calcium to produce strong eggshells, but the higher concentration is unhealthy for growing chicks.

If you buy chicks that have been vaccinated for common diseases and if you practice good hygiene in caring for the birds, you won't need medicated chick food. If you're planning to sell organic eggs, you should start the chickens off on organic feed with free-choice grit (at a size that is made for young birds). Mix in some light greens as the chicks get older to keep their diet varied. As the weather warms and insects become available, drop in some small worms and bugs to supplement their diet. You can also take them outside for short adventures and allow them to peck around in the grass—albeit inside a protected enclosure.

Organic regulations allow for the use of day-old chicks from a conventional hatchery, but after day one, they have to be fed and cared for according to standards set by the National Organic Program.

Cleanliness. There are other things to watch out for throughout the time your chicks are in the brooder. Pay close attention to cleanliness. When chicks are young, they sometimes get "pasties"—waste that collects and dries up on their bottom sides. When the waste material cakes up, it can prevent the chick from doing its duties. If you find it, use a damp towel to wipe and pull it off, clearing the vent.

As the chicks grow larger, they consume more food and produce more waste. And their ability to move around, jump on top of food and water dispensers, and create a mess for their brood mates increases.

Young chicks are adorable, but they can carry diseases. Always buy your birds (or eggs for hatching) from a salmonella-free supplier certified by the National Poultry Improvement Plan (NPIP).

Avoid diseases by keeping their food and water area clean. Be sensitive to any odor that is particularly sour or reeks of ammonia, which means a cleaning is long overdue. Also watch out for aggressive behavior in chicks, especially if the brooder is crowded. Stressed birds can lead to problem pecking and even cannibalism.

Handling. The more you interact with your chicks in the first few days after they arrive, the easier they will be to handle in adult life. Talk to them softly and linger over the brooder when you're feeding, watering, and caring for them. Even if you don't want to make pets of the chickens, the one-to-one familiarity comes in handy when a bird escapes its enclosure and you have to chase it down and pick it up.

Intimacy is nice, but you should use caution when handling young chickens. Handling baby birds can cause stress to the chicks and possibly spread diseases, such as Salmonella. Be sure to wash your hands thoroughly after handling any poultry—young or old. Children, who have young immune systems, should not play with chickens, since they are more susceptible to infection.

Moving Time

If all has gone well, your chicks have grown considerably and feathered out over the past six weeks—and so has the aroma of their brooder. With a full coat of feathers, they should be ready to assume a life outside. If you timed your purchase well and started caring for the chicks in early spring, it should be warm enough to put the birds outside.

Even before the chicks are ready for the outdoor life, it's good to get them accustomed to the world outside their brooder. On warm days, you can scoop up the chicks and place them into a small, fenced-off area to let them peck at the grass and hunt for bugs. Just make sure you sit with them outdoors and that there are no cats, dogs, or predators to harm the vulnerable youngsters.

When it's time to move them outdoors permanently, take one final look at the coop. Make sure it's clean and that there's a fresh supply of food and clean water. Also double-check the surroundings to ensure that both the coop and the run are secure from predators.

At this point the youngsters will still have two to three months of growing before they start laying eggs. They will not get the prime perching spots on roosts inside the chicken coop. And when the groups are gathered for a feeding of greens or kitchen scraps, the youngsters will have to steal a meal and run away with it to avoid being bullied by the older hens.

When Will Eggs Begin?

If you don't have roosters and you bend over to pick up a young pullet, it may assume the breeding position when you reach down. The hen will squat on the ground with its back flattened and wings slightly spread, waiting for the rooster to mount. Pat the girl's back a few times to give her a thrill. It should shake its feathers and dance away contentedly. Regardless of whether you pat its back, this is a sure sign that eggs will soon be arriving.

Protection from Predators

Chickens are easy prey for so many predators that it's hard to keep track. Raccoons will grab at their necks and pull their heads off through a fence. Hawks will swoop down from above to score an easy meal. Coyotes, foxes, fishers, and bobcats will chase down and kill birds that are left outdoors. And weasels, minks, and rats will sneak through the tiniest of cracks and kill chickens in a poorly protected enclosure.

Any and all losses to predation take a toll, not only on your flock but also on your spirits. All the work and responsibility of caring for the chicks and bringing them to egg-laying age to see the flock wiped out by one or two nights of predation can be heartbreaking. It pays to make sure your chickens are safe.

To keep burrowing animals out of the hen house and run, I dig a trench around the entire perimeter, 1 ft. (30 cm) deep x 15 in. (38 cm) wide. I cut a 20- (6 m-) to 30-ft. (9-m) length of 4-ft. (1.2 m) heavy plastic chicken fencing into three 16-in. (40.5 cm) strips, and I place a strip in the trench. I staple one side of the steel chicken wire to the bottom frame of the coop, and then I flatten the fencing into the trench at a 45-degree angle away from the coop.

Chickens are susceptible to multiple diseases, from influenza to salmonella infections.

I secure the fencing with rocks, then I bury it in the trench. When a burrowing animal comes up to the edge of the coop and tries to dig in, it gets a few inches into the ground before running into a wall of fencing. To get around the barrier, it would have to start digging 15 in. (38 cm) away from the edge of the coop, and that's just not likely to happen.

Going Mobile

Traveling poultry shelters called chicken tractors can be moved from place to place around the yard to practice rotational grazing. They're inexpensive to make, often consisting of a simple A-frame covered with chicken wire and a lightweight plywood shelter at one end. The bottom is sometimes open or covered just with fencing, which allows the chickens to graze and scratch on the grass beneath.

Chickens are allowed to graze in a fresh section of the lawn or pasture area for a day or two (depending on the number of birds and size of the tractor), then the entire structure is dragged, wheeled, or carted with handles to a fresh grazing site. Moving the tractor by its width or length one day at a time lets your birds graze on new material each day. It also means they won't pick the pasture clean and create a big mess with their droppings. Depending on the size of the tractor and the flock within it, the birds' manure can be left behind to fertilize the soil and help the lawn or pasture regenerate.

There are numerous plans available online—some with wheels, skids, extended handles, and other methods of achieving mobility. Tractors, also called "arks," are ideal for small flocks in suburban and urban farms, but large-scale producers use the concept, too.

There are variations to the mobile tractor that alter the game a little. An old pop-up camper, for instance, can be converted to a full-sized chicken coop for a larger flock. Once towed into place, it can be left in an open pasture surrounded by portable fencing. The larger grazing area that's available inside a ring of portable fencing means the tractor won't have to be moved as frequently and can remain in place for more than a week. Then, when the grass starts to get thin and covered with droppings, a farmer can pull up the fencing, hitch a real tractor to the rig, and tow it to a new site. Once the portable fencing is reinstalled, the chickens can be released to forage in green pastures once again. This model is often referred to as day ranging. Chicken tractors and day ranging techniques can be used for both egg-laying flocks and for meat birds.

Author and farming pioneer Joel Salatin of Polyface Farm in Virginia's Shenandoah Valley developed a model that mimics nature in allowing free-range chickens to follow grass-fed beef in a rotational grazing pattern. The chickens scatter the cow manure as they graze the pasture, spreading it out and leaving their own droppings, just as wild birds followed buffalo on the plains of North America. When the chickens finished their job, the grass grew back and the cattle returned.

When building a chicken tractor, think of these possibilities:

- A chicken tractor that is used as a permanent shelter—meaning there is no other stationary coop—will need to be sturdy enough to thwart predators. Lightweight tractors with only chicken wire for protection can leave birds vulnerable to larger animals that are strong enough to break through. Consider using sturdy wire fencing for a more reliable shelter.

- Placing a section of fencing across the bottom of the chicken tractor will prevent animals from tunneling underneath the sides and gaining access to your birds.

- Tractors need fairly level ground for best operation. Don't use a big tractor over lawn or pasture area that's riddled with holes and hillocks.

- You'll need a door at the top of the tractor to reach in and change water, fill the feeders, and harvest eggs.

- Tractors can be used for turkeys, ducks, geese, and even organic, grass-fed rabbits.

Pests and Diseases

Chickens can help clear the landscape of slugs, bugs, pests, and weeds, but they can also be subject to their own parasites and diseases.

Lice. Head lice, body lice, and shaft lice (they feed on feathers) are among the external parasites that attack chickens, irritating the birds and causing them to lose sleep and feeding time and eventually weakening the flock and hurting production. To inspect birds for lice, part the feathers, especially under the wings. You may see lice scurrying for cover. To prevent them, avoid introducing new birds to an existing flock. Also keep other birds and mice away from chickens. Some remedies recommend diatomaceous earth in a dust bath, which kills parasites, but it is also irritating to chickens' respiratory systems. A SARE (Sustainable Agricultural Research and Education) grant funded a manual of preventive, homeopathic, physical, and herbal treatments for chicken maladies. Some suggestions from the compendium include adding wood ashes to their dust bath and applying limestone to their run.

Worms. Several types of parasitic worms infect chickens—some transmitted in the droppings of other chickens and others passed along from eating worms, grasshoppers, slugs, and other soil creatures that are food for free-ranging birds. To prevent worms, rotate chicken runs and coops; don't keep chickens of different age groups; and don't spread poultry manure on pastures that birds will use for grazing. Clean infected litter from the coop and run.

Diseases. Marek's disease affects young chickens, causing tumors and paralysis. Newly hatched chicks can be vaccinated before shipping. *Salmonella pullorum* and *S. gallinarum* (fowl typhoid) can kill much of a flock. Other types of salmonella can be harmful to humans. Always buy chicks or hatching eggs from a salmonella-free hatchery participating in the National Poultry Improvement Plan (NPIP).

The best general prevention for diseases is buying healthy birds, keeping them in uncrowded, sanitary conditions, and not introducing disease from one flock to another. If you're visiting another farm with chickens, do not walk through the coop, run, or areas where birds are grazing. Avoid introducing new birds to your flock, especially older birds that may have been exposed to diseases in other flocks. Consult with your organic certifying agent for advice on treating chickens.

Barred Rocks (aka Barred Plymouth Rocks) are a dual-purpose breed. If you want your meat birds to be organic, they must be raised—and processed—organically.

Meat Breeds

Raising chickens specifically for meat involves a much shorter time commitment than caring for layers. Chicks are ordered from a hatchery and arrive the same way as layers, but the most common breeds used for meat production are programmed to do two things: eat and pack on weight. Some are ready for dressing out in about six weeks.

Cornish Cross. The most popular breed for meat chickens is the Cornish Cross, bred from commercial Cornish chicken and a white Plymouth Rock. These chickens grow extremely fast and develop thick breasts and fat legs, with pale yellow skin and white feathers. They grow so quickly that they won't survive much longer than their designated time for dressing out because their legs don't grow strong enough to support their massive body weight.

Cornish Cross chickens are typically not left to free range like other birds but are candidates for tractor culture. They often create so much waste as they reach full weight that they produce a high volume of manure and must be moved frequently to prevent a buildup of droppings.

Red Ranger. Another common meat bird, Red Ranger, has a more normal (but still fast) growth rate. It can reach more than 6 ½ lb. (3 kg) in ten to twelve weeks and is prized for its full flavor. Red Rangers don't have the same size breasts as Cornish Cross chickens, and their drumsticks (legs) are longer and a little less plump. But they have more "real-chicken" flavor and are better suited to being raised free range.

Large Heritage Breeds. A number of large heritage breeds that were intended as dual-purpose chickens can also be raised as meat birds. Plymouth Rock, Rhode Island Red, and Delawares are a few more common names. But don't think that the laying chickens you kept for two years will be tender-tasting roasters. Older chickens are as the name implies, "tough old birds" that are better suited to stewing than stuffing and roasting.

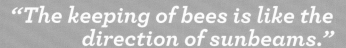

"The keeping of bees is like the direction of sunbeams."

—Henry David Thoreau

When people hear that I run a small-scale farming business, they often ask if I keep animals. "Oh, yes," I answer, "a hundred thousand or so." Well, it's true—but only if you count the bees in our hives. It's also true that the limitations of my land (a single acre), our zoning laws (suburban), and a full-time career (off the farm) all conspire to keep me from the rural joys of raising sheep, goats, dairy cows, horses, or grass-fed beef.

Bees

I raise bees or more precisely, my wife, Valerie, tends to the hives—places where she occasionally reminds me, the women do all the work. And as I think back, it was she who took the classes, and it was she and my youngest daughter who painted the boxes and built the frames that keep our colonies in order. Meanwhile I sometimes jump in for the arduous task of extracting and pouring the honey.

As rewarding as bees are to the spirit, the business venture has been unpredictable. Seven years ago, when we set up the first hive, our summer was long, warm, and productive. We harvested about 70 lb. (31 kg) of dark amber honey that fall.

With such a successful start, we decided to expand and add another hive the following spring, That summer, the bees were finicky, and we collected 30 lb. (13.5 kg) total from the two hives. We added two more hives the following spring and took in a little more than 50 lb. (22 kg) from four hives.

That doesn't mean our efforts were a failure. We lost a few hives, captured a swarm from our cherry tree, and had several years of good production, but nothing like the first. We also had the opportunity to watch one of the world's most miraculous creatures at work—close up and personal, with only a few welts from bee stings.

Bees are fascinating insects, and they are perfect participants in the organic arena. The insects pollinate our fruit trees, berry bushes, and vegetables, then reward us with sweet, golden honey each year. We treat the culture of bees almost as a thank-you to Mother Nature—giving back what has been given to us.

Bees are a fascinating addition to the hobby farm. They can fly more than 2 miles (3 km) to bring back pollen and nectar to the hive.

Getting Started

The best advice for a beginning beekeeper is to take a bee class. Local beekeeping groups in your community (you can find them with an Internet search) often hold classes beginning in winter, teaching newcomers about bees, their habits, how to care for them, and how to prevent pests and diseases. Some classes build a beehive over the course of the lessons, and at the time of graduation, new beekeepers are ready to buy a "package" of bees.

A beekeeper checks one of his deep chamber frames for brood and eggs. The bees will store their surplus honey inside a "super" located above the brood chambers.

When questions arise, former students can rely on the contacts they made during class. It's a great way to be introduced to the business of bees.

Types of Hives

Most commercial and hobby beekeepers use box-style Langstroth hives, which are essentially boxes containing snug-fitting panels that hold honeycomb in a neat and easy-to-remove fashion. Bees use the panels, called frames, to grow brood (bee larvae) and to store pollen and honey.

Boxes filled with frames are generally arranged with the bottom two deep boxes (also called chambers) set up for the bee brooding. It's the section where the queen hangs out and lays eggs into individual hex-shaped cells. The cells are fed by attending bees and ultimately grow into adults. The deep chambers will also contain stored honey and pollen for the growing hive.

Above the brood chambers is where the honey supers are added and is where the bees will store excess honey. This is the honey that the beekeeper harvests.

Bees fill the honeycombs with nectar, then fan it with their wings until it reaches the right thickness and concentration before the bees cover the cells with a wax capping. Once the frames have been filled with capped honey, beekeepers can harvest, removing the frames, slicing off the caps with a hot knife, and spinning the frames in an extractor, which uses centrifugal force to draw out the honey. Honey can be harvested several times a year, depending on how strong and how industrious your bees are. In New England and the Northeast, early summer honey is generally light colored, while later-season honey, which contains nectar from many summer and fall meadow flowers, is deep brown and has a strong wildflower taste.

Honey that has been capped and jarred can last indefinitely on a shelf at room temperature. Sometimes, though, when the honey frames are harvested before all the cells are capped and the material contains a higher percentage of water, the honey can ferment.

A Langstroth beehive will cost about $200— less if it's unpainted or if you build it yourself. Sometimes a club will let you borrow the group's honey extractor, or you could take your frames to an established beekeeper, who will typically spin out your frames for a predetermined share of the honey. Buying a quality extractor will set you back about $500. (You could spin frames and collect honey from other beekeepers to help defray the cost.)

Top Bar Hives

Some organic farmers prefer to use a more natural bee habitat than the highly structured Langstroth beehive. Top-bar hives consist of a long V-shaped box with a removable lid that uses bars (sections of wood) spanned across the box. Bees build natural comb from the bars, and it hangs down into the box.

A honeybee collects nectar from a flower. These fascinating insects have two stomachs—one for digestion and another to store nectar or water for the hive.

Bee Prepared: 10 Tips to Get You Started

1. **Are you sensitive or allergic to honeybee stings?** Bee stings will be an inevitable consequence, even with protective gear. It is a good idea to evaluate beforehand how you may react to stings.

2. **Check local laws.** Find out if there are local regulations regarding beekeeping and registering your hive. Also check with your insurance company to see if your homeowner's policy covers having beehives on your property.

3. **Join a local beekeepers association.** Most states and local counties have established organizations and clubs that organize monthly meetings, bee school classes, and other informational and social events. New members are always welcome.

4. **Study up!** Read books and magazines and attend a bee school. Bee school is an affordable series of introductory classes offered through most beekeeper associations starting in late January to mid-February and running eight or so weeks. Students are taught about honeybees and their life cycle, what to expect the first year, how to buy and build equipment, order bees, inspect hives, and recognize and treat diseases and pests. School usually culminates with a final class that is a hands-on demonstration of a live bee package installation. Often you can combine orders with classmates for bulk deals on equipment and bees.

5. **Be an apprentice.** Consider being an apprentice alongside an established beekeeper for one season. This is an excellent way to see what is involved in keeping a hive without making an equipment investment. All you need is to purchase protective gear.

6. **Keep it simple.** Purchase just the basics to start. Make sure you have a good understanding of what each piece of equipment does and is used for. All you need to get started is a complete hive, bees, a smoker, hive tool, and protective gear.

7. **Purchase new equipment.** Invest in new equipment for your first couple of seasons. As a new beekeeper it is often too easy to miss or overlook diseased or pest-ridden equipment due to inexperience. It is also impossible to truly know the previous beekeeper's hive-housekeeping habits or whether any pesticides may have been used on or near the equipment. The money you might save buying used equipment isn't worth the possible loss and cost of your new hive. If you must purchase used equipment, take along an experienced beekeeper for counsel.

8. **Consider starting with two colonies.** While it may seem like two means double the work, generally it is not, and it will give you a chance to compare hives, possibly spotting problems earlier because you notice a difference. It also gives you a backup in the event you lose one hive, which can be common with novice beekeepers.

9. **Understand you may not get honey the first season.** The goal of your first year is to get a healthy hive up and running. Every hive is different and will build and strengthen at different rates depending on your location and the weather.

10. **Step back and enjoy your bees.** Beekeeping should be fun, so take time to enjoy your hive. Set up an observation area where you can sit and safely watch these amazing insects at work. You will gain a new appreciation for nature, your bees, your garden, and the food that you eat.

Top-bar hives are much easier to build than box-style Langstroth hives because the plans don't have to be as precise. The design is also easier, and there are fewer parts. One downside is that bees will have to build out large amounts of comb—without the structure of frames—so it may take longer to produce usable honey. But access to the bees is easy because a single frame can be removed without disturbing others. (For more information, see "Resources" on page 356.)

Different Bees, Different Traits

There are several different types of bees used by commercial and hobby beekeepers, each with different temperaments, their ability to produce honey, and other traits. Many beekeepers will tell you one group is more docile, more productive, or hardier than the other, but many variations exist among hives of the same group, often depending on the quality of the queen. Again, one more reason to take a lesson and develop a friendship with experts living in your area. They can give you local advice about the best bees to buy.

Italians. Italian bees are good honey producers; they are somewhat disease resistant and are said to be less aggressive in defending the hive (docile). They also build up the hive population throughout the season, which can sometimes result in higher amounts of honey being used to feed the brood. Italians are the most popular bees used in North America and have an attractive golden tan color.

Carniolians. Darker in color than Italians, Carniolians are grayish brown to nearly black. They build populations quickly in the spring, and early in the season; they can appear to be two to three times more productive than other bees. They also have a tendency to feel overcrowded in midsummer, before their brooding or honey frames are even partly filled. This can cause them to swarm (fly from the hive in search of a new home). Our Carniolians swarmed several times, and we were only able to recapture the bees once. They are also characterized as a fairly calm bee.

Bee Stings

Knowing a little about your bees' behavior can help you minimize bee stings. Honeybees will feel threatened and are more likely to sting when they encounter dark colors and certain aromas that mimic pheromones. The scent of banana oil is similar to the alarm pheromone and gets bees agitated. Don't use tropical or banana-scented shampoos before working on your bees. Also avoid black or dark-colored clothing. You'll notice that beekeepers all wear white suits around their hives. Some things to keep in mind:

- Bee stingers are barbed, and they can remain attached to the skin. If you're stung, gently scrape off the stinger, as if you were removing a tiny splinter. After a bee stings you and loses its stinger, it will die.

- Bee stings can sometimes bring on an allergic reaction that can be fatal if you're sensitive to the venom. If you've ever had excessive swelling, redness, or dizziness and nausea after a bee sting, talk to your doctor before taking up beekeeping.

Russians. Russian bees are a newer bee stock introduced by the USDA from eastern Russia. They appear to resist mites and may prove to be a more durable bee in the United States. Russians like to build queen cells (specific brooding chambers where queen bee larvae are raised).

Bee Ailments

Bees can be bothered by numerous diseases and pests, including mites, and the recent phenomenon known as Colony Collapse Disorder. There are ways to check bees and treat for mites, including the natural treatment of sprinkling powdered sugar onto the bees. (When they groom themselves to remove the sugar, they also remove mites from their bodies.)

Bees are under siege these days—threatened by the cumulative stresses of brain-damaging neonicotinoid pesticides, parasitic veroa mites, trachea mites, and the stress of transportation en masse to pollinate crops that are grown 100 miles away. Raising bees in an organic environment is the responsible thing to do if you're in favor of pollination (that is, food production).

Staying Organic

Raising bees without the use of pesticides makes obvious sense, but even if your hives are treated naturally and every square inch of your property is under organic production, your honey won't be likely to qualify as organic. This is because bees can fly more than 2 miles (3 km) away to forage for nectar and pollen, so it's very likely that somewhere in their journeys, they will have come into contact with pesticides from some sprayed farmland, orchard, or residential property.

Still, if you keep your property free of chemicals, you're giving the bees a better environment in which to forage for nectar.

Goats

Goats have long been praised as the ideal animal for small-scale, start-up farmers—easier and more profitable than cows because they can supposedly make a meal out of everything in sight, miraculously turning weeds, briars, old newspapers, and maybe even the seat of your best Sunday britches into nutritious low-cost milk and meat. And legend has it they require so little upkeep that they can practically fend for themselves.

While a little bit of that is true, most of it is rural legend culled from cartoons and not-so-conventional wisdom. The bottom line is beginners need to beware of any promise that goats are easy, low cost, or maintenance free. Like all animals kept by humans, they require a good deal of human care. And it shouldn't be any real surprise that there are start-up costs beyond the price of the animals themselves.

It's true that goats are more efficient grazers than cattle, and they produce more offspring and do so at a younger age than cows. But goats can also be susceptible to diseases and parasites—especially when raised in moist, temperate regions with limited space to roam. And they can become all-too-easy prey for coyotes, feral dogs, and other predators, taking a heartbreaking toll on a herd if you leave them unprotected.

Goats also don't eat everything in sight. Most will shy away from any food that is dirty or spoiled. They will, however, make short work of green twigs and brush, foraging more like deer than sheep or cattle.

In much of the world, where grass and grain are hard to come by, goats live on weeds, underbrush, and other forage materials that provide roughage for their rumen. But in the United States and Canada, grass and grain make up a significant portion of their daily diet. To keep your animals healthy, you'll need to ensure they receive high-quality forage and hay along with feed. That means browse material when it's in season and good-quality hay during the winter months. They will also need a steady supply of clean water, something that you can't overlook in either the heat of summer or in winter, when the water buckets freeze.

Goats also require adequate space—especially if they are being raised naturally. If you want to raise them organically, they need access to pastureland that has not been treated for at least three years with synthetic fertilizers or chemical pesticides. And the animals will need enough room to be comfortable inside a shelter, with access to fresh water all season long.

Goats are intelligent and playful and can provide natural organic fertilizer for your garden. But they have basic needs that must be met for optimum health.

You'll also need to choose a breed that's right for you. There are goats that are suited to different purposes—dairy, meat, pets, and showing. And there are goats with varying degrees of tolerance for heat, cold, dry weather, and moisture. With all the variables, it pays to talk to someone who is experienced with goats and can offer advice that is specific to your location. A cooperative extension office, 4H club, nearby university, or even a fellow goat keeper may be a good place to start. Meanwhile, here are some basics.

Dairy Goats

There's a reason why dairy goats have become popular additions to a growing number of family farms. They're efficient producers of milk, sometimes yielding more than 1 gal. (4 L) per day per animal in a season that lasts about ten months. But those numbers are only estimates, and they can vary greatly depending on the quality of the animal and the specific breed—from tiny Nigerian dwarf goats to full-sized Nubian milkers. Also, most goats will drop their milk production slightly after several months, and some will dwindle to only a few cups a day for the last few months. The point is a single animal can provide more than enough milk for a family of four.

But you'll need more than a single animal. Goats are herding creatures by nature, and it's always best to keep at least two of them—especially if you don't want them working 24-7 to escape their enclosures. And with two or more goats, there will be at least twice as much milk. Some of it can be used to nurse young kids, but once the young'uns are weaned, you'll have to make use of what you and your family can't drink. Thankfully, milk from healthy goats tastes just as good as cow's milk and has even better nutritional value. And it can also be used to make cheese, ice cream, yogurt, kefir, cosmetics, and soap.

Getting Started

Setting up a small-scale commercial goat dairy involves meticulous commitment to cleanliness and strict adherence to state and local regulations. It's a daunting task for even those who are already experienced at raising goats. Beginners would be wise to set modest goals, easing into the process by milking for themselves and their families or trying their skill at making goat-milk soap and goat cheese for themselves and their families. Hobbyists can usually sell goat milk to others for soap but not for human consumption until they have a registered dairy operation. Check the rules in your state.

Healthy, Tasty Goat's Milk

People who say goat's milk tastes different from cow's milk either haven't tasted goat's milk or they've tasted poor-quality goat's milk. Good goat's milk has no perceptible difference in flavor, although it has slightly higher calcium and phosphorus and slightly less cholesterol.

Owners who allow an unneutered male goat to mingle with lactating does risk spoiling the milk; the bucks will invariably impart an unmistakable musky flavor. Bucks should be neutered unless they're being used for breeding purposes, and they should be kept in separate quarters away from milking does.

Another way that goat milk becomes less appetizing is if a lactating doe has too many internal parasites; the milk can develop a capric acid flavor, taking on the taste of goat's milk cheese and making the milk undesirable. Does should be tested for parasites and treated when necessary.

Before purchasing your first goats, you'll need to consider the animals' basic needs—food, water, shelter, and general health care.

Feeding and Care

First of all, your goats will have to be fed year-round with a diet that includes roughage, grasses, and grains. Goats are ruminants, so they need roughage in the form of well-cured hay, pasture material, and browsing food—usually the tender leaves, shoots, and twigs from low-growing trees and shrubs. (Think of the foods that deer would eat in the wild.) During the winter months, when pasture and browsing material is scarce or nonexistent, you'll need to give them more hay. When animals are producing milk, they'll need hay with a higher nutritional value, like alfalfa hay or hay with another legume that will provide extra nitrogen and protein. Animals that are breeding or growing from kids into adults need legume hay as well. Depending on the quality of their pastureland and the amount of legumes in their daily rations of hay, goats will also consume varying amounts of grain. The amount also depends on the age of the animal as well as the breed and your commitment to the grass-fed movement.

Dairy goats reward their owners with sweet, healthy milk. What you and your family can't drink can be sold to others for soap and creams.

Goats also need access to fresh, clean water, free-choice baking soda (to induce belching and prevent occasional bloating), adequate salt, and free-choice mineral supplements in granular form, not in a block.

Space for Grazing

Happy goats like to roam around and forage for food, nibbling leaves, grasses, and other goodies as they wander. In most moderate climates, count on giving each milking goat roughly ½ acre (0.2 ha) to roam. Goats that graze in closer quarters will invariably feed on the same grasses where other goats have passed, spreading parasites and becoming ill. Some internal worms have alternate life stages in snails and slugs that live in the grass, where they eat the goat droppings, then pass the parasites back to the goats. When goats are confined to a small pasture area and feed on the same grass over and over, the parasites have multiple opportunities to complete their life cycles. That can be a problem for organic producers who are restricted in their use of synthetic parasite controls such as Ivermectin. Natural remedies like garlic and diatomaceous earth can be used for worms, but they are not always as effective in eradicating established parasites.

Don't Let Your Goats Eat These!

There's no truth to the notion that goats can eat tin cans, newspapers, and everything from old shoes to new furniture and digest them in their indestructible cast-iron stomachs. Many substances, including common backyard plants, are toxic to goats. Scout your property to remove them before letting the goats run free. The list includes many landscaping favorites such as azaleas, boxwood, rhododendron, and mountain laurel; flowering hellebores and lily-of-the-valley; and ubiquitous wild black cherry, pine trees, and milkweed. One common type of lawn grass, tall fescue, contains endophytes, which are specific types of fungi that make the turf toxic to insects. Unfortunately, endophytes also render the potential grazing material unhealthy and even dangerous to goats and other grazers.

Some other common plants that are toxic or potentially harmful to goats: aconite, black snake root, blue cohosh, crow poison, death camas, dicentra, false hellebore, hemp, horse nettle, jimson weed, larkspur, lobelia, lupines, marijuana, monkshood, nightshade, poison rye grass, spider lily, spotted water hemlock, staggerweed, wild parsnip, wolfsbane, arrow grass, black locust, broomcarn, buckeye, cherry, choke cherry, dogbane, elderberry, ivy, Johnson grass, kafir, laurel, leucothoe, maleberry, oleander, seven bark, sneezewood, sorghum, velvet grass, white snakeroot, wild hydrangea, clover, cocklebur, down broome grass, squirrel tail grass, pokeweed, ponderosa pine needles, baneberry, buttercups, crowfoot, ground ivy, spurge, and white cohosh.

A complete list of toxic plants, along with the types of toxicity each plant causes is available at the Cornell University website: ansci.cornell.edu/plants/goatlist.html

Rotational Grazing. One practical and natural way to help prevent the spread of parasites is by using a pattern of rotational grazing. Use a movable fence to keep your goats in an enclosed area until they consume all the grass in that location. Then move them to another spot where the grass is taller. They will happily munch on the new grass and won't have a chance to go back and eat from the soiled grazing area, breaking the parasites' life cycle. Goats ordinarily don't eat forage that has the smell of another goat's urine or droppings, but when they're crowded and food is limited, they can break that tendency.

Ellen Gould, who raises Saanen and Nigerian dairy goats at Sisters Three Farm in Mendon, Massachusetts, advises that grass should be allowed to grow about knee high before goats are given the opportunity to graze.

For best results, use three or more pasture areas to ensure adequate time for the grass

Goats can be raised on grass and grains, but they also need high-quality hay. Minerals are also important, especially copper and selenium. Free-choice mineral mixes can be placed in plastic cups mounted inside their shelter so they can take them as they need them.

Goat Dairy Breeds

Name of Breed	Description
Alpine	The French Alpine breed is a healthy milk producer in a wide range of climates. Color is varied, except solid white or light brown. Size is medium to large, with does around 135 lb. (61 kg). Their excellent milk production makes them a popular dairy choice.
LaMancha	Cross flop-eared Nubians with short-eared goats from Oregon, and you get LaManchas, a calm and docile breed with strong milk production. Height is at least 28 in. (71 cm) and weight is about 130 lb. (60 kg).
Nigerian dwarf	Let's hear it for the little guy! Nigerian Dwarf is a miniature dairy goat with a gentle disposition, making it great for a pet. Despite its size, the breed is known for high-quality milk, which is often high in butterfat. Production is about one-quarter to one-third of what you might expect from a full-sized dairy goat.
Nubian	If you like size (about 135 lb./61 kg), you'll like Nubians. They don't produce as much milk as some of the smaller breeds, but the milk is high in protein and butterfat. They can be loud, and their hound-dog ears make them comically attractive. They are one of the most popular dairy goat breeds in the United States.
Oberhasli	This Swiss breed used to be called Swiss Alpine and has a striking cinnamon-brown color known as chamois, with a black stripe along its back and black on its underbelly and legs. Size is medium to small, at 28 in. (71 cm) and about 120 lb. (54 kg) for does. Milk production is moderate, and fanciers say the milk is sweeter than other goat milk.
Saanen	Even larger than Nubians, this Swiss breed has grown in popularity in the United States, thanks to its high-quality milk. Color is pure white or cream with short, fine hair. Height is at least 30 in. (76 cm), and weight is about 135 lb. (61 kg). The National Saanen Breeders Association notes that the Saanen doe is "Queen of the Dairy Goats."
Sable	The distinction between Sables and Saanens is one of color. Basically, any Saanen that isn't colored white or light cream is classified as a Sable; all other characteristics are the same as Saanens. Sables are classified as a separate breed but for show purposes only.
Toggenburg	Named for Switzerland's Toggenburg Valley, Toggs have light to dark brown color, with white ears and white on the legs, tail, and face. Toggs are small to moderate in size, at least 26 in. (66 cm) and 120 lb. (54 kg), but they have conspicuously large udders. They're an active breed and moderate milk producers.

SOURCES: Perdue University Dairy Goats information: ansc.purdue.edu/goat/factsheet/breeds.htm; University of Oklahoma, Breeds of Livestock: okstate.edu/breeds/goats/

to regenerate. Note, though, that some fields may require a full year to break the parasites' life cycle and may be better used for cutting hay. Rotational grazing also lets you inspect the pasture before the goats start feeding, which means you can remove poisonous plants.

Space for Indoor Shelter

Dairy goats also need adequate indoor space—about 15 sq. ft. per animal for a bedding area, with 25 sq. ft. of outdoor exercise area. Goats don't require fancy quarters, but a basic shelter that keeps out the rain and wind is essential. Concrete or wooden floors are easier to clean than dirt floors, but a dirt floor is less costly and is actually recommended by the American Dairy Goat Association. However, the shelter should be located on high ground ,and the area must be well drained so that it does not flood during rainy periods.

The exercise pen and any pastureland should be fenced to a height of at least 5 ft. (1.5 m), depending on the size of your goats. Some are good leapers or climbers and can clear a 5-ft. (1.5 m) fence, while others will be contained. If predators are a problem, consider an electrical strand at the top (nose level for the goats), the middle (nose level for coyotes), and bottom (nose level for foxes or belly-crawling coyotes). Also consider guard animals. Dogs that are trained to protect and not prey on the goats are an option, but other choices include guard llamas and guard burros. As you might guess, adding llamas or burros and feeding them along with the herd can substantially increase the cost of keeping animals, but if the alternative is losing your animals to predation, there's really not much choice.

Some goats are better leapers and climbers than others. A 5-ft. (1.5-m) fence should be enough to keep them contained.

Choosing the Right Breed

Here's where you do have some choices. There are many breeds of goats around the world that can be milked, but only eight that are recognized as dairy breeds in the United States (see table on page 308). Goat fanciers all have their favorites, and some have different characteristics—size, coat, color, milk production, and demeanor.

Buying Your Goats

When your fences are up, your shelter is built, your feeding regimen is established, and you understand how to care for animals, it's time to make a purchase. The place to do that is a trusted farm, where you can look at the goat, its herd members and parents, and talk to the owners who can show you records of the animal's health and heritage.

Beware of auctions, where some goat owners might go to dump their unwanted animals. Dairy goats produce offspring, and not all offspring are desirable, so animals of lesser quality may be sold to unsuspecting buyers at auction. If you're in doubt about your ability to judge a goat, bring an expert who can ask the right questions and judge the goat's health. Also ask for information on its heritage, along with records of past medical treatments. If a seller won't tell you that information and or doesn't have up-to-date medical records, don't buy the animal.

In addition to asking for medical records and heritage, check the animal's physical appearance for the following signs of potential problems:

- The animal should have a shiny coat with no wounds and no thin or bald spots. There should be no apparent sores or scars.

- Are her eyes OK, or are they dull and cloudy? There should be no pinkeye, no sores, or ulcers around the mouth, teets, or hooves.

- How is the animal's muscles? She should have tone with a long, well-proportioned torso and not look overly thin. Emaciated appearance may be a sign of parasites.

- Is the udder well formed, with no signs of mastitis (swelling or reddish coloration)?

- Kids should be free of Caprine Arthritis Encephalitis (CAE) syndrome, a viral disease spread via nursing.

- Ask for records on other diseases in the herd, including pinkeye, Coccidiosis, and Johne's disease (pronounced Yo-NEES or yo-NAYS).

Rabbits

It's unlikely that rabbits will ever replace cattle, pork, or poultry as a mainstay in the American diet, but they are well suited to small-scale meat production. They take up little space, are quiet, and capable of generating hundreds of pounds of meat.

Rabbit meat is a delicacy in many parts of the world, with a mild flavor and a high percentage of meat to bone, especially with Californian, New Zealand, and other select breeds that produce the best-quality food. Rabbit is also very high in protein—with less fat than beef, chicken, turkey, or pork. And rabbit meat has an ideal mix of fatty acids, making it heart-healthy as well.

If you find the right market for rabbit meat, the price can range from $5 to $10 per pound, which adds up quickly when you consider that rabbits multiply like, well—rabbits. They also grow rapidly to adult weight. In less than three months, a fryer-sized rabbit (3½ to 5 lb./1.5 to 2.3 kg) is ready for the kitchen. Roasting-sized rabbits are considerably larger—roughly 8 to 11 lb. (3.5 to 5.0 kg) each and are generally ten weeks to six months old. The meat on young fryers is tender and pinkish white, growing darker and more textured as the animals mature into roasters.

Finally, rabbits fit in well with an organic farming operation since their manure is high in nitrogen and phosphorus, making it an ideal fertilizer or addition to the compost pile.

You can create a rabbit operation inside a barn, shed, or another sheltered outbuilding, away from the other activities on the farm, and largely out of sight. Because rabbits are quiet—they don't crow or cluck or moo or bleat—a meat rabbit operation can easily go unnoticed, even by close neighbors. Another plus is the fact that authorities don't consider rabbits a type of livestock or poultry, so production is less regulated than with other farm animals. Still, you should check your community zoning regulations before getting started.

Before jumping into a rabbit operation, get advice from others and read a good book about the potential problems.

Getting Started

John Howcroft, a friend who raises rabbits and beef and grows fruit at his 15-acre (6 ha) Perry Hill Farm in Acushnet, Massachusetts, says there are three important steps to take before purchasing your first rabbit:

1. Get the knowledge.

2. Get the equipment.

3. Get the right breed.

Easy Does It!

John, my friend who raises rabbits, keeps his rabbits cool using ice–filled containers. He fills a dozen or more 2-quart (2–L) plastic juice bottles with water and keeps them in his chest freezer. When the temperature nears 90°F (32°C), he puts one bottle of ice in each cage to keep his rabbits cool. When the danger is over, he cleans and fills the bottles and returns them to the freezer for the next time.

He recommends reading up on the subject and joining a local rabbit club to learn from others who are raising rabbits. Two books are at the top of his list: *Storey's Guide to Raising Rabbits* and *The American Rabbit Breeders Association Official Guide to Raising Better Rabbits and Cavies*. (See "Resources" on page 356 for more information.) Joining the American Rabbit Breeders Association (ARBA) or a regional group will help, too. It can bring you information and put you in touch with other breeders, who may be able to help you find equipment and the right breeding stock.

The second step is getting the right equipment. Most rabbits are raised in individual wire cages with water and food inside and a tray below to catch droppings. Wire cages keep the animals cooler in summer and because the droppings fall below, the animals never sit in manure and have fewer problems with parasites and diseases.

Organic production methods are different and more humane than standard rabbit techniques, offering the animals outdoor space and community time. But there's a good deal more work to manage the operation.

Location and Shelter

Rabbits don't take up much space, but they do have some basic needs that must be met; location and shelter first and second among them. In the wild, rabbits are natural prey for coyotes, foxes, fishers, bobcats, and even owls and hawks. In a suburban backyard, they can become a meal for all of the same carnivores, in addition to neighborhood dogs. To avoid predation, rabbits should be kept in a protected area. If your rabbitry is outdoors, you should fence the area to keeps dogs and coyotes away.

Rabbits also don't tolerate extreme heat, so their shelter should be located in a spot that receives some shade from the summer sun. If you live in the South, where temperatures exceed 90°F (32°C) for much of the summer, rabbits may not be a good option. In far northern regions where temperatures remain below freezing for extended weeks, keeping rabbits can be difficult during the winter months.

Feeding and Care

Rabbits can eat a mixed diet of pellets and fresh vegetables, which you may have in great supply, but there are pitfalls to sudden variations in diet. For instance, many people think rabbits love lettuce, but feeding young rabbits a heaping portion will result in enteritis (diarrhea) and can leave the rabbits weak, exhausted, and dehydrated, possibly even killing the animals, Howcroft said. Some foods that rabbits can eat include carrots, turnips, and greens in small proportions, but not kale or cabbage plants, which can give rabbits excess gas that they can't pass.

Feed Pellets. Rabbit pellets sold at commercial feed stores are the easiest option for feeding, along with a supplement of high-quality alfalfa and timothy hay to provide roughage. Together they will fulfill nutrient requirements for optimum growth. Adult bucks and does should eat about 6 oz. (177 ml) of feed pellets per day. Pregnant and nursing does need more calories. Howcroft says he uses a mix with about 16 percent protein.

Diseases. The best way to keep your animals healthy is to keep them well fed and housed in clean, dry shelters. Rabbits don't suffer as many maladies as large livestock, but Coccidiosis can be a problem. Animals infected with the parasitic protozoa suffer a loss of appetite and diarrhea. If you're raising organic rabbits, check with a veterinarian and your certifying agent for an appropriate approved treatment. Knowing an experienced and trustworthy rabbit producer or a good veterinarian who works with organic rabbits can be a big help.

Meat rabbits are best fed on pellets, but they can handle greens in their diet as long as they are introduced slowly. Vegetables hold much more water than pellets and hay and can cause intestinal distress.

In addition to Coccidiosis, rabbits sometimes suffer from Pasteurellosis, a contagious respiratory infection that causes coughing, sneezing, and wet hair around the eyes and nose. Diseased animals should be removed to a separate area to prevent them from spreading the disease.

Choosing the Right Breed

There are numerous breeds of rabbits, but the best two breeds for meat production are Californian and New Zealand. Both are medium to large in size, they put on weight quickly, and have small bones, resulting in a good meat-to-bone ratio. White-colored New Zealands are preferred, said Howcroft, because they are more tolerant of heat. Some breeds such as the Flemish Giant may grow larger, but they take longer to reach market size and the meat quality is not as good.

Having good lines of parenthood helps in more ways than just the flavor of meat. Some breeders will be better at mating and kindling than others, and some does will do a better job of nursing the kits. You can't choose your parents, but you can choose your rabbits' parents, and you should do so wisely. If a doe doesn't kindle enough kits, the bottom line of your operation can suffer as you spend money on food and shelter without reaping the rewards. In general, a healthy doe can produce about six litters of six to eight kits per year, making a total of up to fifty or so bunnies. At that rate, each doe will produce about 150 to 250 lb. (68 to 113 kg) of meat. You don't need separate sets of parents for each litter; one buck will service up to five to ten does, allowing a steady production of young rabbits throughout the year. A good-sized backyard rabbit operation might consist of twenty does and several bucks—capable of producing up to 500 lb. (227 kg) of meat.

Naturally with the start-up cost of capital equipment—cages, feeders, nesting boxes, and breeding stock, it may take several years before the operation becomes profitable. As a beginner, you can avoid heavy losses by starting out smaller and expanding as you learn the ropes. One buck and two or three does will suffice for someone who is just learning how to care for rabbits.

The Benefits of Pastured Rabbit Meat

The Maine Organic Farmers and Gardeners Association (MOFGA) cites the following benefits to pastured meat rabbit production:

- The pasture time allows the animals to behave naturally, exercising and breathing open air, which improves their quality of life.

- Even though rabbit meat is already healthier than other livestock, rabbits raised on organic pastureland produce more meat and meat of nicer quality, with a higher percentage of omega-3 fatty acids.

- When the animals produce manure, it is applied naturally to the soil in small doses, enriching the grass and improving the land, without the need of a farmer to cart the droppings to a compost bin, make compost, and then spread that around his plants or pasture. (Organic, grass-fed rabbits should not graze on the same area for one year to prevent the spread of diseases, primarily Coccidiosis, a common intestinal malady for rabbits that is caused by ingesting parasitic protozoa.)

- The production rate is slower, since animals take longer to reach their ideal weight because they eat a more varied diet and get more exercise than in a confined cage.

- Rabbits that are best for production from feed may not be best suited for pasturing. The table below lists some recommended breeds for organic care, including grazing.

Best Grass-Fed Breeds

Name of Breed	Weight (lb.)	Weight (kg)
American Chinchilla	16	7.3
Californian	10.5	4.8
Champagne d'Argent	12	5.4
Cinnamon	11	4.9
Crème d'Argent	11	4.9
New Zealand	12	5.4
Palomino	11	4.9
Satin	11	4.9

Doing It the Organic Way

Feeding rabbits with organic feed pellets will cost more money, but the problem may be one of convenience. Local supply stores may not stock certified organic feed in large enough quantities for a meat-rabbit operation.

Management is also an issue. Rabbits do not always do well in mixed company. Bucks can become overly aggressive and can kill other bucks. That makes it difficult to give the animals adequate access to the great outdoors.

One method that organic rabbit ranchers use is a modified chicken tractor—a movable large-sized hutch that can be carted from one place on the lawn to another for rotational

When it's time to mate, the doe is introduced to the buck's cage, where the ritual can be just a few minutes long. When he's finished, the male passes out momentarily.

grazing. As the rabbits exhaust the food source in one area, they can be moved, shelter and all, to an adjacent area where the grasses and clover are growing tall and lush. Rabbits can be returned to a more traditional shelter at night to avoid predators. The bottom of the movable hutch should be covered with slats or caging wire to prevent the rabbits from digging out. Rabbits are burrowing animals by nature, and when the weather gets warm, they want to dig into the cool ground, where they can be closer to their optimum temperature of 50°F (10°C).

Sunshine can also be difficult to manage. A shady area over an improvised rabbit tractor might cover the animals in the morning but then leave them exposed in the afternoon. Also, if you supplement the animals' diet with garden scraps, make sure that everything they eat has been grown to the exacting standards of the National Organic Program, with only materials approved by the Organic Materials Review Institute (OMRI). Rabbits raised in a rotational grazing pattern can get close to half their food from the grass, but it will take them longer to fatten up because the calorie intake isn't as high and they get more exercise outdoors than in a cage.

The last stage in the process is the butchering. If you don't have the heart to do it yourself, you'll want to find a reliable butcher to do the work. Once butchered, rabbit meat can be stored in a freezer until you're ready to use it or sell it. As your operation grows, you may need a large freezer to store the meat until customers can pick it up. All these details are part of the planning process and should be considered before you purchase your first set of breeders.

When I worked as a produce manager, my old boss "Red" used to tell me that "the fresher the produce, the less time you have to spend fussing around with it." By "fussing around," Red meant trimming, primping, and packaging it to look good for the customers. Red knew that if the lettuce was fresh, the celery was crisp, and the leeks were clean and green, the produce would sell itself. His wisdom points to the biggest marketing advantage that local farmers have over grocery chains: freshness.

From Farm to Fork

Because you grow your own vegetables—and because you don't have to pick them two days in advance and then ship them 1,500 miles (2,414 km) away—freshness is pretty much guaranteed. Your local organic produce is also healthier, more flavorful, and longer lasting than the produce grown by conventional farmers. With those selling points on your side, you'll find several available marketing vehicles for getting your goods to the consumer.

The three top marketing ventures for small organic farms are farmers' markets, Community-Supported Agriculture (CSA) programs, and restaurant sales. Explore them in the following pages to decide which one or which combination of the three will work best for your business.

Farmers' Markets

Farmers' markets are one of the oldest venues used by small-scale farmers around the world. Despite being the norm elsewhere, farmers' markets are just now beginning to draw larger numbers of North American buyers.

Farmers' markets bring growers and customers together for commerce, conversation, and the ritual weighing of sumptuous summer squashes.

Buoyed by the Slow Food movement, a culture of weekend gourmets who seek in-season culinary treasures, and customers clamoring for a better connection to their food, farmers' markets are enjoying phenomenal growth. These open-air multi-vendor bazaars are the easiest way for many growers to sell their goods. Chances are you have half a dozen within easy driving distance from your farm—close enough for you to market your fresh-grown produce without the burden of high transportation costs.

The Scene

Small, local farmers' markets spring up in various local venues: a park, library, school, or other public place where, once or twice a week, farmers and market gardeners peddle their produce directly to eager customers. It's an arrangement that eliminates the middleman, bringing more profits to the farmers and providing fresher food to the customer. It also strengthens the bond between buyers and their local food producers.

Market Management

A community group, city or town agency, merchants association, agricultural partnership, or other sponsor usually requires farmers to sign a contract—charging either a weekly or seasonal fee. The agreement may have specific requirements for farmers who participate and may possibly state that. . .

- farmers sell only fruits, vegetables, berries, eggs, or meat items that were raised on their farm.

- processed food items be locally packaged with the producer's name, address, and contact information included on the label.

- farmers adhere to state or city rules regarding the sale of farm goods.

The Simple Farm Stand

Don't overlook one of the simplest ways to sell your farm-fresh goods—an old-fashioned roadside stand. An improvised shed or even a large table under the shade of a pop-up canopy can become your store. Fresh produce and ample parking on a well-traversed road are important factors, says Myrna Greenfield, owner of Good Egg Marketing, a Boston-area business that specializes in helping small farmers build branding and sell their goods.

Another key is signage, she said. A hand-painted sign that says "Fresh-picked sweet corn" makes it tempting for drivers to pull over.

Below are some additional things to consider:

- Check with local officials before getting started. There may be an issue with zoning or traffic.

- Make sure perishable food is stored properly to prevent spoilage.

- Ask yourself whether you have the time to work at the stand or will you run it on the honor system? If using an honor system, make sure the cash is secure. Mount a strong box securely and leave a slot for cash. A thief isn't likely to steal heads of lettuce, but a box of money may be all too tempting.

• farm owners themselves be present to sell their wares.

Contracts often contain an indemnity clause to protect organizers against lawsuits. You may not need a lawyer to review the document if there's one required for a local farmers' market, but you should read the document very carefully before signing.

Government Assistance

Federal and state programs now provide assistance to farmers' market customers. The USDA awards grants to states, allowing them to fund the Senior Farmers' Market Nutrition Program (SFMNP), which provides low-income senior citizens with SFMNP coupons redeemable at farmers' markets. The program encourages seniors to gain access to healthier, fresh foods, and it draws more customers to local farmers' markets. Similar programs are available for Women, Infants, and Children (WIC), providing Farmers' Market Nutrition Program (FMNP) coupons, Supplemental Food and Nutrition Program (food stamps), and even fruit and vegetable prescriptions in some states.

If you're thinking of selling your goods at a farmers' market, spend some time scouting out the territory. Ask for advice from vendors, and settle on a venue where you feel comfortable.

The Pros

Because the markets attract farmers with different specialties, they lure varied customers. Some markets feature dairies with locally crafted cheeses, ranchers with grass-fed beef,

Farmers' markets bring growers and customers together for commerce, conversation, and the ritual weighing of sumptuous summer squashes.

Easy Does It!

Markets are a great place to experiment, too. Use your imagination to merchandise your material in a way that makes it more appealing. One of my favorite CSA creations is a fresh salsa verde kit. I had originally prepared the recipe at home and brought it to work with a bag of chips. All of my coworkers wanted to know how I made it and where they could get the ingredients. So, the next week, I brought in some kits: I put a dozen green and purple tomatillos, one giant sweet onion, five serrano chilies, a few jalapeño peppers, and a big bunch of cilantro into a plastic container. I also included one purchased organic lime and a recipe for the salsa. At $8 each, I sold more than I could produce. And every season since, people have asked me to let them know when I would have the kits available again.

I've been to enough farmers' markets to know that this same technique would work for customers in that venue—especially if the market were scheduled just before the weekend, when buyers might be looking for a dish to bring to a summer party or barbecue. Consider the fruits and vegetables you grow that will be in season for the market, and get creative! You can craft any number of recipes, bundles, and kits.

beekeepers selling local honey, and artisan bakers offering homemade breads and muffins. All of this diversity combines to create a significant food-buying attraction that brings more customers to the markets. Larger markets can sometimes take on a festival atmosphere by hosting bluegrass bands, face-painting booths for children, and other entertainment, making the market a weekend destination for families.

The beauty of local farmers' markets is that buyers can pick and choose the best of what local growers are offering—produce harvested fresh from the earth and in peak season. For farmers, the setting provides a place to meet customers, learn more about what people want in fresh food, and share ideas with their earth-wise colleagues. With so many knowledgeable growers assembled, the conversation often revolves around food trends and gardening techniques, sprinkled with a dash of social activism: *Have you tried raw Lacinato kale drizzled with a lemon vinaigrette dressing? Has the cool summer slowed down your melon crop? What about fungal diseases on your tomatoes? Have you signed the letter yet on statewide labeling of GMO foods?*

Farmers' markets are a great place to get started in the business of selling your produce. Unlike with CSAs and restaurant sales, there's no pressure to supply a specific amount of any one commodity at a market. You can bring what's in season and sell your produce based on supply and demand.

The Cons

When manning a booth, you'll spend a good deal of time talking to customers, many of whom just want to look. If you're not outgoing and you don't like interacting with strangers, you might not enjoy the experience of people handling and prodding your precious

Baked goods, jams and jellies, and tangy pickles are among the value-added farm goods sold at farmers' markets. Consider selling your honey, eggs, or fresh-made relishes, too.

produce—like the lady who pinches the tomatoes, pulls husks off the sweet corn, and asks if you can mark down the lettuce because it's late in the day.

A more serious drawback is that farmers' markets can take time away from your busy growing schedule. That could mean that the five hours during which you might have been weeding, watering, planting, or transplanting will be spent selling goods and counting change. Most markets are timed to catch the busy after-work crowd looking to buy fresh produce for their evening meal. But weekend events can last for much of the day, effectively consuming what might be a precious part of your farming workweek. If you have a full-time job in addition to your part-time farming operation, you might not have the time to pack all your goods into boxes, truck them to the market, set up, and then wait behind a table to sell them. (This is where having a partner or farmhand can help.)

The Road to Success

When selling at a market, be observant of all laws and regulations, including some unwritten codes. One of these is to bring only your best items to sell. In other words, don't show up with scrawny green beans, overgrown zucchini, and bitter, half-bolted lettuce. The quality of your produce contributes significantly to the overall customer experience at a farmers' market. You don't want to be skipped over by customers, and you certainly don't want to drag down their impressions of the market as a whole.

Speaking of things that will annoy your fellow farmers, don't sell your produce at below-market prices. If you're a hobby farmer or a grower who doesn't depend on the market to provide an income, that's great for you, but it's bad business and bad taste to undercut the livelihoods of the farmers who need their market income to help pay the bills.

Finally, keep your food fresh. If it's a hot, sunny day, keep your crates in the shade and store highly perishable produce in coolers, topped with ice. Many growers keep their lettuce, greens, and tender produce in shallow wooden or plastic trays covered with wet burlap: Store the burlap in a cooler filled with ice before spreading it over the produce.

The Plan

As we've discussed, selling at a farmers' market is as simple as renting a stall and showing up with as much of your freshly harvested high-quality produce as you like. But you don't have to stop there. If action is consistently slow at one market, consider another location that has more customers, better hours, or less competition. A number of small-scale growers sell their goods at three or four different farmers' markets throughout the week.

Don't like the idea of traveling to a market? Then let customers come to you by selling CSA (Community–Supported Agriculture) shares. Buyers will pay a fee for the full season, then visit your farm every week to pick up their boxes or bags of fresh produce.

Community-Supported Agriculture (CSAs)

One of the newest and most exciting ways to sell produce is through CSAs. While participation in CSAs is a fast-growing trend in farming, it's not something that you should enter into lightly.

The Scene

A CSA works like a pay-in-advance cooperative. Customers buy a share in your farm's anticipated harvest, paying up front for a season- or year-long supply of fruits, vegetables, and other goods. The farmer then packs these shares into baskets, boxes, or grocery bags weekly for customers to pick up.

The Pros

For small- and medium-sized farming operations, the up-front money helps pay for seeds, supplies, and labor before the harvest, thereby guaranteeing a source of income when it's needed most. The arrangement also allows farmers to do their marketing in advance of the season—usually during the winter when there are no fields to plow. In many ways, it's a perfect arrangement. But it is by no means easy.

The Cons

Once you've collected money from shareholders, you're left with the solemn responsibility of upholding your end of the bargain—not just for yourself but also for all the other farmers who sell their produce through CSAs. A failed attempt could leave your customers with a bitter taste about ever going to another CSA, and that would be a great disservice to fellow farmers who are making the business model work successfully across the country.

Many growers sell extensions to their CSA (Community–Supported Agriculture) season, offering pumpkins, squashes, cool–weather greens, and root crops well into fall. Others offer flowers, fruits, berries, or eggs as an add–on throughout the season.

In my travels to workshops and seminars, I've attended multiple lectures by farmers who are experienced at planning and growing vegetables for a CSA. Almost universally they warn that it's not a wise undertaking for inexperienced growers. The pressure of providing weekly shares to faithful customers can be taxing—especially when an unusually hot and dry summer forces lettuce to bolt or a devastating plague of late blight destroys entire tomato fields before a single fruit can be harvested. Growers who aren't skilled or experienced enough to provide a variety of high-quality vegetables all season long or to plant successive crops of lettuce, radishes, scallions, and produce that customers are looking for week after week should consider another marketing option.

That said, the risk of failure is part of the bargain in a CSA. Customers should be advised that their investment in your operation is subject to the vagaries of weather, pests, and other unforeseen events. Still, customers are putting their trust in your ability to cope with some emergencies and make the most of smooth stretches in the growing season. You should decide whether you're up for that challenge before starting a CSA.

The Road to Success

The most important thing to remember when planning and executing a CSA is that customers want variety. That variety should also take into account what customers will want to eat. For example, include lettuce or other greens during the summer months when customers might want fresh, cool salads. Don't think for a minute that you can fill a share with 5 lb. (2 kg) of zucchini every week, just because that's what seems to grow best in your fields.

In addition to variety, customers want fresher, crisper, tastier, and healthier produce than they can find at the supermarket—that's why they're coming to you. You have to harvest your produce at its peak value and keep it fresh for as long as you can. And while packaging appearance has no bearing on the freshness of your tomatoes, it does have a bearing on the customers' perception of your produce and your business. The bottom line is, always do right by your customers and your business will thrive.

The Plan

All those caveats aside, I market the bulk of my produce through a small backyard CSA with six to eight customers per season. I find it fits well with my diverse organic crops, my regimen of successive plantings, and the techniques I've used for more than two decades as a large-scale vegetable gardener. My market garden also keeps my family fed, as I plan two extra shares for our table.

If you decide to market your vegetables through a CSA, you'll need to think about the following five questions:

1. How long should your season last?

2. How many different kinds of vegetables should you grow?

My CSA Growing Schedule

My CSA shares last twenty weeks, beginning in late May and ending sometime in mid-October. For my customers—and most of the northern half of the United States—the first vegetables of the season include radishes, scallions, cabbage, radicchio, mustard greens, perennial rhubarb, salad turnips, spinach, mesclun, and an assortment of Asian greens. Other early-season offerings could include kohlrabi, cauliflower, and perennial asparagus, along with items that had been root-cellared over the winter (think onions, garlic, turnips, parsnips, and other root crops).

The middle of June brings peas, broccoli, cauliflower, carrots, beets, early lettuce, Swiss chard, kohlrabi, green garlic, and onions, along with more radishes and greens. Midsummer will supply the first tomatoes, along with kale, cucumbers, carrots, green onions, successive crops of heat-tolerant greens, bush beans, pole beans, summer squashes, and possibly the first eggplants and peppers of the season.

Late summer and early fall constitute peak harvest time. Baskets will be brimming with a full variety of heirloom tomatoes, peppers, eggplants, and squashes, along with successive mesclun and lettuce, pole beans, and cucumbers. Throughout the summer and fall, kale, Swiss chard, carrots, and a few other greens are weekly staples.

3. How many members can you provide for?

4. How big should your shares be?

5. How do you determine the price?

The answers to these questions are as varied as the people who ask them. You might benefit from talking to other farmers who run CSA programs. Or you could attend a regional farming conference to catch a lecture or seminar taught by someone with experience in CSAs. Here is some information to get you started.

Season Length

The length of your CSA season depends entirely on your own skills for stretching the growing season and making sure you produce a steady and varied supply of vegetables. Typically, CSAs begin in late spring and end in early fall. The growing season will naturally vary by geographical location and the types of vegetables you choose.

Many CSA managers sell a late-season share in addition to their regular summer program, providing winter squashes, late fall greens, turnips, fall broccoli, cabbages, lettuces, and Asian greens. Depending on your location, the fall CSA might tack on another six weeks, bringing the season to a full close around late November.

Harvest Timing

Because the most important part of running a CSA is providing customers with the freshest, highest-quality produce, you'll need to carefully consider the timing of your harvests.

Some vegetables need to be harvested earlier than you might think. For best value and customer satisfaction, harvest cucumbers while the fruit are still thin and crisp, not when the cukes are swelling and have thick skin, big seeds, and bitter flavor. Broccoli needs to be cut when the heads are tight and firm, not after the buds begin to separate and open into tiny yellow blossoms. Pick zucchini and summer squashes when they're 6 to 8 in. (15 to 20.5 cm) long and still thin. (Sure they'll reach the size of baseball bats if you let them, but the skin gets tough, the seeds grow large, and the culinary quality of your fresh produce is lost.) Harvest lettuces, mesclun, and greens before the plants start to go to seed and the leaves get bitter. When the crop begins to fade, pull it and sow another.

On the other hand, don't be tempted to pick vegetables too early. Tomatoes taste best when they're allowed to ripen on the vine. Don't pick a beefsteak before it's ready, cheating it out of those last few days under the late-summer sun. The same goes for peppers, which grow thicker walls just before the fruits ripen from green to red, yellow, brown, or purple. (For information on harvesting specific crops, refer to the vegetable descriptions in chapter 5.)

The time of day when you choose to harvest will also affect the quality of your vegetables. In almost all cases, the best time to pick is early in the morning, while the temperature is cool and the sun isn't beating down on the crops. After the cool evening, roots have drawn in moisture and the bounty is at its best. Of course if you're heading to a farmers' market and you want to set up your stand by 7:00 A.M., you probably won't be getting up to pick lettuce at 2:00 A.M so it can be washed, packed, and then trucked to the market. Crops that are picked the day before sale should be kept cool and fresh.

Harvesting your beets, cauliflowers, salad turnips, and greens in the early morning and selling them to CSA (Community–Supported Agriculture) customers the same day keeps the food at peak freshness.

Keeping Produce Fresh

You'll have to develop ways of keeping your produce fresh from the time you harvest it to the time the customer picks it up. To keep the produce fresh after picking—especially during the summer months—remember that water is your vegetables' best friend. I always pick vegetables with a large plastic barrel of cold water nearby. After cutting a head of lettuce, for example, I plunge it into cool water to keep the fluids from draining from the stems and causing the leaves to wilt. The result: lettuce stays crisper for longer. The same goes for fresh-cut leaves of Swiss chard, kale, and other greens. Spinach—roots and all—gets water treatment, too.

Even the root crops—beets, turnips, radishes, and carrots—go into the water. Turnips, radishes, and beets all have edible greens, which I try to keep crisp and appetizing. Feathery green tops on a bunch of carrots not only provide a great handle for carrying the bunch, but the unwilted greens also tell customers that their carrots are as fresh as they can be.

Looking As Good As It Tastes

I sometimes think of my old boss Red when I'm fussing too much with the vegetables I've picked, but I think appearance is part of the allure of buying a bundle of fresh organic produce.

Once I harvest crops at my farm, I rush them to the cleaning station: a double-well outdoor sink with running cold water. I then bag and portion the dry produce—beans, peas, tomatoes, peppers, squashes, eggplants, and similar items. Then I give lettuce, cabbages, and greens a good washing; rinse green garlics, green onions, leeks, and root crops well to remove any soil; evenly trim the stems of Swiss chard, spinach, broccoli, and kale; and inspect all leaves for insects, slugs, or foreign material. When I've finished with that, I bundle individual portions with rubber bands and place them in buckets of fresh water until customers come to pick up their shares.

I made this improvised cleaning station from a CSA (Community-Supported Agriculture) member's unwanted kitchen sink. Here my wife, Valerie, washes fresh beets, while kales, Swiss chard; other greens soak inside the sink's other well.

Large-scale farmers—and even many small-scale commercial farmers—can't provide the same tender loving care when harvesting their crops, and that gives us an edge. The extra effort takes a toll, sometimes doubling the time necessary to pick and prepare our produce, but we think it's worthwhile. After spending so many hours, days, weeks, and months growing our vegetables, it would seem shortsighted to just pick the crops and stuff them in a bag—or even worse, to let the precious vegetables linger too long in the sun, where they could wilt in a few short minutes. So I make sure the produce in my CSA shares looks as good as it tastes. My customers repeatedly tell me they're impressed by the quality and freshness of what I grow.

Fresh organic lettuce, greens, scallions, and herbs look as good as they taste. Insist that your customers arrive on time to pick up their goods before things start to wilt.

Share Sizing and Pricing

Share sizes can vary, but consider that you're providing a family of two adults and maybe two children with all the vegetables they might eat in a week. Your customers have jobs and busy schedules, so not every day will end with a full meal that includes grilled eggplants, fresh salads, or sautéed greens. But, as a general rule of thumb, you should provide enough produce so that each shareholder can eat something new every day.

I shoot for about a dozen different vegetables totaling 8 lb. (3 kg) per week (heavier when zucchinis, summer squash, and tomatoes are in season). A twenty-week season will run my customers just under $500, which works out to about $25 per week. For additional fees, customers can get a dozen eggs per week or opt into the six-week fall CSA.

To get an idea of how to structure your prices, visit other farms in your region that provide CSA programs. Another good resource is the Local Harvest website: localharvest.com which provides links to farms, farmers' markets, and CSA programs across North America. The majority of the shares I've seen at other CSAs in the suburban Boston and Providence areas are priced similarly or slightly higher than mine for about the same variety and quantity of produce. Most weeks, if you compare the cost of each item in one of my shares to their organic supermarket equivalents, the bags are a definite bargain. So I feel good about the prices I charge, and my customers are satisfied enough to return the next season.

Subscription Variations

CSA programs can really be tailored to whatever suits you and your customers. Farms routinely sell varied sizes of shares—sometimes creating half-sized shares for smaller families and larger shares for big families. The policy can become a problem if you're forced to give someone half a watermelon or half a head of Chinese cabbage. A simpler option is to keep all shares the same size but allow customers to buy them every other week.

I also like to spread my shares throughout the week so that I don't have to pick for everyone on the same day. I provide one set of shares in the evening on a weeknight and the other set on Sunday afternoon. For larger CSAs (more than 100 customers), some farmers I know spread their shares evenly across the week or split them among three days—say, Monday, Wednesday, and Friday. The schedule is up to you and your customers, but you should be firm about pickup days. If someone doesn't pick up their goods, you can't be expected to keep them fresh and cool for an extra day. Let them know up front that uncollected shares will be donated to a food pantry.

Some CSAs run by larger farms allow buyers to choose their own vegetables from the items harvested that day. For example, Brookfield Farm in Amherst, Massachusetts—a nonprofit CSA that's one of the oldest and longest-running programs in the country—has a big barn with tables where customers can pick over the produce and box their own shares. Signs and chalkboards advise customers of rules and weight limitations for certain items. (No, you can't just fill your bag with 8 lb. (3 kg) of heirloom tomatoes.)

Added Value

Some CSAs sell a variety of farm goods in addition to fresh vegetables. You make the rules—as long as your customers agree to them before the season begins. If you have chickens, you might add half a dozen eggs each week. Maybe you could include fruits, berries, flowers, herbs, or even fresh-baked bread for an additional price.

Personally, I like to include a few bonus items at no additional cost. I love growing kitchen herbs, and I include a bunch in every week's share. I also grow edible flowers and include nasturtium blossoms with our lettuce mixes. Because edible flowers are so fragile, they are rare and extremely expensive; it's a small gesture, but customers truly appreciate the novelty. Depending on how successful my honey harvest has been, I sometimes include a jar of honey with the final CSA as a sign of appreciation for sticking with me all year long.

You don't have the time or inclination to produce bonus items? I know some innovative CSA managers in southeastern New England who contract with berry growers, bakers, and even a few local fishermen to include nonfarm items through their CSA programs, but that

Fresh-picked tomatoes, eggplants, peppers, and other summer vegetables cover the kitchen counter on one of our pick-up days.

takes a tremendous amount of organization. For the most part, you should keep your CSA simple: Limit it to the items you grow on your property. Just set the price accordingly and be consistent about providing fresh, high-quality organic produce.

CSA Marketing 101

Once you've established your CSA, you should never have a shortage of customers: Fulfill your end of the bargain and people will tell others about their culinary experiences. You may even develop a waiting list. When it comes to marketing CSAs, nothing's better than word-of-mouth advertising. But finding the first customers often takes a little bit of work.

Once I decided to start a CSA, I approached customers with a one-time, act-now opportunity to participate in my "experimental" CSA. I charged half price—only $250 per share—and told my potential customers that I was confident the same share would double in price next season. I got eight customers. The year went well, and several customers stayed with me when I started again. Meanwhile, neighbors who wandered into my expanded and highly productive growing area wanted to know what I was doing. After admiring the vegetables I produced, they and some friends joined the following year. Other CSA shares went to coworkers and friends of friends who heard about what I was doing. I suspect that's common for many CSA producers, but it's not exactly an organized plan for finding a large customer base.

A farmer who runs a CSA in Upstate New York—and who took a more organized approach—said she began her first marketing drive by placing flyers at places frequented by

A Facebook page is an easy way to keep in touch with customers. Also think about starting your own farm web page for better control of content and branding.

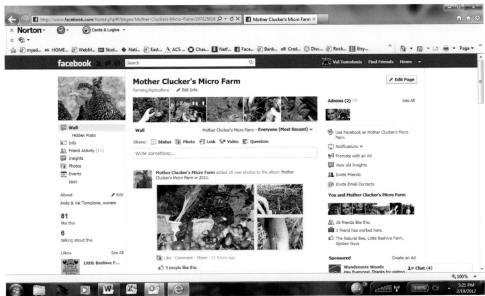

There are multiple ways to attract customers to your fledgling CSA (Community–Supported Agriculture) program. Consider posting notices on bulletin boards or using social media websites.

health-conscious people. She targeted gyms, yoga studios, spas, daycare providers, and other locations. She also offered a discount price for her first year and attracted more than fifty customers.

Given the growth of social media, you might also post advertisements for your farm on the networks you already use. Ask your online friends if they are interested and to spread the word. Maybe even create a Facebook business page for your farm where you can post a description of your farming operation, photos, and notices. I have a Facebook page for my farm (Mother Clucker's Micro Farm), and our customers post comments about how much they like the produce we grow. The site has brought us several inquiries from potential customers, some of whom will likely be on board and commenting in the season to come.

Restaurant Sales

Many growers have developed profitable businesses by selling their fresh produce to local restaurants. If you can find a chef with a steady need for something that you grow and if you are a consistent enough grower to provide the specific ingredients in the right quantities, the relationship can be financially rewarding. Restaurants pay a much better price than wholesalers, and they buy in larger quantities than customers at a local market.

But breaking in is not always easy. Chain restaurants usually operate commissaries and buy their food items in bulk from large-scale distributors, so they're not a good option. Locally owned restaurants—which are essentially small businesses like your farm—are a much better bet. And the more discriminating the restaurant, the more likely it is that its chef will have an appetite for local food. Consider the following points:

Convince Chefs. Before you begin a relationship, you will have to convince a chef that your produce is of the highest quality and that you are a dependable grower. You may have to provide samples over several visits and show convincing evidence that you can produce the right quantity of ingredients at the right time.

Be Flexible. Your courtship with a chef might take months to complete, and his or her needs will invariably determine at least part of the next season's growing plan.

Specialize. Upscale restaurants look for eclectic menu items that aren't available from other sources. If you specialize in hard-to-get produce, you might be a natural fit. Think of baby fennel, wild arugula, purple opal basil, Bull's Blood beet greens, or heirloom Black Krim tomatoes. Follow food trends and keep in contact with chefs to ensure that you are growing things that are in high demand.

Develop Client List. Because one restaurant isn't likely to buy everything you grow, you will have to contact several chefs to develop a client list. If the restaurants are located in a nearby city, you can save time by making all your deliveries on the same morning.

Don't Be Discouraged. Some chefs might not want to take chances with a new and somewhat inexperienced grower. Don't be discouraged by an early rejection.

Green Is the New Gold

One Massachusetts grower I know recently switched from mixed farming to growing lettuce greens for restaurants in Boston. He sows successive crops of arugula and other fast-growing lettuces, cuts the greens with sharp scissors, then sorts, washes, dries, and bags before selling it to restaurant owners in the Boston area. He makes deliveries on the same day to maximize his driving time and makes more than $30,000 a year.

It took him time to establish a system, find the right customers, and meet chefs. Then he needed to convince them that his greens were superior. But after doing the legwork, he developed a reliable customer base for a business that can continue to expand. He now grows greens under protection in the winter months and charges a premium price for them because they are out of season.

While some might not have the growing acumen or business savvy, others can certainly earn $10,000 or more runnng a part-time business growing only specialty salad greens for the restaurant trade.

Be Reliable. Most important, remember that if a chef takes the time to develop a menu that includes your ingredients, he and all of his diners are counting on you to deliver them.

Talking to other growers who have supplied restaurants is always a good idea. Not only can they show you the ropes but they also might have customers they had to drop or a waiting list of restaurants that they would be willing to share.

Other Marketing Opportunities

If you talk to other farmers and food buyers, you may find even more ideas for marketing your farm goods. Some larger growers, CSA organizers, or food distributors may need specific ingredients that you could provide.

Food marketing specialist Myrna Greenfield, founder of Good Egg Marketing in Boston, advises farmers to consider selling to schools, hospitals, nursing homes, and other institutions. The price may not be as attractive as selling directly to retail customers, but you don't have to spend time at the market selling your items. A good place to start is the website of the National Farm to School Network (farmtoschool.org), which works to pair farmers with schools for education and marketing. The national site provides links to state contacts.

"Start small. Add things slowly as you learn how to do them well. Don't go into debt."

—Julie Rawson, Executive Director, Northeast Organic Farming Association, Massachusetts Chapter

Think for a minute about your own organic farm. Is it a peach grove in Georgia with sweet pick-your-own freestones? Maybe it's a ranch in Ontario, where grass-fed Hereford cattle snort steam under a gray November sky? Or is it a classic organic vegetable patch in the Connecticut River Valley, with soldierly rows of broccoli sprouting from crumbly chocolate-cake soil? Your farm can be whatever you want it to be, but you need a plan to keep it on track.

Why Plan?

Your farm's operation can be as simple or complex as you decide to make it, but if you want to turn your hobby farm into an income-producing venture or expand and take on new customers beyond your friends and neighbors, you may need to create a more formal business plan. By discussing the possibilities and working through the mathematics, you should be able to prevent your ambitions from overtaking your means of fulfilling them.

Farming commercially without a business plan is a little like launching a raft into the ocean without a compass, map, or provisions and hoping you'll bump into some nice sunny island to live on. Simply put, it can be an invitation to disaster. You might be an amazing grower, work very hard, and still be unsuccessful if you don't prepare for the needs—both current and eventual—of your farm as well as your vision for it.

Here's one example of how easily you could run astray: Let's say you're running a successful microscale organic farm and earning close to $5,000 per year by selling ten CSA (Community-Supported Agriculture) shares plus pastured eggs and raw honey. Extrapolating from your original business model to create a larger one might seem logical. You do some quick math and posit that if you signed up one hundred subscribers, you could count on an income of $50,000 per year. But you don't account for the expense of additional land to raise the crops. You'll also need larger and more expensive farm equipment to cultivate, harvest, clean, and package the produce. And you'll need a packing shed for prepping the produce, a walk-in cooler to keep the greens fresh, and a barn or shed to store all your expensive

Will your vegetable business bring in enough to pay for a new hoop house, or will the income barely cover the cost of seeds? Plan before you plant.

Take an Oath for Growth

When you become a personal farmer, growing food for yourself and your family, you take on a certain level of responsibility—striving to make sure that everything you put on the table is healthy, fresh, and of the highest quality. That responsibility increases at the moment you decide to sell food to the public.

As a food supplier, your customers will depend on you for the health and happiness of their family members. This trust should never be taken lightly. So before you get down to the business of planning for your farm, you need to take an honest look at yourself, your abilities, and your ambitions.

The success of even the smallest organic farming operation depends on standards that are essential for you, your customers, and the other farmers who work the land for a living. Consider taking a simple pledge before delving into the world of small-scale agribusiness.

Know Your Limitations. To be successful, you'll need to have patience, time, and the ability to raise high-quality foods. Don't attempt to sell produce unless your vegetables and fruits are market worthy (no tadpole beans, swollen cukes, or lettuce that has turned bitter and started to bolt).

Make the Most of Your Land. Choose crops and livestock wisely to strike a balance in what you can do. If you try to do too much and start out too quickly, you can go into debt and become one of the agriculture trade's casualties. That won't do you, your customers, or farming any good.

Strive to Get Better with Each Season. Become a student of agriculture by attending workshops, trying new fruit and vegetable varieties, and treating the science of organic growing as a serious endeavor. Seek solutions to problems, and accept occasional failures as learning experiences. Don't settle for the second-rate crops that are wilted, old, underdeveloped, or overripe.

Recognize That Meat, Eggs, Dairy, and Produce Have Value. Don't undercut other local farmers by selling your food for less than it's worth. In addition, don't apologize if your prices are higher than the grocery store chain that sells lettuce at a loss so customers will purchase other goods at higher prices. Grow good food, and price it for what it's worth. To learn about local prices, check out other farmers' markets and farm stands. If you can afford it, consider a sliding scale for low-income customers. And donate unused produce to the local food pantry.

Be Organic. Even if your business earns less than $5,000 per year and is exempt from USDA organic certification inspections by law, adhere to the principles of organic farming. The standards are in place for a reason; they protect the principles and livelihoods of organic farmers who have made the decision to protect the food supply and the planet from those who would take shortcuts for profits.

machinery. Finally, because you can't possibly do everything yourself, you'll need employees. Suddenly, your $50,000 in earnings looks more like a serious loss.

This example may be oversimplified, but clearly, things can get complicated in a hurry. And if you haven't considered and prepared for all the possible pitfalls—crop failure, machinery breakdowns, drought, or some other unforeseen catastrophe—you might join the ranks of those who were forced to give up their dream after it careened out of control.

Unfortunately, you won't find a one-size-fits-all template for an agribusiness plan. Farmers are as varied as the products and services they produce, so a pick-your-own blueberry farm will have different needs from a wholesale dairy operation, a dry-harvest organic cranberry bog, or a roadside vegetable stand. What you will find at university websites and elsewhere are multiple templates for farm budget plans that may be similar to yours. And local farmers and cooperative extension services can help you adapt an existing model to fit your ideas.

Once you have guidance and a model, you and your partner(s) will be the best people to develop the specific working plan for your small agribusiness. If you do the planning before the planting, you can make the most of your money, time, and effort.

Is your goal to market fresh, naturally grown, chemical-free vegetables and sell them at weekly farmers' markets? If so, write it down.

Plan the Flight and Then Fly the Plan

This chapter provides you with a basic overview of the things to consider before creating a business plan, which takes a lot of work, research, and introspection. Once you've created goals, written down and answered all your questions, considered your own skills and shortcomings, and learned what business-specific finances entail, you'll be better prepared to tackle business-plan templates.

Thankfully, many schools, cooperative extension services, organizations, farm cooperatives, and industry support groups are willing to help walk new farmers through the minefield of agribusiness decisions. They run workshops, maintain websites, and hold conferences to stress the importance of hatching out a business plan for your small farm. To find an organization offering farm planning workshops near you, contact local farmers, your local university cooperative extension service, your community's Agricultural Committee, or your state's Department of Agriculture, or the regional organic farming organization. (For specific programs and companies that can help, see "Resources" on page 356.)

Business Plan versus Planting Calendar

The information discussed in this chapter outlines a business plan that incorporates financial goals, personal expectations, and budgeting. The farm production plan described in chapter 4, however, is a set of instructions for managing the farm—when to plant, when to harvest, when to test the soil, which seeds to buy, and where to rotate the Brassicas next year.

Create a Mission Statement

Every good farm plan begins with a vision or mission statement, sometimes called an executive summary, which is a simple collection of thoughts that describes the essence of your business. The point of this is to hone the big picture; the more specific goals follow. Your mission statement should answer some basic questions without getting too technical:

· Who are you?

· What are you trying to accomplish?

· Why are you choosing this particular type of farm business?

· How are you going to succeed?

Your mission statement doesn't have to be a twenty-page report. Just write down a few detailed, well thought-out sentences that answer the questions above and incorporate your personal beliefs. Here's an example:

> *We are small-scale family farmers who raise premium vegetables using chemical-free and sustainable methods. We sell directly to local customers at reasonable prices.*
>
> *Our success is based on our ability to grow a wide variety of produce, harvest it at the peak of ripeness, and deliver it while it's still fresh and delicious.*

Whether you're growing citrus or herbs in a hoop house, if you plan to make a business out of your operation, create a mission statement.

Skills Assessment

One exercise that you can do to help the planning process along is a self-assessment of your skills. A small-scale organic farm involves so many different talents—planting, growing, sales and marketing, website development, bookkeeping, and management, just to name a few. This is where it comes in handy to involve a spouse or partner whose skills differ from yours; you can decide which person is best suited for each of the varied tasks needed to reach your goals. Without a partner, you'll have to figure out workarounds where your skills are lacking.

For example, a farmer who may be a gifted vegetable grower and business manager could be completely out of place and uncomfortable when dealing with people. Unless he or she has a partner who enjoys talking to the public, it might be wise to avoid farmers' markets, where customers sometimes come to gab about their gardens and squeeze the tomatoes without buying anything. This farmer might, instead, design a business plan for less-chatty wholesale farming.

Thankfully, each new farmer brings a lifetime of skills to the business. According to Judith Gillan, director of the New England Small Farm Institute in Belchertown, Massachusetts, experience in sales and marketing, bookkeeping, management, engineering, or construction all translate to different aspects of farming. The institute runs numerous training programs for new farmers, including some that help farmers assess and rank their particular skills. Gillan asserts that when you match tasks to your skills and the activities that keep you motivated, you'll be more successful and derive more satisfaction from your work. Take a look at what your business needs—from bookkeeping to composting—and decide honestly whether you'll be able to fill those needs. If you can, how? If you can't, who can?

Set Goals

Once you've established your mission statement, you can set goals that work directly to fulfill the mission. The S-M-A-R-T guideline is a widely used goal-setting tool for successful businesses that is perfectly suited to a small-scale farm. To make your goals meaningful, they should be Specific, Measurable, Attainable, Rewarding, and Timed.

Specific. Make your goals clear and to the point. Don't write that you want to grow a diverse range of several kinds of produce, for example. Instead, write that you want to grow at least twenty varieties of heirloom tomatoes and what those varieties are. Or set a goal of providing at least a dozen different items in your CSA share bag each and every week, adding and subtracting crops as the season progresses.

Measurable. If your goals can't be measured, you won't be able to mark your progress in achieving them. Instead of setting a goal of earning "enough money to be comfortable," aim for an actual amount, based on how many farmers' markets you'll attend each year and what kinds of crops you'll bring, for example. Or base it on a specific number of CSA subscribers and the estimated profit from each share. Then set related goals, such as how many CSA members you want to have, the length of the season, or the number of meat chickens you'll raise and process during the year.

Attainable. Even if your goals are specific and measurable, they will be worthless if they aren't realistic. So if you're running a small roadside vegetable stand, don't set a goal of earning $100,000 a year. Be reasonable and don't put an impossible amount of pressure on yourself.

Rewarding. Your goals should be satisfying to achieve. I'm guessing you didn't get into small-scale farming just to become rich, so base some of your goals on personal

fulfillment—helping your customers discover, each month, one new vegetable variety that they'll like, for example.

Timed. Don't use the words *eventually* or *whenever* in your goals. The idea behind a written goal is to attach a time element to a specific task or set of tasks. So instead of saying you want to produce tomatoes early in the growing season, set a target date of July 1. That's a tall order, but it's attainable if you take the proper steps, such as choosing early-fruiting varieties, starting them indoors under a light table, and growing them in a greenhouse.

Prepare for Action

After you've written down your mission statement and goals, it's time to plan the process by which you will actually "git-er-done." This is the place where pie-in-the-sky ideals need to meet a cold dose of reality. Look at each of your goals and create a practical plan for meeting them. For instance, if you set a goal of signing up ten CSA subscribers in your first year, how will you reach that goal? Can you find enough customers in your pool of coworkers, neighbors, and friends? If not, where and how will you do your marketing? Will you advertise in the local newspaper? Will you create a Facebook page, write a blog, or develop a website? (See chapter 9 for more marketing ideas to help you reach your goals.) How will the cost of marketing impact your profits? Consider all the variables, all the related questions, and all the possible solutions, and then sketch out a plan.

Each goal that you set needs to be accompanied by a strategy to fulfill it. If you can't figure out ways to accomplish the tasks associated with the goal, you need to go back and redefine your goals to make them more attainable.

Review, Revise, and Maintain Control

Once you've settled on a mission, determined your goals, and figured out a plan to make everything come together, you should be done with planning, right? Not so fast. Your plan will need constant tweaking and fine-tuning—both up front and over time—to meet your mission. If you run into a softer-than-expected market, or competition reduces your customer base, you may need to raise prices or find a new facet of your business to replace the lost money. Even if things are working smoothly, the results may not bring you satisfaction—either financially or personally—which can affect your motivation. Maybe you'll find that you don't like working on weekends in the summer or that you'd rather spend your Saturdays out in the field than behind the table at a local farmers' market. If so, you'll need to rethink your goals to stay motivated.

If you're turning your pastime passion into a moneymaking business, plan the operation with a budget and record-keeping system separate from your family finances.

Thinking Long-Term

When you're excited about starting your farm, it's hard to think about its last years. But the best business plans account for all phases of the business. The beginning influences the end, and vice versa. Consider these questions now and you'll be better prepared when they really start looming.

- How long will you run the farm? Are you planning to retire and move away in five to ten years? If so, you'll want to keep long-term debt to a minimum, and your plan should reflect that reality.

- How is your health? Farming can be strenuous work, putting physical demands on your body. If you're not in the best shape for hewing trees, clearing land, digging up boulders, and shoveling compost, consider how much time and effort those tasks really take. Is your plan going to include less physical labor as you get older?

- What will you do with your farm property when you get too old for the rigors of raking, planting, plowing, and picking? Will your children take over the business? Will you sell the farm and use the money to pay for your retirement?

Managing Farm Finances

Some part-time farming operations are so small that they can run their finances using cash in an old tin box. Others use computer spreadsheets to track every transaction and manage all their operations through a business bank account. The size of your farm operation will play some role in the scope of your financial planning, but once you've made a commitment to run a business instead of a hobby, one rule is universally applicable: Don't mix your personal finances with your business finances. Randomly feeding a business endeavor with money from the family budget can drain your personal bank account and allow you to operate a poorly managed, money-losing operation.

The only way to really know if your business can stand on its own is to fund it with a reasonable start-up account and then keep track of income and expenses. Even if your part-time business is earning only a few thousand dollars a year or barely breaking even, it's important to know precisely how much money is coming in and going out. Accurate records will help you gauge your farm's profitability, allow you to make informed decisions about major purchases, and help you predict your earnings and expenses in the coming years. Without getting too complicated, you'll need to follow a few basic bookkeeping rules.

One of the best ways to keep records is through a dedicated checking account that keeps track of most of the money flowing through the business via checks, deposits, and debit and credit charges. Such a bank account will have the added benefit of clearly separating the farm business from your personal finances.

The Rodale Institute, a research nonprofit in Emmaus, Pennsylvania, recommends other record-keeping tips for farm businesses:

- Whenever you buy or sell anything for the farm, write down the name of the buyer or seller, the date of the sale, the quantity, the unit price, and the total price.

- Avoid cash transactions. Checks and charges are easier to record, and charges (for expenses) can also boost your credit rating. When you have to use cash, always get a receipt.

- Keep all receipts in the same easy-to-access place. Make it a part of your daily or weekly routine to file receipts (or, if using a computer, scan them).

- If you're keeping written records in a ledger and not on a computer, use a pencil instead of a pen. This makes correcting mistakes simpler.

Two Kinds of Expenses

A key step in record keeping and budgeting is distinguishing between operating expenses and capital expenses. Everything that you spend money on to actually run the farm during the year—fixed costs like rent or mortgage payments and variable costs like gasoline and fertilizers—constitutes an operating expense. Any long-term purchases that you make for your farm are considered capital expenses. Larger capital expenses are usually financed over multiple years, and the payments are added to the operating budget as fixed operating expenses throughout the course of the loan.

Some common operating expenses:
- Advertising costs
- Business travel expenses
- Dues (in a cooperative, for example)
- Fertilizers
- Gasoline for motorized equipment
- Insurance premiums (fixed operating costs)
- Miscellaneous supplies (plastic bags, egg cartons, berry boxes, rubber bands, and so on)
- Mortgage payments (fixed operating costs)
- Organic insect controls
- Phone and Internet service
- Seeds
- Tractor loan payments (fixed operating cost)
- Wages and benefits for hired help
- Water and electricity

Some common capital expenses:
- Cold frame
- Computer
- Garden shed
- Greenhouse
- Lawn mower
- Rotary tiller
- Building materials (for a chicken coop, for example)
- Fencing
- Hand tools
- Tractor attachments

Farm operating expenses are allowed as business deductions at tax time, so having a complete record of all transactions will make things easier on tax-filing day. You can't simply deduct the cost of a major capital expense against your income, but you can write off depreciation on a shed, tractor, chicken coop, or other large purchase over several years as it declines in value. For specific advice on tax matters, consult a tax accountant with experience in farm properties.

Creating a Budget (or Budgets)

Once you've separated family and farm finances and kept accurate records over the course of a season, you should be able to do some meaningful analysis. Subtracting expenses from your farm's total income will determine how profitable your business was for the year. And using those numbers will help you set a budget with reasonable targets for the coming year's business.

Farming businesses big and small use four basic types of budgeting:

1. **Enterprise budgets** track a full season of operating expenses and revenues for a single farm enterprise—beehives, free-range eggs, or pick-your-own strawberries, for example.

2. **Whole-farm budgets** tally all the numbers from multiple enterprise budgets to calculate the entire farm's finances. They often include off-farm income as well.

3. **Partial budgets** track small segments of the farming operation to make specific decisions, such as purchasing a tractor attachment or switching to a drip irrigation system for saving time and conserving water.

4. **Cash-flow budgets** track cash receipts and expenses over time (usually over a year) to determine whether finances are on target. They are helpful for determining the feasibility of a major purchase or loan.

What has worked best for my small farming business is an enterprise budget, which is useful for planning purposes and can be easily tailored to suit any small-scale operation. Taking stock of individual enterprises within your multi-dimensional farm can help you spot profitable areas and increase overall farm income. This kind of worksheet will also point out money-losing operations that might otherwise go unnoticed.

Let's use a small-scale free-range organic egg business as an example. You create an enterprise budget by listing all the receipts and expenses for the most recent full year. In this kind of worksheet, $160 of poultry netting for portable fencing is divided over its four-year projected life at $40 per year. Small capital purchases such as feeders and watering containers would be listed as miscellaneous expenses in the year they were

Hiring Finance Help

There are more complexities in finances than there are varieties of winter squash, and there may be more facets to business planning than you want to deal with. A good solution is to hire a competent bookkeeper.

Look for someone with expertise in farming or farm-related businesses. Some accountants work part-time as consultants and can help you set up the financial parts of your business plan.

Dan Kaplan, who manages the successful nonprofit CSA Brookfield Farm in Amherst, Massachusetts, suggests that you find a bookkeeper or accountant who can set up your financial planning, explain all the intricacies, train you in the use of QuickBooks and other software, and then check back with you periodically to make sure the records are in order. Obviously, you'll save money if you can use outside services such as these to become more adept at the financial aspects of your business. In addition, being more hands-on in all areas of your business may help you notice problems sooner and find solutions more readily.

purchased (especially if they are low-cost common expenses that tend to even out year after year). You'll simply weigh your sales of eggs against your variable costs (such as organic chicken feed, utilities, bedding, and egg cartons) and your fixed costs (such as land rent and financing for your chicken coop). Do what makes sense to you, but try to be as honest and accurate as possible when listing expenses. A quick online search will give you plenty of sample enterprise budgets to work from.

When I created this kind of worksheet, I found that the rising prices for my organic chicken feed had increased our operating expenses too quickly. After one of my customers told me my egg prices were much lower than a competitor's, I hiked them up a bit to offset the operating expenses and boost my bottom line.

An enterprise budget is also handy for getting a commonsense glimpse at whether it's worth pursuing a larger-scale capital expense. If you see that your chicken business is generating $500 in yearly profits, for example, you might think twice about spending $2,000 on a new chicken coop and run. Unless you raise prices significantly, you'll have to sacrifice your profits for the next four years to pay off that one-time investment. And if you borrow money to finance the coop, you also have to consider the interest payments.

Record your business expenses—from gasoline and truck repairs to feed, hay, and new livestock.

Foundation for a Business Plan

A new chicken coop along with its adjoining fenced-in run is a capital expense. But the birds, bedding, feed, grit, and egg cartons are all operating expenses, as are low-cost items like feeders and waterers, which need to be replaced every few years.

Do You Need Additional Insurance?

What would happen if someone became ill or had a bad reaction to food that was purchased on your farm? What if unwashed eggs caused a stomach ailment? How about if the children next door tried to take the top off your beehive? Small business owners working from their home may not be covered for losses and liability on a farm business—especially if they are relying on a home insurance policy that was issued before the farm business began.

During a call to my insurance agent to discuss plans for a vegetable CSA and backyard egg sales, I mentioned that I had been keeping bees on the property for the past six years. I was quickly informed that my insurance was being canceled and that I had thirty days to find another insurer. Fortunately, I found a farm insurance agent and bought a new policy that was only slightly more expensive than my old insurance.

If you're intending to sell eggs, honey, organic vegetables, or value-added products made at home such as soap, you might consider a similar insurance policy that covers your home and farm property for multiple types of risks; the price varies depending on the ceiling of the liability coverage. You might also consider forming a limited liability company (LLC). An LLC can protect personal assets, leaving only the assets of the farming business itself exposed to judgments in lawsuits. It can also provide a range of tax advantages. (Consult a lawyer or tax advisor for more information.)

Borrowing Money for Expansion

This is another area where a business plan comes in handy. If you ever have to borrow money for expansion, equipment, or some other capital expense, banks will want to see that you're working with a financially feasible plan. Lenders want to see the five Cs: **C**haracter, **C**apacity, **C**apital, **C**ollateral, and **C**onditions (or **C**ovenants). A detailed business plan should address all these elements to the lender's satisfaction.

Character. Your integrity and business savvy are both considered when a lender makes a gut call to accept or reject your application. The mission statement for your business—if it's original and made from the heart—will help a lender understand what you're trying to accomplish and provide some insight into your character. (Of course, your credit history will play a pretty big role, too.)

Capacity. You have to prove to lenders that your business is going to generate enough money to pay all the bills and still pay back their loan. Your operating budget and cash flow, along with any steady off-farm income, will be compared with your expenses and overall debt.

Capital. The percentage of the business that you actually own—that is, your equity—is key in the eyes of a lender. If you've personally backed your business, you'll have a stronger stake in working to make it successful.

Collateral. If your farm's operating budget isn't as strong as it should be and you don't have an adequate stake in your own business, you probably won't qualify for a loan unless someone else backs you. But you still might be able to borrow money if you have valuable property that you can offer as insurance to repay the loan. However, your business plan should demonstrate that you are in control of your finances and able to pay back money without defaulting.

Conditions. A lender will also want to know what the money is being used for. And that may be part of your overall plan. General economic trends and the outlook for farming income may also influence lenders.

Check into extra liability insurance if you turn your hobby into a business. You might think raising vegetables is harmless, but visitors or customers could injure themselves on your property and take legal action.

Your Business Plan and You

Your farm business plan can be as ambitious as you decide to make your operation. But if you want to be satisfied with the results, it will have to suit the lifestyle you want to lead. Ultimately, formulating a detailed plan can help you realize the time commitment and effort it will actually take to run your business, along with unforeseen expenses and risks that you might have missed had you not gone through the exercise.

Because the business will involve your family, the planning should be something you do with your husband, wife, or farming partner, making sure that everyone is in agreement before any major decisions are made.

Above all, remember to leave some time to enjoy your farm. Too many growers joke that they work all day and late into the evening to grow fresh, healthy fruits and vegetables for their CSA members and then go home exhausted—with barely enough time to eat a frozen pizza before bedtime.

Never forget the simple pleasures that drew you to farming, and don't let lofty ambitions and the push for profits get in the way of pursuing them.

Don't put all your eggs in one basket. If you're applying for financing to expand the farm, lenders will want to see that you have enough capital to cover the loan.

The 5 Components of a Business Plan

1. **Executive Summary.** What is your vision or mission statement? What are you doing and why do you think it will be successful?
2. **Business Description.** What are your specific costs and objectives? You can't manage what you can't measure. What are your goals and when do you want to accomplish them?
3. **Operation.** What will your workplace and business look like? How will you do what you plan to do?
4. **Marketing.** What are you selling? How do you price it? How do you promote it?
5. **Finance.** What is the cash flow? Are the profits substantially greater than your losses? Will it work?

Certified Organic

Deciding to give up chemicals and use organic practices is a personal choice you make for yourself, your family, your neighbors, and anyone else who shares in the bounty of your harvest. Once you decide to sell your farm-raised goods to the public, however, it's time to make another decision—whether to become USDA "certified organic." Being certified means you can use the USDA Certified Organic label. It is the gold standard of natural growing, recognized by customers around the world.

Without certification, you can call your food naturally grown, chemical-free, earth-friendly, produced without pesticides, sustainably raised, or another term that honestly describes your overriding philosophy and practice. But if you want to use the "organic" label, you'll need to be certified, and that means meeting the National Organic Standards.

Farmers who knowingly misrepresent their operations by calling their goods organic without following the rules can be fined up to $11,000 per violation. Farms are also subject to inspections, and at least 5 percent of certified organic operations' products are tested every year for pesticide residues and other prohibited substances.

Requirements for Certification

Getting certified is a process that requires an application, fees, detailed record keeping, site inspections, and developing a relationship with a third-party certifier, whose judgment will ultimately determine whether your operation becomes "organic." Certifiers can be independent agents, nonprofit organizations, or members' state agriculture departments. In California, the State Organic Program administers organic certification.

Wherever you're growing, being certified organic is not simply a matter of switching from chemical fertilizers and pesticides to natural materials. For farmers who grow vegetables, fruits, berries, grains, and other crops, requirements include the following.

Pesticides. No use of chemical pesticides is allowed. That also means using organically approved pesticides only when less-invasive ways to control pests—such as handpicking, physical barriers, and natural predators—are either unsuccessful or not possible.

Fertilizers. No synthetic fertilizers or prohibited soil inputs are allowed. Fertilizers derived from sewage sludge are also prohibited, and even natural amendments are regulated. Compost must be made according to standards that ensure all ingredients are natural and that potentially harmful pathogens are killed in the composting process. Applying manures must be timed to prevent contamination of food crops.

Seeds. Seeds cannot be treated with chemical fungicides. In addition, growers must choose organically grown seeds when they are available. Non-organic seeds are allowed when an equivalent variety is unavailable from an organic supplier, as long as the seeds have not been treated with non-approved materials. Seeds cannot be genetically engineered, meaning they can't be the result of manipulation by science to include traits from another species. Hybrids created by deliberate cross-pollination are allowed. If you use annual seedlings instead of seeds, they must also be certified organic.

Planting Stock. Perennial planting stock such as berry bushes and fruit trees must come from an organic supplier or have been raised organically under your care for at least one year before you can market their fruit as organic.

Farm Plans. Plans for your farm must include improving the soil and promoting biological activity through crop rotation, manure, compost, cover crops, and other natural amendments.

Weeds. Weeds must be controlled by physical or mechanical methods—hand-pulling, hoeing, cultivation, tilling, flaming, and mulching—and not by spraying herbicides.

Preventive Methods. Preventive methods must be used to discourage plant diseases through cultural practices, including cover crops, crop rotation, adequate spacing (air drainage), choosing resistant varieties for your region, proper irrigation, pruning, and cleaning up dead and sick plant material to prevent diseases from spreading.

Farm Practices. Overall farm practices that conserve water, promote biodiversity, and benefit the overall ecology must be employed.

Growing Area. If your growing area was used in non-organic production—with chemical pesticides, synthetic fertilizers, or any other unapproved materials—you'll have to wait three years before marketing your crops as organic.

Organic Systems Plan

The rules are strict, but they are subject to some interpretation, according to the specific conditions and practices on your particular farm. That's where your organic certifier plays a key role. You'll have to develop an Organic Systems Plan for your farm and update that plan every year. This detailed document becomes a working record of activities on your property. The plan requires you to keep records of nearly everything you do on the farm to demonstrate compliance to the National Organic Standards. Part of keeping the records is saving receipts for seeds, seed-starting materials, fertilizers, transplants, organically approved controls for pests and diseases, and other supplies. Your records will also include maps of your fields.

Poultry and livestock producers will need to keep records and save receipts for organic feed, bedding materials, and medications for treating parasites and diseases.

The rules for animals also restrict the use of some medications; however, sick animals must be treated—even if it requires non-approved treatments. After treating, the animals may no

longer be marketed as organic. Vaccinations are allowed, but growth hormones are prohibited, along with antibiotics (except as therapy). Livestock must be raised organically from the third trimester of gestation for mammals and from the second day of their lives for poultry.

Animals must also be raised on organic feed and treated humanely—with enough room to move around; adequate food, water, and shelter; and access to the outdoors. Any pastureland where the animals graze must be free of chemical fertilizers, pesticides, and other prohibited materials for at least three years.

Getting Started

To begin the certification process, first get acquainted with the rules. The website of the USDA's Agricultural Marketing Service contains numerous links to information about the National Organic Program and the National Organic Standards Board (NOSB), which makes recommendations to the USDA. Go to ams. usda.gov/nop/ and follow the links. Among the headings, you'll find a link to USDA accredited certifying agents. The listing includes the certifiers' specialties, along with their names, phone contacts, email addresses, and websites.

After you've made the decision to go organic and you've researched the rules for your particular operation, contact a certifier. It makes sense to choose a certifier that is close to your location because he or she will be familiar with soil types, geographic conditions, regional pests and diseases, and other specific challenges you may face. Your certifier will be able to get you information on the application process, forms, instructions, and a listing of what's allowed or prohibited for your farm.

You can also download a free introduction to organic certification from the National Center for Appropriate Technologies (NCAT) and Appropriate Technology Transfer for Rural Areas (ATTRA) at attra.ncat.org/attra-pub/organcert. html

The fee for certification varies, depending on the size and type of your organic farm and your individual certifier. The guidebook from NCAT/ATTRA says costs can "range from a few hundred dollars to a few thousand dollars." (A similar guide from Pennsylvania State University cites a low price of $750 for farm certification.) There are rebates available under a cost-share program.

Once you have filed an organic application, a certified inspector will visit the farm to make sure it is in compliance. Schedule the inspection at a time when your farm is in operation. (For instance, you wouldn't schedule a December inspection for an outdoor tomato farm in Michigan.) After the inspection, the certifier will review your submitted material, along with information from the inspection and make a decision. Sometimes the decision will be delayed until more questions have been answered, especially for a farmer's first application. Once an application has been filed, the process can take six weeks or more before your farm becomes certified. After certification, you'll have to file a new Organic System Plan and pay a renewal fee every year.

Exemption for Small-Scale Farmers

If you're a small-scale producer making less than $5,000 in gross organic sales per year, you are exempt from two key parts of the program:

You don't need to be certified to sell, label, or market your products as organic. However, you can't use the USDA organic seal or call your goods "certified" organic. And you're not allowed to sell your products as ingredients in another producer's certified organic product. Customers may require you to sign an affidavit stating that you adhere to USDA organic regulations.

You also don't need to fill out an Organic System Plan to document your practices. But you still must keep records and follow other USDA requirements, including rules for production or handling.

If you choose, you can still apply for optional organic certification. In California, all growers who call their goods organic must register with the State Organic Program—even growers who earn less than $5,000 per year.

Certified Naturally Grown

Some farmers and producers choose not to become certified by the USDA. They follow the environmentally friendly practices of organic growing but eschew the official designation because the application process, record keeping, and inspections are too time-consuming, especially when added to the regular tasks of farming the soil, caring for poultry or livestock, developing a business plan, and marketing and delivering your goods. Other small-scale food producers argue that the application and renewal fees are too high. Many simply say they don't want to be controlled by a large-scale federal organization like the USDA. Finally, some assert that the "certified organic" designation benefits larger-scale growers who ship their goods to faraway locations, because buyers need proof in the form of the USDA Certified Organic label. Meanwhile, small-scale local farmers who market directly to customers can speak about their philosophy and practices—farmer to buyer.

Since the National Organic Program was enacted in 2002, a group of organic farmers in New York developed an alternative certification system: Certified Naturally Grown (CNG). The "organic" term they had been using for years to describe their operations was suddenly forbidden—unless they took on extra paperwork and paid fees to the USDA. Organizers and members say the CNG designation is tailored to small-scale, independent farmers who market their goods directly to customers or to restaurants and small-scale grocers. It's based on the National Organic Program, follows the same guidelines, and is just as strict. But certification and compliance relies on the Participatory Guarantee System favored by

Weighing the Pros and Cons of Certification

Pennsylvania State University lists the following benefits and challenges of becoming a certified organic farmer:

The Benefits

- Learn new, ecologically based practices.
- Some costs of production may be lower.
- Improved record-keeping practices.
- Certification that your practices meet organic standards.
- Access to new, expanding markets and price premiums.
- Reduced exposure to synthetic agricultural chemicals.

The Challenges

- Certification costs money.
- Some costs of production may be higher.
- Increased record-keeping requirements.
- Restricted use of synthetic chemicals and fertilizers.
- Increased management intensity.
- Organic price premiums not available during transition.

IFOAM, the International Federation of Organic Agriculture Movements.

Benefits cited by CNG include the following:

CNG is a standardized alternative designation for small-scale growers who might otherwise choose not to become certified.

The fees are lower. Using unpaid farmers as peer inspectors saves the additional fee for farm inspections.

Because it reduces paperwork, CNG is more appealing to growers who produce diversified crops and use multiple systems on their farms. Crop diversity improves the environment, encourages beneficial insect populations, and is more in line with traditional "organic" philosophies than large monoculture farms that qualify for USDA Organic Certification.

Peer inspections and direct involvement with both consumers and small-scale businesses improve the local farming community by opening lines of communication. When peer farmers conduct inspections, they can make helpful suggestions on organic techniques for improving the soil, controlling pests, eradicating weeds, or fighting diseases naturally. Conversely, USDA Certified Organic inspectors are prohibited from making suggestions during their visits.

CNG has standards for beekeepers that include a list of prohibited and allowed materials.

Some growers who pay the price for organic certification and have fought to maintain organic standards argue against Certified Naturally Grown participation, saying it can confuse consumers, erode participation in the National Organic Program, and make it easier for some producers to cut corners because inspections are conducted by peers instead of independent USDA accredited certifiers. More information on CNS certification is available at the nonprofit group's website: naturallygrown.org

Acknowledgments

I owe an unending debt of gratitude to all who helped produce this book, beginning with the editors and publishers of *Hobby Farms* magazine and I-5 Publishing. Thank you especially to Andrew DePrisco, whose vision brought the opportunity to a first-time author and whose guidance provided support throughout the project. Thanks also to the keen eyes of my editor, Dolores York, an experienced, well-organized professional whose knowledge of her craft gave shape to these pages. I also thank photo researcher Elizabeth Tunnicliffe and art director Mary Ann Kahn for giving life to the pages through careful photo selection and graphic design.

Beyond those who worked to create this book, there are acres of others deserving recognition—most notably, the organic and sustainable farmers who walked me through their fields or answered questions in between customers at the local farmers' market. The men and women who dedicate their lives to producing our food perform one of the most underappreciated jobs on the planet, and those who do so while striving to maintain our natural resources deserve appreciation from all. When busy, hardworking farmers can also find time to answer questions from a wide-eyed wannabe, they merit not only thanks but also awards for tolerance.

Sincere gratitude to: Julie Rawson and Jack Kittredge of Many Hands Organic Farm; Weston Lant of Lucky Field Organics; John Mitchell of Heirloom Harvest CSA; Pete Lowy and Jennifer Hashley of Pete and Jen's Backyard Birds; Ryan Voiland of Red Fire Farm; Jim Gerritsen of Wood Prairie Farm; Derek Christianson of Brix Bounty Farm; Lynda Simkins of the Natick Community Organic Farm; Ellen Gould of Sisters Three Farm; and Dan Kittredge of the Bionutrient Food Association.

Thanks also to the many experts who provided information through interviews, lectures, seminars, workshops, and correspondences: Eliot Coleman, Lee Reich, Michael Darre, Marianne Sarrantonio, Dan Kaplan, Jon Frank, Don Franczyk, Al Averil, Cliff Hatch, John Howcroft, Ruth Hazzard, and the staff and volunteers of the Northeast Organic Farming Association's Massachusetts Chapter.

Resources

The best place to turn to for information on organic farming is to others who are in the farming community. They are familiar with all issues related to organic practices—from soil development to pest control to marketing and business. I strongly recommend joining a regional organic farming organization and attending workshops. Not only will you gain knowledge from the courses you take but you'll also make valuable contacts that you can reach out to for assistance with specific problems. Organic farmers are by nature generous with their time, even though they often don't have much of it to spare. Joining an organic farming group may also allow you to place bulk orders on fertilizers, seeds, and other supplies that will likely make up for your annual dues and more.

Organic and Sustainable Farmers Groups

In the United States

Beginning Farmers: beginningfarmers.org
Comprehensive listing of resources covering all aspects of farming and providing content relevant to new, experienced, and aspiring farmers, as well as farm educators, activists, and policymakers.

Midwest Organic and Sustainable Education Service (MOSES): mosesorganic.org
Provides a diverse listing of free and affordable resources for farmers interested in learning more about organic farming.

National Sustainable Agriculture Information Service (ATTRA): attra.ncat.org/index.php
Extensive searchable database on a multitude of topics, providing information and technical assistance to farmers, ranchers, cooperative extension agents, educators, and others involved in sustainable agriculture in the United States.

Northeast Organic Farming Association (NOFA): nofa.org
A loosely coordinated group of state NOFA chapters whose purpose is to advocate for and educate about organic and sustainable agriculture, family-scale farming, homesteading both rural and urban, agricultural justice, and related issues.

State Groups

Alabama: Alabama Sustainable Agriculture Network: asanonline.org

Alaska: University of Alaska Cooperative Extension Services: uaf.edu/ces/ah/sare

Arizona: University of Arizona Cooperative Extension: cals.arizona.edu/extension/sustainableag/related.html

California:
Community Alliance with Family Farmers: caff.org
California Certified Organic Farmers (CCOF): ccof.org

Connecticut: Northeast Organic Farming Association (NOFA—CT): ctnofa.org

Florida: Florida Organic Growers: foginfo.org

Georgia: Georgia Organics: georgiaorganics.org

Hawaii: Hawaii Organic Farming Association (HOFA): hawaiiorganic.org

Illinois: Illinois Organic Growers Association (IOGA): illinoisorganicgrowers.org

Indiana: Indiana Organic Gardeners Association: gardeningnaturally.org

Iowa: Iowa Organic Association (IOA): iowaorganic.org

Kansas: None available at the time of this printing.

Kentucky: Organic Association of Kentucky (OAK): oak-ky.org/home.shtml

Maine: Maine Organic Farmers and Gardeners Association (MOFGA): mofga.org

Maryland: Maryland Organic Food and Farming Association (MOFFA): marylandorganic.org

Massachusetts: Northeast Organic Farming Association—MA (NOFA—MA): nofamass.org

Michigan: Michigan Organic Food and Farming Alliance (MOFFA): moffa.net/home.html

Missouri: Missouri Organic Association (MOA): missouriorganic.org

Montana: Montana Organic Association (MOA): montanaorganicassociation.org

Nebraska: Nebraska Sustainable Agriculture Society (NSAS): nebsusag.org

New Hampshire: Northeast Organic Farming Association—NH (NOFA—NH): nofanh.org

New Jersey: Northeast Organic Farming Association—NJ (NOFA—NJ): nofanj.org

New York: Northeast Organic Farming Association—NY (NOFA—NY): nofany.org

North Carolina: Carolina Farm Stewardship Association: carolinafarmstewards.org

North Dakota: North Dakota Organic Advisory Board: ndorganics.nd.gov

Ohio: Ohio Ecological Food and Farm Association: oeffa.org

Oregon: Oregon Tilth: tilth.org

Pennsylvania:
Pennsylvania Certified Organic (PCO): paorganic.org
Pennsylvania Association for Sustainable Agriculture (PASA): pasafarming.org

Rhode Island: Northeast Organic Farming Association—RI (NOFA-RI): nofari.org

South Carolina: Carolina Farm Stewardship Association: carolinafarmstewards.org

Tennessee: Tennessee Organic Growers Association (TOGA): tennesseeorganics.org

Texas: Texas Organic Farmers and Gardeners Association (TOFGA): tofga.org

Vermont: Northeast Organic Farming Association—VT (NOFA – VT): nofavt.org

Virginia: Virginia Association for Biological Farming (VABF): vabf.org

Washington: Tilth Producers of Washington: tilthproducers.org

In Canada

Canadian Organic Growers (COG): cog.ca
Diverse membership-based charitable organization with chapters throughout Canada; supplies resources and support for organic farming and backyard gardening.

Farm Start: farmstart.ca
Resources, support, and startup programs for aspiring, new, and experienced farmers in Canada.

Organizations Promoting and Advocating Organic Farming

In the United States

Center for Food Safety (CFS): centerforfoodsafety.org
A public interest and environmental advocacy organization working to protect human health and the environment by curbing the use of harmful food production technologies and by promoting organic and other forms of sustainable agriculture.

Meetup.com: organic-farming.meetup.com/
Social networking websites for organic farmers.

Organic Consumers Association: organicconsumers.org/organicgroups.cfm
Nonprofit advocacy organization that campaigns for food safety, organic agriculture, and sustainability. Searchable database contains an extensive listing of organic farming sites in the United States, Canada, as well as international listings.

Organic Farming Research Foundation: ofrf.org
A support structure for farmers focusing on policy, education, grant making, and community; has a searchable database.

Organic Trade Association (OTA): ota.com/index.html
A membership-based business association that represents North American businesses across the organic supply chain with a mission to promote and protect organic trade to benefit the environment, farmers, the public, and the economy. The focus is on all things organic in both food and nonfood industries.

USDA: National Organic Program (NOP): ams.usda.gov/nop
Information from the USDA's Agricultural Marketing Service about U.S. organic food standards.

In Canada

Atlantic Canadian Organic Regional Network (ACORN): acornorganic.org
A membership-based nonprofit organization that promotes organic agriculture in the Canadian Atlantic provinces. Website features links to information on certification, transitioning to organics, farmers' markets, ACORN conferences, and more.

Government of Canada: canada.gc.ca/home.html
Search "General Principles and Management Standards" for the latest information about Canadian organic product standards.

Organic Consumers Association: organicconsumers.org/organicgroups.cfm
Nonprofit advocacy organization that campaigns for food safety, organic agriculture, and sustainability. Searchable database contains an extensive listing of organic farming sites in the United States, Canada, as well as international listings.

Publications

Journals and Magazines

Acres U.S.A.: acresusa.com
A comprehensive guide to sustainable agriculture; highlights techniques and tips for growing healthy crops and livestock.

Farming Magazine: farmingmagazine.net
A quarterly publication that strives to educate farmers and to celebrate farming on an ecologically conscious scale; created in the spirit of stewardship for our planet and for those who live on it.

GRIT Magazine: grit.com
A rural lifestyle resource website and publication; includes informative articles, DIY projects, and practical advice for more sustainable living.

Growing for Market: growingformarket.com
A journal for market farmers; full of ideas, analysis, and news that will help make your farm business profitable and enjoyable.

Hobby Farm Home Magazine: hobbyfarms.com
A home magazine for those living and/or loving the rural country life; highlights farmhouse activities such as cooking, crafting, collecting, pet care, and home arts and skills.

Hobby Farms: hobbyfarms.com
Great website and publication resource geared to the rural country enthusiast, hobby farmer, small production farmer, and everyday gardener; provides relevant information on all aspects of rural life—from small farm equipment to livestock and crops.

Modern Farmer: modernfarmer.com
A website and quarterly publication for window-herb growers, career farmers, people who have chickens, people who want to have chickens, and for anyone who wants to know more about how food reaches his or her plate.

Mother Earth News: motherearthnews.com
A website and magazine packed to the max with great reads and extensive resources.

Books

Belanger, Jerry. *Storey's Guide to Raising Dairy Goats: Breeds, Care, Dairying, Marketing.* 4th ed. North Adams, MA: Storey Publishing, 2010.
No-nonsense basics on all dimensions of goat care.

Bennett, Bob. *Storey's Guide to Raising Rabbits.* North Adams, MA: Storey Publishing, 2009.
Essential information on raising rabbits.

Bradley, Fern M., and Barbara W. Ellis, eds. *Rodale's All-New Encyclopedia of Organic Gardening.* Emmaus, PA: Rodale Press, Inc., 1992.
Over 400 entries of practical, up-to-date gardening information from garden experts.

Coleman, Eliot. *Four-Season Harvest: Organic Vegetables from Your Home Garden All Year Long.* 2nd ed. White River Junction, VT: Chelsea Green Publishing, 1999. Great advice for extending the growing season from organic guru Eliot Coleman.

———. *The New Organic Grower: A Master's Manual of Tools and Techniques for the Home and Market Gardener.* 2nd ed. White River Junction, VT: Chelsea Green Publishing, 1995.
Practical information for market gardeners and small-scale sustainable farmers.

———. *The Winter Harvest Handbook: Year Round Vegetable Production Using Deep Organic Techniques and Unheated Greenhouses.* White River Junction, VT: Chelsea Green Publishing, 2009.
Useful and practical advice on producing vegetables throughout the year.

Damerow, Gail. *Storey's Guide to Raising Chickens.* 3rd ed. North Adams, MA: Storey Publishing, 2010.
A detailed and informative resource on raising chickens.

Ekarius, Carol. *Small-Scale Livestock Farming: A Grass-Based Approach for Health, Sustainability, and Profit.* North Adams, MA: Storey Publishing, 1999.
Includes case studies of successful farmers, focusing on the details of livestock farming.

———. *Hobby Farm: Living Your Rural Dream for Pleasure and Profit.* Irvine, CA: Hobby Farm Press, 2005.
Honest and informative advice on running your dream farm.

Fox, Thomas J. *Urban Farming: Sustainable City Living in Your Backyard, in Your Community, and in the World.* Irvine, CA: Hobby Farm Press, 2011.
Lots of information on raising food in urban settings.

Gershuny, Grace. *Start with the Soil: The Organic Gardener's Guide to Improving Soil for Higher Yields, More Beautiful Flowers, and a Healthy, Easy-Care Garden.* Emmaus, PA: Rodale Press, Inc., 1997.
Great resource on everything related to soil.

Jeavons, John. *How to Grow More Vegetables (and Fruits, Nuts, Berries, Grains, and Other Crops).* 8th ed. Berkeley, CA: Ten Speed Press, 2012.
Biodynamic vegetable farming using intensive growing techniques in small spaces.

Kains, Maurice G. *Five Acres and Independence: A Handbook for Small Farm Management.* New York, NY: Dover Publications, 1973.
A Depression-era classic with wisdom that still survives.

MacKenzie, David. *Goat Husbandry.* 5th ed. London: Faber and Faber Ltd., 1996.
Filled with drawings, diagrams, charts, photos, and easy-to-understand information.

Perry, Jill, and Scott Franzblau. *Local Harvest, A Multifarm CSA Handbook.* Gaithersburg, MD: Signature Book Printing, 2010. Available free online: agmarketing.extension.psu.edu/ComFarmMkt/PDFs/local_harvest_csa.pdf
Details the how-tos and nuts and bolts of developing, implementing, and running a multifarm (cooperative) CSA program.

Salatin, Joel. *Pastured Poultry Profit$.* Swoope, VA: Polyface, Inc., 1993.
Plain talk on raising pastured chickens.

———. *Raising Better Rabbits and Cavies.* Bloomington, IL. American Rabbit Breeders Association. 2000.
A complete guidebook with practical advice.

Sayer, Maggie. *Storey's Guide to Raising Meat Goats: Managing, Breeding, Marketing.* 2nd ed. North Adams, MA: Storey Publishing, 2010.

Smith, Edward C. *The Vegetable Gardener's Bible.* North Adams, MA: Storey Publishing, 2009.
A classic book for large-scale home gardeners.

Wiswall, Richard. *The Organic Farmer's Business Handbook.* White River Junction, VT: Chelsea Green Publishing, 2009.
Step-by-step guide on managing a farm as a business (comes with a companion CD that offers additional business tools).

Certification

In the United States

Certified Naturally Grown (CNG): naturallygrown.org/
Nonprofit organization that offers certification targeted at small-scale, direct-market farmers and beekeepers using natural methods. Works with peer reviews by other local CNG farmers. Also accepts memberships from nonfarmers.

Midwest Organic Services Association (MOSA): mosaorganic.org
Nonprofit organic certification agency, rooted in the Midwest, with growth extending throughout the United States. Certifies crops, livestock, processors, and handlers (a private certifier).

Organic Materials Review Institute (OMRI): omri.org
National nonprofit organization that maintains a database on which input products are allowed for use in organic production and processing.

USDA National Organic Program (NOP): usda.gov (click on "Laws & Regulations" and then "Organic Certification")
Provides links to information on the USDA's National Organic Program in the United States, including its regulations, policies, and accredited certifying agents.

In Canada

Canadian Food Inspection Agency: inspection. gc.ca (click on "Food-Organic Products" and then "Certification and Verification")
Provides links to information on the organic program in Canada, including its regulations, policies, and accredited certifying agents.

Ecocert: ecocertcanada.com
Certification body for sustainable development in Canada (a private certifier).

Resources by Chapter

Chapter 1—The Good Earth

AgriSupply: agrisupply.com
Large-scale supplier of farm and garden equipment and accessories.

Beginning Farmers: beginningfarmers.org/finding-land-to-farm
Provides an extensive state-by-state listing of Land Link programs in the United States.

Center for Rural Affairs: cfra.org/landlink
Brings together beginning farmers and established landowners in the United States and offers consulting services to help farm transitions through its Land Link program; Land Link services provided at no charge to participants.

FarmLink: farmlink.net
Matches new farmers who are looking for land or mentorship with farm owners who have land available or expertise to share throughout Canada.

Lands of America: landsofamerica.com/america
Large rural listing service specializing in land for sale.

World Wide Opportunities on Organic Farms (WWOOF): wwoof.net
Worldwide organization linking people who want to volunteer on organic farms with people who are looking for volunteer help.

Chapter 2—Tools That Rock the Farm

FarmTek: farmtek.com
Offers greenhouses, high tunnels, and agricultural and livestock supplies.

Gempler's: gemplers.com
All-purpose tool and farming online supply catalog.

Small Farm Tools: smallfarmtools.com
The engineers and builders behind Eliot Coleman's many inventions, including movable greenhouses; offer innovative farming tools and ergonomic, efficient hand tools.

Valley Oak Tools: valleyoaktools.com
Offers broadforks, wheel hoes, accessories, and other items.

CHAPTER 3—The Living Soil

Soil and Compost

National Institute of Food and Agriculture (NIFA): csrees.usda.gov
Provides links to state cooperative extension services, where you can find local soil test labs.

USDA Natural Resources Conservation Service (NRCS) Web Soil Survey: websoilsurvey.sc.egov.usda.gov
An online interactive map service that allows the user to view the soils of the area of interest; has soil maps and data available online for more than 95 percent of the nation's counties.

Public Soil-Testing Labs

Colorado State University: soiltestinglab.colostate.edu

Cornell University—Soil Health: soilhealth.cals.cornell.edu

Iowa State University: soiltesting.agron.iastate.edu

Texas A&M—AgriLife Extension: soiltesting.tamu.edu

UMass-Amherst Soil and Plant Tissue Testing Laboratory: soiltest.umass.edu

University of Guelph (Canada): guelphlabservices.com

Utah State University: usual.usu.edu

Private Soil-Testing Labs

International Ag Labs: aglabs.com

Logan Labs: loganlabs.com

Nutrient-Dense Farming

Bionutrient Food Association (BFA): bionutrient.org
Nonprofit member-based association dedicated to increasing quality in the food supply and advocating for vital soils, nutritious food, and healthy people.

Remineralize the Earth: remineralize.org
Nonprofit membership organization that promotes the regeneration of soils and forest with finely ground gravel dust to create fertile soil rather than by using chemical fertilizers and pesticides.

Vermicomposting
University of Wisconsin Extension: 4.uwm.edu/shwec/publications/cabinet/composting/Vermicomposting%20Supplies%20Source%20List.pdf
List of sources for worms and worm-composting supplies.

Biochar
International Biochar Initiative: biochar-international.org
Nonprofit membership organization committed to sustainable biochar production and use.

CHAPTER 4—We Plow the Field . . .

Online Publications
University of Missouri Extension: extension.missouri.edu
Provides online access to informational agricultural publications that include organic and sustainable pest and weed management methods.

Drip Irrigation
Growers Supply: growerssupply.com
Offers large selection of ground fabrics, row covers, drip irrigation systems, and other growing supplies.

Biodynamic Farming
Biodynamic Farm and Garden Association: biodynamics.com
Based on the work of Rudolf Steiner, biodynamic farming is a spiritual-ethical-ecological approach to agriculture, food production, and nutrition.

CHAPTER 5—Vegetables and Herbs: The Spice of Life and Farming

Gourmet Garlic Gardens: gourmetgarlicgardens.com
A garlic information center, where you can learn about garlics of all kinds and buy many kinds of gourmet garlics directly from growers.

Woodchucks in Rhode Island, Rhode Island Department of Environment: www.dem.ri.gov/programs/bnatres/fishwild/pdf/woodchuc.pdf
General information about woodchucks, including damage identification and methods of control.

Seed Resources
Certified Organic Associations of BC: certifiedorganic.bc.ca/rcbtoa/training/organic-seeds.htm
Listings of suppliers of organic seeds in Canada.

Organic Seed Alliance: seedalliance.org
Advocacy, research, and educational organization founded to advance the ethical development and stewardship of the genetic resources of agricultural seed.

Seed Suppliers
Albert Lea Seed: alseed.com
Wide selection of field and cover crop seeds with many choices of organic corn, grains, legumes, and cover crops.

Baker Creek Heirloom Seeds: rareseeds.com
Broad range of hard-to-find open-pollinated and heirloom vegetables.

Fedco Seeds: fedcoseeds.com
Offers untreated and organic seeds; good prices for organic seeds in sizes for small or large growers.

Johnny's Selected Seeds: johnnyseeds.com
Extensive selection of organic certified and conventional seed for vegetables, herbs, and flowers; informative catalog (carries farm and garden supplies, along with innovative tools).

Peaceful Valley Farm & Garden Supply: groworganic.com/
Selection of organic growing supplies and tools for growers at every level, from home gardeners to commercial farmers.

Seed Savers Exchange: seedsavers.org
Large selection of heirloom seeds, seed potatoes, and seedlings, including organic choices.

Territorial Seed Company: territorialseed.com
Selection of supplies, plants, and seeds, including organic and heirloom varieties.

Wood Prairie Farm: woodprairie.com
Wide variety of organic seed potatoes and organic seeds.

CHAPTER 6—Fruits of the Earth

Adam's County Nursery: acnursery.com
Producer and supplier of fruit trees.

Cornell University: fruit.cornell.edu/berry/nurseries
Comprehensive listing of berry and small fruit crop cultivars and the nurseries that sell them.

Nourse Farms: noursefarms.com
Producer and supplier of bare-root berry plants.

Stark Bro's: starkbros.com
Producer and supplier of fruit and nut trees, berries, and other garden plants.

CHAPTER 7—Chickens in Your Backyard

Meyer Hatchery: meyerhatchery.com
Supplier of chicks and other poultry; also carries a large selection of poultry supplies.

Murray McMurray Hatchery: mcmurrayhatchery.com
Supplier of chicks, pullets, and other fowl.

My Pet Chicken: mypetchicken.com
Online source for chicks and supplies; includes resource link to a help forum.

Stromberg's Chicks and Game Birds Unlimited: strombergschickens.com
Supplier of chickens and other game birds.

CHAPTER 8—Branching Out: Honeybees, Rabbits, and Goats

American Dairy Goat Association (ADGA): adga.org
Membership association organized to collect, record, and preserve the pedigree of dairy goats; website also offers information for dairy goat breeders and herders.

Beginning Farmers: beginningfarmers.org/goat-farming
Offers information on farming goats.

Fias Co Farm: fiascofarm.com
Provides information and resources on goat care, health, and husbandry.

National Goat Handbook. University of Maryland: outlands.tripod.com/farm/national_goat_handbook.pdf

CHAPTER 9—To Market, to Market

Agricultural Marketing—Penn State Extension, College of Agricultural Sciences: agmarketing. extension.psu.edu
Extensive compilation of online topics and free publications available for farmers of all levels (includes business management, financial tools, and marketing).

Farmer's Market Coalition: farmersmarketcoalition. org
Member-based information, resource, and advocacy association whose purpose is to strengthen farmers' markets for the benefit of farmers, consumers, and communities.

Growing Small Farms—North Carolina State University Cooperative Extension: growingsmallfarms.ces.ncsu.edu/
Extensive direct marketing guide and information about how to start a Community-Supported Agriculture (CSA) program on your farm.

Local Harvest: localharvest.org
Online national database for searching for area farmers' markets, CSAs, family farms, and other sources of sustainably grown food; good place to list your farm stand or CSA.

Rodale Institute Farm Finder: newfarm. rodaleinstitute.org/embedfarmlocator
Interactive website to locate or list a farm.

USDA Agricultural Marketing Service: ams.usda. gov
Maintains information, resources, and farmers' market listings under the "Farmers Markets and Local Food Marketing" link.

Regional Farmers' Market Resources
(also check with individual agricultural departments in your state or province)
Alabama: fma.alabama.gov

Alaska: alaskafarmersmarkets.org

Connecticut: ctnofa.org/FarmersMarkets.htm

Delaware: dda.delaware.gov/marketing/delaware_ farmers_markets.shtml

Oklahoma: okfarmandfood.org/ oklahomafarmersmarkets

Wyoming: wyomingfarmersmarkets.org

Financial and Business Resources

AgPlan: agplan.umn.edu
Free online step-by-step guided instructions to help rural business owners develop a business plan (through the resources of the Center for Farm Financial Management at the University of Minnesota).

The Carrot Project: thecarrotproject.org
Listings of financing and business resources and options for small farms.

Certified Naturally Grown, The Grassroots Alternative to Certified Organic: naturallygrown.org
A grassroots alternative to the USDA's National Organic Program; meant primarily for small farmers distributing through local channels such as farmers' markets and CSA programs.

Cornell University—Northeast Beginning Farmers Project: nebeginningfarmers.org
"New Farmer Hub" section contains essential resources for the business side of running your farm (includes business templates).

Farmer's Pledge—Northeast Organic Farming Association of New York: nofany.org/sites/default/files/Generic%20digital%20pledge%202013_0.pdf
Copy of farmer's pledge established by NOFA—New York.

New Entry Sustainable Farming Project (NESFP): nesfp.org
Organization works locally, regionally, and across the country with new farmers to help build successful businesses, find farmland, and improve farming skills; also offers business planning workshops.

Ontario Ministry of Agriculture and Food: omafra.gov.on.ca
Website maintains an extensive agricultural business management section (includes how to build a business plan).

Rodale Institute: rodaleinstitute.org
Extensive online business planning section for farmers.

USDA: National Organic Program (NOP): ams.usda.gov/nop
Information from the USDA's Agricultural Marketing Service about U.S. organic food standards.

Bibliography

There were numerous sources used in researching information for this book. In addition to interviews, workshops, seminars, and observations from experiences, I consulted multiple books, websites, university publications, agricultural fact sheets, seed and organic product catalogs, agricultural organizations, and home gardening resources. To keep this listing concise, I have eliminated some sources of information that were readily confirmed through multiple sources.

Introduction

Balfour, Lady Eve. *The Living Soil*. London: Faber and Faber Ltd., 1943.

Carson, Rachel. *Silent Spring*. Boston: Houghton Mifflin Co., 1962.

Essential Elements for Plant Growth, Law of the Minimum. University of Wisconsin: soils.wisc.edu/~barak/soilscience326/lawofmin.htm

Federal Organic Foods Production Act of 1990. Organic Consumers Association: organicconsumers.org/Organic/ofpa1990.html

A History of Organic Farming: Transitions from Sir Albert Howard's War in the Soil to the USDA National Organic Program. Weston A. Price Foundation: westonaprice.org/farm-a-ranch/history-of-organic-farming

A History of Pesticide Use. Oregon State University: people.oregonstate.edu/~muirp/pesthist.htm

Howard, Albert. *An Agricultural Testament*. London: Oxford University Press, 1943.

———. *The Soil and Health*. New York: Schocken Books, 1947.

The Organic Center: organic-center.org/organic-fact-sheets/top-12-reasons-to-go-organic/

Parr, J. F., and S. B. Hornick. The Evolution of Sustainable Agriculture in the United States: A Recent Historical Perspective. Agricultural Research Service, U.S. Department of Agriculture: infrc.or.jp/english/KNF_Data_Base_Web/PDF%20KNF%20Conf%20Data/C3-8-102.pdf

Post War Fertilizer Explodes, Farming in the 1940s. Wessels Living History Farm: livinghistoryfarm.org/farminginthe40s/crops_04.html

Steiner, Rudolf. *Spiritual Foundations for Renewal of Agriculture*. Milwaukee: WI: Biodynamic Farming and Gardening Association, 1933.

Chapter 1—The Good Earth

Rosen, Carl J. *Lead in the Home Garden and Urban Soil Environment*, St. Paul, MN: Department of Soil, Water and Climate, University of Minnesota Extension.

"Water Sources," Northeast Beginning Farmers Project. Cornell University: nebeginningfarmers.org/farmers/land/water-sources/

Soil Survey Staff. Web Soil Survey. Natural Resources Conservation Service, USDA. Available online: websoilsurvey.nrcs.usda.gov/ [Accessed 11/02/2013].

Integrated Pest Management (IPM) Principles. U.S. Environmental Protection Agency: epa.gov/pesticides/factsheets/ipm.htm

An Introduction to Wind Erosion Control. Alberta Agriculture and Rural Development: agric.gov.ab.ca/$department/deptdocs.nsf/all/agdex3524

McKay, Erin. "Buying the Farm House," *Hobby Farms* magazine, February/March 2003: hobbyfarms.com/home-and-barn/farm-house-buying-14824.aspx

Foster, Ruth S. *Landscaping That Saves Energy and Dollars*. Guilford, CT: Globe Pequot Press, 1994.

Chapter 2—Tools That Rock the Farm

Balter, Michael. "Farming Was So Nice, It Was Invented at Least Twice," news.sciencemag.org (news site of the journal *Science*), July 4, 2013: news.sciencemag.org/2013/07/farming-was-so-nice-it-was-invented-least-twice

Earthway Precision Garden Seeder, product description, Earthway Outlet, online catalog for Earthway tools. Owned and operated by Reed Hill General Store: earthway-outlet.com/Earthway-Precision-Garden-Seeder.htm

Ekarius, Carol. "Tractor Shopping Guide," *Hobby Farms* magazine, December/January 2004: hobbyfarms.com/farm-equipment-and-tools/tractor-shopping-14977.aspx

Fishel, Frederick M. Using Your Hand-Held Lawn and Garden Sprayer. Institute of Food and Agricultural Sciences, University of Florida: edis.ifas.ufl.edu/pi211

Learn Everything about 3 Point Hitch Attachments, website with information, product descriptions, and catalog of tractor attachments: 3pointtractorattachments.com/

Logan, William Bryant. *Smith & Hawken: The Tool Book.* New York: Workman Publishing Company, Inc., 1997.

Pettis, Stephen D. Take Care of Garden Tools. Georgia Center for Urban Agriculture, College of Agricultural and Environmental Sciences, University of Georgia: apps.caes.uga.edu/urbanag/Home&Garden/indexFS.cfm?storyid=2566

Productive Tools for Garden and Farm, Easy Digging, blog site with online tool catalog: easydigging.com/hoss-seeder-planter.html#one

Thomson, Bob, with James Tabor. *The New Victory Garden.* Boston/Toronto: Little, Brown and Company, 1987.

Chapter 3—The Living Soil

Bionutrient Rich Food & Health. Bionutrient Food Association: bionutrient.org/bionutrient-rich-food/bionutrient-rich-food-health

Cahilly, Wayne. "How Is Your Soil Texture?" *Fine Gardening:* finegardening.com/how-to/articles/hows-your-soil-texture.aspx?id=81680

Canola Council of Canada: canolacouncil.org/crop-production/canola-grower's-manual-contents/chapter-6-acidity,-salinity,-solonetzic/chapter-6

Gershuny, Grace. *Start with the Soil: The Organic Gardener's Guide to Improving Soil for Higher Yields, More Beautiful Flowers and a Healthy, Easy-Care Garden.* Emmaus, PA: Rodale Press, Inc., 1993.

Gilman, Steve. *Organic Soil Fertility and Weed Management.* White River Junction, VT: Chelsea Green Publishing, 2002.

Gruver, Joel, Western Illinois University and Michelle Wander, University of Illinois, Extension. "Use of Tillage in Organic Farming Systems: The Basics." Aug. 15, 2013: extension.org/pages/18634/use-of-tillage-in-organic-farming-systems:-the-basics#.Ulxc-VOY9HA

Harvesting Hope. 2013 USBI North American Biochar Symposium, Amherst Center for Agriculture, University of Massachusetts: ag.umass.edu/news/harvesting-hope

Ingham, Elaine R. "Brewing Compost Tea," *Fine Gardening:* finegardening.com/how-to/articles/brewing-compost-tea.aspx

Ingham, Elaine R.; Andrew R. Moldenke, Oregon State University; and Clive A. Edwards, The Ohio State University. Soil Biology Primer. Natural Resources Conservation Service, USDA: soils.usda.gov/sqi/concepts/soil_biology/biology.html

Jenkins, Joe. Thermophilic Bacteria, Composting Stages, and the Sanitization of Compost, A Growing Culture: agrowingculture.org

McDowell, Dr. C. Forrest, and Tricia Clark-McDowell. Home Composting Made Easy, Step-by-Step Instructions, Products & Resources for Backyard Composting: homecompostingmadeeasy.com/

Menjoulet, Brie. Urban Chicken Manure Management. University of Missouri Extension: extension.missouri.edu/webster/backyardchickens/UrbanChickenManureManagement-BrieMenjoulet-%28ScreenVersion%29.pdf

Raymond, Dick. *Garden Way's Joy of Gardening,* North Adams, MA: Storey Publishing, 1982.

The Soil Management Series. University of Minnesota Extension: http://www1.extension.umn.edu/agriculture/tillage/soil-management/soil-management-series/

Soil Taxonomy: Classifying Soils and the Soil Orders. The Cooperative Soil Survey, University of Missouri: soils.missouri.edu/tutorial/page4.asp

Soil Testing. Agricultural Service Laboratory, Clemson University: clemson.edu/public/regulatory/ag_svc_lab/soil_testing/

Stell, Elizabeth P. *Secrets to Great Soil, A Grower's Guide to Composting, Mulching and Creating Healthy, Fertile Soil for Your Garden and Lawn.* (Storey's Gardening Skills Illustrated): North Adams, MA: Storey Publishing, 1998.

Taking a Soil Sample. Michigan State University Extension: msusoiltest.com/get-your-soil-test/

Trautmann, Nancy, and Elaina Olynciw. Compost Microorganisms. Cornell Composting, Cornell University: compost.css.cornell.edu/microorg.html

Tucker, M. Ray. Essential Plant Nutrients: Their Presence in North Carolina Soils and Role in Plant Nutrition. North Carolina Department of Agriculture, 1999: ncagr.gov/agronomi/pdffiles/essnutr.pdf

What is Biochar? International Biochar Initiative: biochar-international.org/

Chapter 4—We Plow the Field . . .

Camp, Danya. Beneficials in the Garden. Galveston County Master Gardeners: aggie-horticulture.tamu.edu/galveston/beneficials/index.htm

Cox, Bonnie. Oregon Tilth. Training Systems and Pruning in Organic Tomato Production. eXtension, America's Research-Based Learning System: extension.org/pages/18647/training-systems-and-pruning-in-organic-tomato-production#.UoDJVSeY9HA

Cranshaw, W. S. Lady Beetles. Colorado State University Extension: ext.colostate.edu/pubs/insect/05594.html

Eaton, Alan T., PhD, Dealing with Woodchuck Damage. Integrated Pest Management, University of New Hampshire Cooperative Extension: extension.unh.edu/resources/files/resource000562_rep584.pdf

Growing Fresh Market Tomatoes. University of Wisconsin Extension: datcp.wi.gov/uploads/Food/pdf/mk_fc_80_web.pdf

Grow Organic, the website for Peaceful Valley Farm and Garden Supply: groworganic.com

Identification of Common Landscape Pests and Beneficial Organisms in Nevada. University of Nevada Cooperative Extension: www.unce.unr.edu/publications/files/ag/2006/sp0608.pdf

McDonald, John E., Jr., Massachusetts Division of Fisheries and Wildlife, and Craig S. Hollingsworth, Department of Entomology, University of Massachusetts. Preventing Deer Damage: extension.umass.edu/vegetable/articles/preventing-deer-damage

Merrill, Richard. Attracting Beneficial Insects to the Garden with Beneficial Flowers. Renee's Garden: reneesgarden.com/articles/Beneficial%20Insects%20Guide.pdf

Natural Enemies Gallery. University of California Agricultural & Natural Resources: ipm.ucdavis.edu/PMG/NE/

Perry, Dr. Leonard. Effective Deer Fencing. University of Vermont: pss.uvm.edu/ppp/articles/deerfences.html

Pests in Gardens and Landscapes, Voles (Meadow Mice). University of California Agriculture & Natural Resources: ipm.ucdavis.edu/PMG/PESTNOTES/pn7439.html

Sullivan, Kristi L., and Paul Curtis. Voles, Wildlife Management Fact Sheets. Cornell University: wildlifecontrol.info/pubs/Documents/Voles/Voles.pdf

"Top Ten Garden Insect Pests," *Organic Gardening* magazine: organicgardening.com/learn-and-grow/top-ten-garden-insect-pests?page=0,1

University of Georgia, Ft. Valley State College, and USDA: caes.uga.edu/extension/cobb/anr/Documents/VEGETABLESEEDREQUIREMENTS2.pdf

Woodchucks in Rhode Island. Rhode Island Department of Environment: www.dem.ri.gov/programs/bnatres/fishwild/pdf/woodchuc.pdf

Chapter 5—Vegetables and Herbs: The Spice of Life and Farming

Bird, Richard. *Companion Planting.* New York: Sterling Publishing Company, Inc., 1998.

Cahill, Betty Jo. Growing Basil. Colorado State University Cooperative Extension, Denver County: coopext.colostate.edu/4DMG/VegFruit/basil.htm

Cornell University Vegetable Growing Guides. Cornell University: gardening.cornell.edu/homegardening/scene0391.html

Corn Insect Pests. CropWatch: Insect Management. University of Nebraska, Lincoln: cropwatch.unl.edu/web/insect/cornpestmgt

Essential Facts for Parsley. The Herb Society for America: herbsociety.org/herbs/documents/Parsleyfactsheet_000.pdf

Euser, Barbara. Make Time for Thyme in Your 2007 Garden. Marin Master Gardeners, University of California: ucanr.org/sites/MarinMG/Marin_Master_Gardener_Independent_Journal_Articles/?uid=213&ds=275

Genetically Modified Foods (Biotech Foods) Pros and Cons: webmd.com/food-recipes/features/are-biotech-foods-safe-to-eat

Golden Harvest Organics: ghorganics.com

Gourmet Garlic Gardens: gourmetgarlicgardens.com/

How to Grow Dill. herbgardening.com: herbgardening.com/growingdill.htm

Icebox Watermelons and Watermelon Variety Descriptions. Mount Vernon Northwestern Washington Research and Extension Center: agsyst.wsu.edu/watermelonphotos.html and vegetables.wsu.edu/watermelon.html

Jett, John W. That Devilish Parsley. West Virginia University Extension Service: wvu.edu/~agexten/hortcult/herbs/parsley.htm

Johnny's Selected Seeds: johnnyseeds.com

Kitazawa Seed Co.: kitazawaseed.com/

Insert new entry above Loha:

Kuepper, George, and Mardi Dodson. Companion Planting: Basic Concepts & Resources, Horticulture Technical Note, July 2001. Appropriate Technology Transfer for Rural Areas (ATTRA): attra.ncat.org

Loha-unchit, Kasma. Cilantro—Pak Chee. Thai Food and Travel: thaifoodandtravel.com/ingredients/cilantro.html

McClure, Susan and Sally Roth. Rodale's Successful Organic Gardening: Companion Planting. Emmaus, PA: Rodale Press, Inc., 1994.

Micro Greens Production #8079. Johnny's Selected Seeds: johnnyseeds.com/Assets/Information/MicroGreensTechSheet.pdf

New England Vegetable Management Guide. University of Massachusetts: nevegetable.org/crops

Organic Corn Seed Coat Treatments. Neely-Kinyon Memorial Research and Demonstration Farm Trial, 2006: extension.agron.iastate.edu/organicag/researchreports/nk05cornseedcoating.pdf

Relf, Diane, and Alan McDaniel. Seeds for the Garden. Virginia Cooperative Extension: pubs.ext.vt.edu/426/426-316/426-316.html

Rosmarinus officinalis cv. Madalene Hill Rosemary. Mountain Valley Growers: mountainvalleygrowers.com/rosoffmadelinehill.htm

Rotating Vegetables by Family. Cornell University Extension: ccetompkins.org/garden/food-gardening/rotating-vegetables-family

The Seed Savers Exchange: seedsavers.org

Sikora, Edward J., and Joseph M. Kemble. Common Diseases of Leafy Greens. Auburn University, Alabama Cooperative Extension Service: aces.edu/pubs/docs/A/ANR-1189/ANR-1189.pdf

Smith, Edward C. *The Vegetable Gardener's Bible*. North Adams, MA: Storey Publishing, 2009.

Southern Exposure Seed Exchange: southernexposure.com

Strom, Stephanie. "Disease cuts corn yields." *New York Times,* Sept. 30, 2013: nytimes.com/2013/10/01/science/earth/a-disease-cuts-corn-yields.html?_r=0

Terra Organics, Certified Organic Seeds: terraorganics.com/

Vegetable Garden: Selected Vegetable Crops. A–Z Master Gardener Manual, University of Arizona: ag.arizona.edu/pubs/garden/mg/vegetable/crops.html#top

Chapter 6—Fruits of the Earth

Asian or Oriental Pears. Penn State Extension, College of Agricultural Sciences: extension.psu.edu/plants/gardening/fphg/pome/variety-selection/asian-or-oriental-pears

Asian Pears. University of Maryland: extension.umd.edu/anmp/asian-pears

Beckerman, Janna. Using Organic Fungicides. Perdue University Extension: perdue.edu

Blueberry Maggot. North Carolina State University: *ipm.ncsu.edu/small_fruit/maggot.html*

Bordeaux Mixture. UC IPM Online, University of California Agriculture and Natural Resources: ipm.ucdavis.edu

Cranberry Fruitworm. Penn State Extension, College of Agricultural Sciences: extension.psu.edu/plants/gardening/fphg/blueberries/insect-pests/cranberry-fruitworm

Fruit Trees: Planting and Care of Young Fruit Trees. University of California Agriculture and Natural Resources: homeorchard.ucdavis.edu/8048.pdf

Growing Peaches. The University of New Hampshire Cooperative Extension: extension.unh.edu/resources/files/Resource000586_Rep608.pdf

Parker, Michael L. Training and Pruning Fruit Trees. North Carolina Cooperative Extension: ces.ncsu.edu/depts/hort/hil/ag29.html#pruning

Peaches and Nectarines. Clemson University Extension: clemson.edu/extension/hgic/plants/vegetables/tree_fruits_nuts/hgic1354.html

Sciarappa, William J., PhD, and Gary C. Pavlis, PhD. *Selecting Blueberry Varieties for the Home Garden*. New Brunswick, NJ: Rutgers Cooperative Research and Extension.

Strawberry Selection and Establishment. University of Nebraska-Lincoln: byf.unl.edu/StrawberryNewPlanting

2013 Organic Production Guide for Organic Blueberries. Cornell University Cooperative Extension, New York State Department of Agriculture & Markets and New York State Integrated Pest Management.

Chapter 7—Chickens in Your Backyard

Barbe, Derek L. Basic Guide for the Backyard Chicken Flock. Florida Cooperative Extension Service, Institute of Food and Agricultural Sciences, University of Florida: edis.ifas.ufl.edu/an239

Best Chicken Breeds for Egg Laying and Hatching: egglayingchickens.com/best-egg-laying-chicken-breeds.html

Brown Egg Layers and White Egg Layers. Cackle Hatchery, Lebanon, MO: cacklehatchery.com/page2.html

Glos, Karma E., ed. and comp. Remedies for Health Problems of the Organic Laying Flock, A Compendium and Workbook of Management, Nutritional, Herbal, and Homeopathic Remedies. SARE Project FNE02-412: kingbirdfarm.com/Layerhealthcompendium.pdf

Poultry Breeds. Department of Animal Science, Oklahoma State University: ansi.okstate.edu/breeds/poultry/index.htm

Small-Scale Egg Production, (Organic and Non-Organic). Penn State Extension, College of Agricultural Sciences: extension.psu.edu/business/ag-alternatives/livestock/poultry-and-game-birds/small-scale-egg-production-organic-and-non-organic

Chapter 8—Branching Out: Honeybees, Rabbits, and Goats

About Dairy Goats. American Dairy Goat Association: adga.org/index.php?option=com_content&view=category&layout=blog&id=86&Itemid=87

Bartels, Mary-Frances R. A Primer on Backyard Meat Rabbit Raising Practices. Rudolph's Rabbit Ranch and Waterfowl Farm: rudolphsrabbitranch.com/rrr.htm

Bennett, Bob. *Storey's Guide to Raising Rabbits*. North Adams, MA: Storey Publishing, 2009.

Breeds of Dairy Goats—Nigerian Dwarf. Purdue University: www.ansc.purdue.edu/goat/factsheet/NigerianDwarf.htm

Breeds of Livestock. Department of Animal Sciences, Oklahoma State University: ansi.okstate.edu/breeds/goats/

Buying a Dairy Goat. New York State 4H Dairy Goat Fact Sheet No. 3: ansci.cornell.edu/4H/dairygoats/dairygoatfs3.html

Common Breeds of Dairy Goats in the U.S. Perdue University Dairy Goats Information: ansc.purdue.edu/goat/factsheet/breeds.htm

The Different Types of Honey Bees. North Carolina Cooperative Extension Service: beesource.com/resources/usda/the-different-types-of-honey-bees/

Fanatico, Anne, and Camille Green. National Center for Appropriate Technologies: https://attra.ncat.org/

Kirchofer, Kevin. Plants Poisonous to Livestock—Toxic Plants and the Common Caprine. Cornell University College of Agriculture and Life Sciences: ansci.cornell.edu/plants/goatlist.html

Morelle, Rebecca. Neonicotinoid Pesticides "Damage Brains of Bees." BBC News. March 27, 2013: bbc.co.uk/news/science-environment-21958547

Neat Facts About Bees, All About Bees. Ontario Beekeepers Association: ontariohoney.ca/kids-zone/bee-facts

Rabbit Production. Penn State Extension, College of Agricultural Sciences: extension.psu.edu/business/ag-alternatives/livestock/additional-livestockoptions/rabbit-production

Schivera, Diane. Raising Rabbits on Pasture. Maine Organic Farming and Gardeners Association: mofga.org/Publications/MaineOrganicFarmerGardener/Winter20092010/Rabbits/tabid/1392/Default.aspx

Small-Scale Sustainable Rabbit Production. National Sustainable Agriculture Information Service: attra.ncat.org

Subspecies: The Place of Honey Bees in the World. eXtension, America's Research-Based Learning Network: extension.org/pages/21742/subspecies:-the-place-of-honey-bees-in-the-world#.UpxdGieY9HB

Transitioning to Organic Sheep or Goat Dairy Production. NOFA—NY Organic Fact Sheet: nofany.org/sites/default/fi les/Transitioning%20to%20Organic%20Sheep%20or%20Goat%20Dairy%20Production_0.pdf

University of Maryland National Goat Handbook: outlands.tripod.com/farm/national_goat_handbook.pdf

Chapter 9—To Market, to Market

Community Supported Agriculture. Local Harvest: *localharvest.org/csa/*

Farmers Markets Increase Access to Fresh, Nutritious Food. Farmers Market Coalition: farmersmarketcoalition.org/education/increase-access-to-fresh-nutritious-food/

Finn, Emily (posting). Trending at the Market: WIC Cash Value Vouchers, November 7, 2013. Farmers Market Coalition: farmersmarketcoalition.org/trending-at-the-market-wic-cash-value-vouchers/

Selling Directly to Restaurants, Tips from Alternative Technology Transfer for Rural Areas (attra.org). May 16, 2012. Penn State Extension, College of Agricultural Sciences: extension.psu.edu/plants/sustainable/production-tips/selling-directly-to-restaurants

Chapter 10—Foundation for a Business Plan

Certified Naturally Grown: The Grassroots Alternative to Certified Organic, http://www.naturallygrown.org/

Farmer's Pledge, Northeast Organic Farming: nofany.org/sites/default/files/Generic%20digital%20pledge%202013_0.pdf

Frain, Michelle, Marketing Coordinator for the Rodale Institute, and Christine Ziegler, Editor, FACT SHEET: Part 3—Recording and tracking your costs: newfarm.rodaleinstitute.org/depts/NFfield_trials/factsheets/pricing/index3.shtml

Introduction to Organic Farming: A Growing Opportunity for Pennsylvania Farmers. Penn State University: pubs.cas.psu.edu/FreePubs/PDFs/uf024.pdf

Jon Jaffe and First Pioneer Farm Credit with Dartmouth, Mass., Agricultural Commission, Nov. 21, 2009. Farm Business Planning Workshop, Dartmouth Cable TV.

The National Center for Appropriate Technologies (NCAT) and Appropriate Technology Transfer for Rural Areas (ATTRA): attra.ncat.org/attra-pub/organcert.html

National Organic Program, Agricultural Marketing Service of the United States Department of Agriculture: ams.usda.gov/nop/

Norman Bender, Extension Educator, Economic Development, University of Connecticut, and Joseph Bonnelli, Extension Educator, Resource Economics, University of Connecticut, August, 2009. Loans, Grants, and other Financial Assistance for NE Organic Farms.

Index

Photo Credits

Author Andy Tomolonis provided the images that appear on pages 19, 22, 31, 41, 42, 43, 46, 74, 98, 99, 105, 106, 109, 110, 112, 113, 115, 117, 120, 121, 122, 123, 127, 130, 134 (middle), 135 (second from top), 137, 143, 147, 148, 149, 153, 154, 156, 157, 222, 230, 233, 236, 265, 285, 328, and 330.

Other images provided by the following: R. Campbell, 62; Department of Environmental Protection, 84; Jack Dyking, 272; Red Garner, 86; Peggy Greb, 24, 266; Tim McCabe, 26; Mother Cluckers, 331; Kristine Nichols, 63; Lynda Richardson, 26, 225, 318; Seed Savers Exchange, 104; USDA, 70; Jeff Vanuga USDA/NRCS, 71.

All other images provided by Shutterstock.

About the Author

Andy Tomolonis is a longtime garden writer, a former produce manager, and an award-winning Boston journalist. He has been an organic gardener and part-time hobby farmer for more than 20 years. Along with his wife, Valerie, he runs a community-supported agricultural operation, selling vegetables, fruits, and berries to subscribers each season. They also sell eggs, honey, homemade soaps, and other farm-raised goods. He lives in Bellingham, Massachusetts.

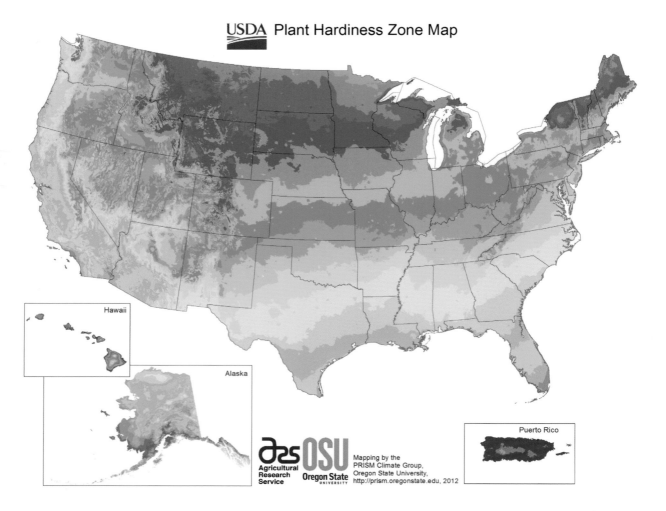

USDA Plant Hardiness Zone Map

Mapping by the
PRISM Climate Group,
Oregon State University,
http://prism.oregonstate.edu, 2012

Average Annual Extreme Minimum Temperature 1976-2005

Temp (F)	Zone	Temp (C)	Temp (F)	Zone	Temp (C)
-60 to -55	1a	-51.1 to -48.3	5 to 10	7b	-15 to -12.2
-55 to -50	1b	-48.3 to -45.6	10 to 15	8a	-12.2 to -9.4
-50 to -45	2a	-45.6 to -42.8	15 to 20	8b	-9.4 to -6.7
-45 to -40	2b	-42.8 to -40	20 to 25	9a	-6.7 to -3.9
-40 to -35	3a	-40 to -37.2	25 to 30	9b	-3.9 to -1.1
-35 to -30	3b	-37.2 to -34.4	30 to 35	10a	-1.1 to 1.7
-30 to -25	4a	-34.4 to -31.7	35 to 40	10b	1.7 to 4.4
-25 to -20	4b	-31.7 to -28.9	40 to 45	11a	4.4 to 7.2
-20 to -15	5a	-28.9 to -26.1	45 to 50	11b	7.2 to 10
-15 to -10	5b	-26.1 to -23.3	50 to 55	12a	10 to 12.8
-10 to -5	6a	-23.3 to -20.6	55 to 60	12b	12.8 to 15.6
-5 to 0	6b	-20.6 to -17.8	60 to 65	13a	15.6 to 18.3
0 to 5	7a	-17.8 to -15	65 to 70	13b	18.3 to 21.1

Canada Plant Hardiness Zone Map

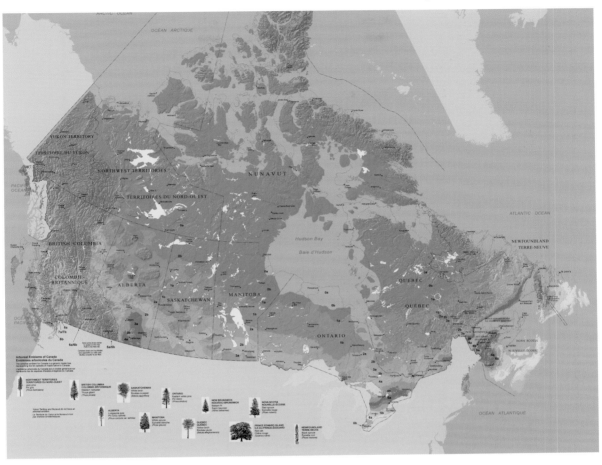

Plant hardiness zones

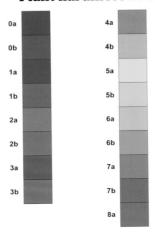

0a	4a
0b	4b
1a	5a
1b	5b
2a	6a
2b	6b
3a	7a
3b	7b
	8a